TECHNOLOGY, ETHICS, AND SOCIETY

Navigating Innovation and Its Impacts

Viktor Wang

innovativeink
PUBLISHING

Cover images Shutterstock.com
Images AI-generated by Viktor Wang

innovativeink PUBLISHING

www.innovativeinkpublishing.com
Send all inquiries to:
4050 Westmark Drive
Dubuque, IA 52004-1840

Print ISBN: 979-8-3197-1505-0
Ebook ISBN: 979-8-3197-1506-7

Published in the United States of America

CONTENTS

Abstract

This chapter examines the evolution of technology and its profound impact on human development, highlighting the dynamic interplay between innovation and societal change. By exploring key technological eras—from the invention of basic tools in prehistoric times to the transformative effects of the Industrial Revolution and the Information Age—the chapter underscores how advancements have shaped social structures, economic systems, and cultural norms. It also delves into the ethical considerations that have accompanied technological progress, providing lessons from history to inform responsible innovation. Through case studies and critical analysis, readers gain a comprehensive understanding of technology's role as both a driver of progress and a source of societal challenges.

Abstract

Chapter 2, *Technological Innovation and Cultural Shifts*, explores the dynamic relationship between technological advancements and cultural transformation. It examines how innovations, from the printing press to the internet, have reshaped societal norms, communication practices, and social structures. The chapter discusses the profound impacts of technology on family dynamics, identity formation, and global connectivity, while also addressing the challenges of inequality and cultural homogenization that arise from technological progress. Through a critical lens, it highlights both the opportunities and complexities technology presents in influencing cultural evolution, and how these shifts shape contemporary life and societal organization.

Abstract

Chapter 3 explores the profound economic impacts of technology, focusing on its role in driving growth, exacerbating inequality, and disrupting traditional industries and labor markets. While technological advancements have fueled economic expansion by increasing productivity, creating new markets, and enhancing business efficiency, they have also contributed to rising economic disparities. Automation and AI have displaced low- and middle-skilled jobs, while rewarding high-skill workers with greater opportunities, leading to labor market polarization. Additionally, the concentration of wealth in tech giants and geographic disparities in access to technology further deepen inequality. The chapter examines how technology is reshaping industries and labor, highlighting both the challenges of job displacement and the emergence of new employment models, such as the gig economy and remote work, and emphasizes the need for policies that ensure equitable growth and worker support in an ever-evolving economy.

Abstract

Chapter 4 explores the profound impact of technology on political structures, examining its role in governance, policymaking, and civic engagement. As digital tools such as big data, artificial intelligence, and social media reshape political power dynamics, governments are increasingly leveraging these technologies to enhance decision-making, increase transparency, and optimize public services. However, the chapter also highlights challenges, including privacy concerns, algorithmic bias, and the digital divide, which may limit equitable participation in political processes. Furthermore, while technology has empowered citizens to engage in political discourse and activism, it has also contributed to issues like misinformation, political polarization, and the concentration of power in the hands of a few corporations. The chapter ultimately underscores the need for a balanced approach to technological integration in politics, ensuring that innovation serves democratic values while mitigating its potential risks.

Abstract

Chapter 5 explores the vital role of ethical frameworks in navigating the complexities of emerging technologies. It introduces foundational ethical theories, such as deontology, consequentialism, virtue ethics, and care ethics, and demonstrates their practical application to contemporary technological dilemmas like AI bias, autonomous weapons, and data privacy. The chapter emphasizes the importance of ethical guidelines, global cooperation, and leadership in fostering responsible innovation. Through case studies and theoretical insights, it highlights the need for adaptable, inclusive, and forward-thinking approaches to ethical decision-making, ensuring technological advancements align with societal values and human dignity.

Abstract

Chapter 6 explores the complex intersections of privacy, surveillance, and security in the digital age, examining the ethical, legal, and practical challenges posed by emerging technologies. It delves into the balance between technological innovation and the protection of individual rights, highlighting the growing concerns surrounding data collection, surveillance practices, and the implementation of security measures. The chapter critically analyzes how advancements such as facial recognition, big data analytics, and cybersecurity protocols affect personal freedoms, public safety, and societal norms. It also considers the global implications of these issues, emphasizing the need for ethical frameworks, transparency, and international cooperation to ensure privacy and security are safeguarded in an increasingly interconnected world.

Abstract

Artificial Intelligence (AI) is reshaping society by offering transformative opportunities while posing significant ethical challenges. Its integration into decision-making processes across sectors such as healthcare, education, and criminal justice highlights the immense potential of AI to enhance efficiency, reduce human error, and address global challenges like climate change. However, the rapid adoption of AI also raises concerns about bias, transparency, accountability, and privacy. Ethical dilemmas emerge as AI systems influence critical decisions, often blurring lines of human responsibility and agency. Addressing these challenges requires a balanced approach, emphasizing fairness, inclusivity, and human oversight. Comprehensive regulatory frameworks, ethical design principles, and collaborative governance are essential to mitigate risks and ensure AI aligns with societal values. By fostering innovation responsibly, society can harness AI as a force for good, tackling global issues while safeguarding individual rights and moral integrity.

Abstract

Chapter 8, *Bridging the Digital Divide: Equity and Inclusion in Technology Access*, explores the growing disparities in access to technology, particularly the internet and digital devices, and the urgent need to address these gaps to ensure equitable opportunities for all. The chapter examines how unequal access to technology exacerbates social, economic, and educational inequalities, particularly among rural, low-income, and minority communities. It discusses the importance of expanding broadband infrastructure, improving digital literacy, and providing targeted programs for underserved populations. By emphasizing the role of governments, businesses, and educational institutions in promoting digital inclusion, the chapter advocates for a collaborative approach to ensure that technology is accessible, affordable, and beneficial for everyone, fostering a more inclusive digital future.

Abstract

Chapter 9, *Sustainability and Technology: Innovation for the Planet*, explores the critical role of technology in addressing environmental challenges and advancing sustainability. It examines technological solutions such as renewable energy, smart agriculture, and digital tools for environmental monitoring, highlighting their transformative potential in combating climate change, resource depletion, and pollution. The chapter also delves into industry applications, consumer-focused innovations, and the integration of green technologies into daily life. While emphasizing success stories and emerging trends, it addresses barriers such as cost, policy gaps, and ethical considerations, advocating for a collaborative, equitable approach to sustainable innovation. This chapter provides a comprehensive analysis of how technology can drive global efforts toward a greener, more sustainable future.

Abstract

Chapter 10 explores the profound impact of globalization and technology on cultural exchange, examining how technological advancements facilitate cross-cultural interactions and global connectivity. The chapter highlights how digital platforms, social media, and emerging technologies such as virtual reality and artificial intelligence enable the rapid dissemination and hybridization of cultures worldwide. It delves into the dual role of technology in both promoting cultural diversity and raising concerns about cultural homogenization. The chapter also addresses the challenges and ethical considerations associated with global cultural exchange, including equity in access, preservation of local cultures, and the digital divide. Ultimately, it emphasizes the need for a balanced approach to ensure that globalization fosters mutual respect, understanding, and inclusivity across cultures.

Abstract

Chapter 11 explores the ethical implications of automation in the future of work, examining its impact on employment, human dignity, and societal well-being. As automation technologies like artificial intelligence and robotics continue to evolve, concerns about job displacement, the erosion of human purpose, and the widening of economic inequalities are becoming increasingly pressing. The chapter delves into how these technologies challenge traditional notions of work and human value, while also addressing the ethical issues surrounding privacy, bias, and accountability in automated systems. It emphasizes the need for thoughtful policy, regulation, and global cooperation to ensure that automation benefits society as a whole, fostering a future where technology enhances human potential and promotes equitable opportunities.

Abstract

Chapter 12, "Responsible Innovation and Global Challenges," explores the critical role of ethical practices in the development and deployment of emerging technologies. It emphasizes the need for responsible innovation that balances technological advancement with social, environmental, and economic considerations. The chapter highlights the importance of a multidisciplinary approach, involving experts from diverse fields to address the ethical challenges posed by technologies like AI, automation, and biotechnology. It also discusses the role of governments, businesses, and international organizations in creating regulatory frameworks and policies that ensure the responsible use of technology, promote global cooperation, and mitigate the risks of inequality, bias, and environmental harm. Ultimately, the chapter argues for a future where innovation benefits society equitably, fosters sustainability, and prioritizes human well-being.

PREFACE

In an era of relentless technological progress, society stands at a crossroads where innovation offers both unprecedented opportunities and profound ethical challenges. Technology has become an integral force shaping every facet of our lives, from the way we communicate and work to the policies that govern our societies and the values that define our humanity. As we grapple with this transformative force, understanding its complexities is essential to ensuring that technological advancement aligns with human flourishing. *Technology, Ethics, and Society: Navigating Innovation and Its Impacts* explores this dynamic intersection, offering a comprehensive analysis of how technology reshapes our lives, redefines societal norms, and raises complex moral questions. This book seeks to equip readers with the intellectual tools to critically evaluate the ethical implications of technological advancement, fostering a deeper understanding of how innovation can be guided to serve humanity responsibly.

The chapters in this volume are thoughtfully curated to traverse the historical, cultural, economic, and political dimensions of technology, setting the stage for nuanced discussions of contemporary and future issues. The journey begins with a historical lens, tracing the evolution of technology and its transformative influence on human civilization. By examining the ways innovation has driven societal change, we can better understand the enduring interplay between technological progress and the social fabric of humanity. This foundational perspective serves as a reminder that technological developments are never neutral; they carry with them the values, priorities, and ambitions of the societies that create them.

From this foundation, the book delves into the cultural shifts, economic disruptions, and political dynamics that define our modern era. The analysis highlights how technological advancements not only accelerate progress but also create ripple effects across industries, cultural norms, and governance structures. The interconnectedness of these dimensions is evident in the way digital technologies reshape communication, redefine work-life balance, and alter the nature of civic engagement. Each chapter aims to provide readers with the insights needed to navigate these shifts, balancing optimism about technology's potential with a cautious awareness of its unintended consequences.

A central focus of this book is its exploration of ethical frameworks and their application to emerging technologies. As advancements in artificial intelligence, data science, and automation redefine the possibilities of human achievement, they also raise critical ethical dilemmas. How do we ensure that privacy and security are not compromised in the name of progress? How can we bridge the digital divide to create equitable access to technological benefits? And how can innovations address global challenges such as sustainability and inclusion while upholding human dignity? By addressing these pressing questions, this book fosters a sense of moral responsibility among innovators, policymakers, educators, and global citizens alike.

The exploration of ethics extends beyond abstract considerations. Each chapter offers practical examples and case studies that illuminate the real-world consequences of technological decisions. Whether discussing the implications of surveillance technologies, the ethical challenges of artificial intelligence, or the disparities in access to digital resources, this book emphasizes the importance of context, critical thinking, and ethical deliberation in shaping the trajectory of innovation.

Technological advancements are also driving profound transformations in globalization, the future of work, and environmental stewardship. These transformations demand urgent attention to the values that underpin responsible innovation. This book synthesizes insights from earlier chapters to propose actionable strategies for addressing these complex challenges. By doing so, it positions technology not merely as a tool for progress but as a means to achieve equity, sustainability, and a shared global vision for humanity's future.

One of the unique aspects of this book is its dual purpose. Not only does it serve as a rigorous academic resource for exploring the intersection of technology, ethics, and society, but it also supports a broader educational mission. This text has been carefully designed to align with the requirements of the Social Science (D4) General Education (GE) curriculum, making it an ideal resource for undergraduate students seeking to fulfill their GE requirements. Additionally, it serves as a foundational text for Career and Technical Education (CTE) program majors, bridging the gap between theoretical exploration and practical application. By addressing topics relevant to both GE and CTE, this book encourages interdisciplinary learning and fosters critical engagement across diverse fields of study.

As technological advancements continue to redefine education, work, and society, the insights presented in this book are more relevant than ever. For educators, this text provides a roadmap for integrating ethical considerations into classroom discussions and professional practice. For students, it offers a framework for understanding the broader implications of the technologies they encounter in their personal and professional lives. For policymakers and industry leaders, it serves as a guide for navigating the complex ethical terrain of innovation with responsibility and foresight.

Technology has the power to drive equity or deepen existing divides, depending on how it is wielded. This dual-edged nature underscores the importance of informed decision-making grounded in a commitment to ethical principles. From developing policies that govern technological deployment to designing systems that are inclusive and fair, this book provides readers with the conceptual tools necessary to act responsibly in a world shaped by innovation.

Equally significant is the book's emphasis on collaboration and interdisciplinary approaches. It acknowledges that addressing the ethical challenges of technology requires dialogue among technologists, social scientists, ethicists, educators, and policymakers. Through case studies, research findings, and reflective questions, the book fosters a holistic understanding of the issues, inspiring readers to seek solutions that transcend disciplinary silos.

As we look toward the future, this book encourages readers to embrace innovation with a critical yet hopeful mindset. By considering the societal and ethical dimensions of technological progress, we can ensure that advances in AI, data science, automation, and other domains are harnessed to create a more equitable and sustainable world. Each chapter invites readers to imagine new possibilities while remaining grounded in the realities and challenges of the present.

Technology, Ethics, and Society: Navigating Innovation and Its Impacts is not merely an academic exploration but a call to action. It challenges readers to think critically, act ethically, and collaborate across disciplines to shape a future where technology enhances human flourishing. It underscores the importance of embracing innovation with a sense of responsibility, ensuring that the digital age is a force for good that upholds equity, sustainability, and dignity for all.

This work is dedicated to students, educators, policymakers, and practitioners who seek to understand and influence the evolving relationship between technology and society. By engaging with the insights presented in this book, readers will be better equipped to navigate the ethical terrain of technological innovation and contribute to a more just and equitable world. As we stand on the brink of new discoveries and challenges, this book invites readers to join a global conversation about the role of technology in shaping our shared future.

Dr. Viktor Wang, Professor
California State University, San Bernardino, USA

FOREWORD

We live in an age when the phrase "technology changes everything" has become almost cliché—and yet, in many ways, it is still an understatement. From the devices in our pockets to the invisible infrastructures that sustain our economies and institutions, technological innovation now shapes nearly every aspect of human experience. We navigate a world in which algorithms influence what we see, whom we hear, and how we decide; in which automation reorganizes work and labor; and in which digital networks carry both solidarity and harm at the speed of light. In such a context, the question is no longer whether technology matters for ethics and society, but whether we are prepared to think about that relationship with the depth, historical awareness, and critical humility it demands. Technology, Ethics, and Society: Navigating Innovation and Its Impacts enters this conversation at exactly the right moment.

One of the most important strengths of this book is its refusal to treat technology as a recent or purely technical phenomenon. The opening chapters take a historical and cultural lens, reminding readers that every major technological shift—printing, industrialization, electrification, computing—has upended social norms, power relationships, and human self-understanding. By tracing the evolution of technology and its transformative impact on human development, the book encourages us to see today's digital, algorithmic, and networked systems as part of a longer story rather than as isolated novelties. That historical grounding matters. It tempers both the breathless enthusiasm that insists "this time is completely different" and the fatalism that shrugs, "we have seen it all before." Instead, it helps readers recognize patterns: who tends to benefit first, who is left out or pushed aside, and how societies have successfully—or unsuccessfully—responded to disruption in the past.

From that foundation, the text turns to culture, economics, and politics with a rare combination of clarity and nuance. Technological innovation never arrives in a vacuum; it is taken up, resisted, reshaped, and contested within specific cultural contexts. The chapters on cultural shifts and global exchange invite readers to examine how new tools and platforms influence identity, community, and meaning-making, sometimes amplifying marginalized voices and sometimes reproducing or intensifying old prejudices. At the same time, the book's analysis of economic impacts—growth, inequality, and disruption—confronts a hard reality: technology can generate extraordinary wealth and efficiency while also exacerbating divides between those who design and control systems and those who are subject to them. By situating technology within labor markets, global supply chains, and financial structures, the book helps readers see why ethics cannot be limited to questions of individual use; it must also address structural questions about ownership, power, and distribution.

The political dimensions of technology are just as carefully explored. In an era of targeted advertising, data-driven campaigns, and algorithmic amplification, questions of governance, policy, and civic engagement have become inseparable from questions about digital infrastructures. The chapter on technology and political structures looks at how tools meant to increase participation can also be used for manipulation or surveillance, and how states and private actors negotiate control over information, platforms, and public space. By examining power, policy, and participation together, the book avoids simplistic narratives that cast technology either as the savior of democracy or as its inevitable undoing. Instead, it invites readers to think concretely about design choices, regulatory frameworks, and social movements that can bend technological systems toward accountability and the common good.

Ethics, in this text, is not treated as an abstract add-on, but as a practical, lived discipline threaded through every domain. The chapters that introduce ethical frameworks for emerging technologies, and that explore privacy, surveillance, and security in the digital age, give students a toolkit of concepts and questions they can carry into real-world dilemmas. Rather than simply listing risks, the book shows how different ethical theories—duty-based, consequence-based, virtue-focused, care-oriented—might approach the same technological problem in distinct ways. This plural perspective is crucial in a world where stakeholders bring different moral languages and priorities to the table. In the discussion of privacy and surveillance, for example, readers are invited to weigh trade-offs between innovation, safety, and individual rights, and to recognize when familiar metaphors ("if you have nothing to hide…") fail to capture the full human stakes of data collection and monitoring.

Nowhere is the need for such careful ethical reflection more evident than in the chapters on artificial intelligence and automation. AI is not simply another software tool; it is a family of techniques that can perform tasks once reserved for human judgment and perception, from medical diagnosis to credit scoring to language generation. By examining AI's opportunities and risks side by side, and by linking decision-making systems to questions of bias, transparency, and responsibility, the book equips readers to ask the right questions: Who designs and trains these systems? On what data? For whose benefit? With what recourse when things go wrong? The later exploration of the future of work extends these concerns into workplaces and labor markets, prompting us to consider not only how many jobs might change or disappear, but what kinds of work we consider dignified and worth preserving.

Crucially, the text refuses to treat access to technology as a neutral backdrop. The chapter on bridging the digital divide foregrounds issues of equity and inclusion, reminding us that the benefits of innovation are not evenly distributed. Gaps in connectivity, hardware, skills, and institutional support often map onto existing lines of inequality—by income, geography, race, gender, or disability. By addressing these disparities directly, the book encourages educators, policymakers, and practitioners to think of technology not merely as a set of tools, but as part of a broader justice project: who gets to participate fully in digital life, and on what terms. The sections on sustainability and planetary impacts extend that justice lens in another direction, asking how technological systems consume energy, resources, and attention, and what it would mean to innovate in ways that are genuinely compatible with the long-term flourishing of the planet.

Taken together, these chapters do more than introduce topics; they model a way of thinking. Throughout the book, readers are invited to move back and forth between big-picture frameworks and concrete cases, between global patterns and local experiences. The final chapter on responsible innovation and global challenges serves as a kind of synthesis, drawing together insights from history, culture, economics, politics, ethics, and environmental studies. It asks what it means to act responsibly when no single actor—no government, no corporation, no researcher, no user—controls the full trajectory of technological change. It suggests that responsibility today is as much about collaboration, transparency, and humility as it is about individual intent. That is a difficult lesson, but an essential one.

As someone who has watched multiple waves of technological enthusiasm and anxiety unfold, I am struck by how well this book balances urgency with patience. It does not deny the speed at which innovations are arriving, nor the intensity of their impacts. But it refuses to be rushed into shallow conclusions. Instead, it offers students and other readers a set of lenses, languages, and questions that they can carry into their own disciplines, communities, and decisions. In classrooms, it will support not only learning about technology, but also learning how to deliberate together about contested futures. In policy and practice settings, it can serve as a shared reference point for conversations that too often happen in silos.

Technology will continue to advance, in ways both foreseeable and surprising. What is less certain is whether our ethical and social imagination will keep up. Technology, Ethics, and Society does not claim to have all the answers—that would be impossible, and dishonest. What it does offer is something more valuable: a thoughtful, structured invitation to think, question, and act with greater responsibility in a tech-driven world. My hope is that readers will not only absorb its content, but also practice the habits of reflection it encourages. If they do, we may yet learn to navigate innovation in ways that honor both human dignity and the fragile world we share.

William J. Rothwell, Ph.D., DBA, SPHR, SHRM-SCP, RODC, FLMI, CPTD Fellow
Distinguished Professor Emeritus and Academy Professor
The Pennsylvania State University, USA

AUTHOR BIOGRAPHY

Dr. Viktor Wang received the 2016 Presidential Award for Exceptional and Innovative Leadership from the American Association for Adult and Continuing Education and is an institutional recipient of the Distinguished Faculty and Scholarly Achievement Award. A devoted mentor and globally recognized leader in education, he has an accomplished career marked by innovation, scholarship, and program-building. Dr. Wang is the author or coauthor of more than 260 peer-reviewed publications and has edited volumes and journals that include contributions from over 1,500 chapters and 200 articles by scholars worldwide. His work is held in major academic collections, including Stanford University Libraries, and he has chaired and successfully guided more than 80 doctoral students to completion. Under his leadership departments have seen enrollment growth, the launch of new academic programs, and expanded international partnerships. He currently serves as editor of the International Journal of AI in Pedagogy, Innovation, and Learning Futures (IJAIPIL), an open-access, peer-reviewed journal publishing research on the educational applications and implications of artificial intelligence.

Technology and Society: A Historical Lens

1

ABSTRACT

This chapter examines the evolution of technology and its profound impact on human development, highlighting the dynamic interplay between innovation and societal change. By exploring key technological eras—from the invention of basic tools in prehistoric times to the transformative effects of the Industrial Revolution and the Information Age—the chapter underscores how advancements have shaped social structures, economic systems, and cultural norms. It also delves into the ethical considerations that have accompanied technological progress, providing lessons from history to inform responsible innovation. Through case studies and critical analysis, readers gain a comprehensive understanding of technology's role as both a driver of progress and a source of societal challenges.

INTRODUCTION

The intricate relationship between technology and human society is a cornerstone of understanding human progress (Arthur, 2009; Smith & Marx, 1994). Technology has not only driven the evolution of civilizations but has also responded to the needs, challenges, and aspirations of society (Bijker et al., 1987; Feenberg, 2010). This chapter delves into this symbiotic relationship, tracing how technology has simultaneously enabled growth and reshaped societal structures (Castells, 2010).

From the invention of the wheel to the rise of digital platforms, technology has been pivotal in redefining human possibilities (Pacey, 1983; Arthur, 2009). Each major technological advancement—be it agricultural tools, industrial machinery, or information networks—has created ripples across social, economic, and cultural domains (Perez, 2002; Castells, 2010). By reflecting on these developments, we can appreciate the nuanced ways in which technology impacts every facet of life (Mokyr, 1990).

The historical lens provides a framework for understanding the trajectory of technological innovation (Mokyr, 1990). Examining key milestones, such as the Agricultural Revolution, Industrial Revolution, and the advent of the Digital Age, reveals recurring patterns of innovation, adaptation, and transformation (Allen, 2009; Castells, 2010). These patterns illuminate how humanity has consistently leveraged technology to solve problems and expand its horizons (Gordon, 2016).

The chapter emphasizes that technology has been both a driver of progress and a disruptor of established norms (Kranzberg, 1986). While technological innovations have brought efficiency, connectivity, and convenience, they have also introduced challenges, including societal inequities, ethical dilemmas, and environmental consequences (Floridi, 2014; Intergovernmental Panel on Climate Change, 2023). Understanding this duality is critical for navigating the complexities of modern technological advancements (Stilgoe et al., 2013).

By exploring the historical context, students will grasp how societal needs and challenges have influenced technological innovation (Rogers, 2003). For example, the need for agricultural efficiency led to the development of tools such as the plow, while the quest for faster communication birthed the printing press and, later, the internet (Eisenstein, 1980; Castells, 2010). These case studies underscore the reciprocal nature of technology and society (Bijker et al., 1987).

The Agricultural Revolution serves as a foundational example of how technology reshaped human development (Diamond, 1997). The shift from hunting and gathering to farming technologies enabled the establishment of settled communities, surplus food production, and the rise of early civilizations (Scott, 2017). These advancements marked a turning point in human history, with profound social and economic implications (Mokyr, 1990).

The Industrial Revolution further highlights the transformative power of technology (Allen, 2009). Machines and mechanized systems revolutionized manufacturing, transportation, and labor practices (Mokyr, 1990; Gordon, 2016). This period also brought significant societal changes, including urbanization, shifts in labor dynamics, and the rise of new economic paradigms (Schumpeter, 2008/1942).

As society entered the Digital Age, the pace of technological change accelerated exponentially (Castells, 2010). The development of computers, the internet, and mobile technologies has fundamentally altered how people interact, work, and access information (Arthur, 2009). This digital transformation has fostered unprecedented connectivity while also introducing challenges such as digital divides and privacy concerns (ITU, 2023; Zuboff, 2019).

Cultural forces have also played a pivotal role in shaping technological development (Bijker et al., 1987). Societal values, beliefs, and priorities influence the adoption and adaptation of technologies (Rogers, 2003). For instance, the cultural emphasis on efficiency and innovation in industrialized nations fueled rapid technological progress, while other societies adapted these technologies in ways that aligned with their traditions and needs (Pacey, 1983).

Conversely, technology has reshaped cultural norms and practices (Castells, 2010). The proliferation of social media, for example, has redefined communication, self-expression, and community building (boyd, 2014; Pew Research Center, 2023). Similarly, technological advancements in art and entertainment have expanded creative possibilities, enabling new forms of storytelling and artistic expression (Castells, 2010).

Economic forces are another critical driver of technological advancement (Schumpeter, 2008/1942). Market demands, competition, and the pursuit of economic growth have consistently spurred innovation (Perez, 2002). However, these same forces can also create ethical dilemmas, such as labor exploitation, economic inequality, and the environmental costs of production (Gordon, 2016; Intergovernmental Panel on Climate Change, 2023).

Political systems and governance structures have historically influenced and been influenced by technology (Jasanoff, 2016). Innovations in transportation, communication, and weaponry have reshaped geopolitics and power dynamics (Castells, 2010). Moreover, regulatory frameworks and public policies have played crucial roles in guiding the ethical development and deployment of technology (UNESCO, 2021).

Environmental considerations form an integral part of the relationship between technology and society (Intergovernmental Panel on Climate Change, 2023). While technological advancements have enabled resource extraction, energy production, and climate monitoring, they have also contributed to environmental degradation (Gordon, 2016). The ongoing challenge lies in harnessing technology to address these issues sustainably (Stilgoe et al., 2013).

Ethical considerations have accompanied technological progress throughout history (Kranzberg, 1986). From debates over the use of early industrial machinery to modern discussions about artificial intelligence and genetic engineering, society has grappled with questions about the moral implications of technological innovation (Floridi, 2014; UNESCO, 2021). Reflecting on historical examples helps inform contemporary ethical decision-making (Jasanoff, 2016).

This chapter encourages students to critically analyze the lessons from history to better understand the role of technology in shaping the future (Stilgoe et al., 2013). By studying past successes and failures, they can identify strategies for promoting responsible innovation that aligns with societal values and goals (Floridi, 2014).

Ultimately, this chapter lays the foundation for exploring the broader themes of the book, providing students with the tools to critically engage with the complex interplay between technology and society (Arthur, 2009; Castells, 2010). Understanding this historical lens is essential for navigating the challenges and opportunities of an increasingly tech-driven world (UNESCO, 2021).

DEFINING TECHNOLOGY AND ITS ROLE IN SOCIETY

Technology is a multifaceted concept that extends far beyond the devices and systems commonly associated with it (Arthur, 2009; Mitcham, 1994). At its core, technology encompasses the tools, techniques, and systems humans develop to interact with and transform their environment (Feenberg, 2010). It spans everything from simple implements like the wheel and lever to sophisticated systems such as artificial intelligence and quantum computing (Castells, 2010). The breadth of this definition highlights its integral role in human development, making it a cornerstone of societal evolution (Bijker et al., 1987).

Historically, technology has been defined as any application of knowledge to solve problems or achieve specific objectives (Rogers, 2003). This broad definition underscores its ubiquity in human life, as technology is present in virtually every aspect of our daily existence—from the food we consume to the communication methods we employ (Arthur, 2009). As a means to address challenges and improve quality of life, technology reflects the boundless potential of human creativity and adaptability (Mokyr, 1990).

Beyond its physical manifestations, technology can also be conceptualized as a dynamic system that evolves in response to human needs and challenges (Arthur, 2009). It is an adaptive mechanism through which societies innovate to overcome obstacles, enhance efficiency, and expand their potential (Stilgoe et al., 2013). For instance, the development of irrigation systems in ancient civilizations addressed agricultural challenges, while modern-day renewable energy technologies aim to tackle climate change (Scott, 2017; Intergovernmental Panel on Climate Change, 2023; IEA, 2023).

Technology, in many ways, acts as a reflection of human ingenuity and creativity (Feenberg, 2010). Its development is often driven by the desire to address fundamental needs, such as survival, security, and comfort, as well as more complex aspirations, including exploration, self-expression, and connectivity (Arthur, 2009; Castells, 2010). These motivations have led to transformative breakthroughs, such as space exploration, genetic engineering, and the advent of the internet (Floridi, 2014).

Technological eras—from the plow and printing press to the steam engine and smartphone—illustrate the bidirectional influence between innovation and social change

Read as a wave pattern, the image compresses centuries of S-curves—agricultural, industrial, and digital—stacking rather than replacing one another. It suggests that technologies arise from social priorities and then feed back to reshape those priorities, turning tools into infrastructures that set new norms, skills, and institutions. What looks linear is really cumulative path-dependence: today's platforms stand on yesterday's inventions. The ethical throughline is a recurring bargain—efficiency and reach traded against new risks and inequities—which is why the image reads as both a history lesson and a policy prompt about how to distribute benefits and steer the next wave.

As a human adaptive system, technology does not exist in isolation; it operates within the context of societal, cultural, and environmental factors (Bijker et al., 1987; Jasanoff & Kim, 2015). For instance, the tools developed by early civilizations to cultivate crops were shaped by their environmental conditions and cultural practices (Scott, 2017). Similarly, contemporary technologies are influenced by global trends, market demands, and ethical considerations (Perez, 2002; UNESCO, 2021). This interplay highlights the reciprocal relationship between technology and society (Pacey, 1983).

Cultural values often play a significant role in shaping technological development (Pacey, 1983; Jasanoff & Kim, 2015). For example, societies with a strong emphasis on innovation and competition may prioritize advancements in fields like artificial intelligence and biotechnology. In contrast, cultures with a focus on sustainability and harmony may channel technological efforts toward renewable energy and environmental conservation (IEA, 2023; Intergovernmental Panel on Climate Change, 2023).

The evolution of technology is also marked by its capacity to reshape societal structures and norms. Historically, the invention of the printing press revolutionized knowledge dissemination, democratizing access to information and enabling social and political reforms (Eisenstein, 1980). Similarly, the digital revolution has redefined communication, commerce, and governance, creating both opportunities and challenges for modern societies (Castells, 2010; ITU, 2023).

However, technological progress is not without its ethical and societal dilemmas. The same innovations that bring convenience and efficiency can also generate inequality, environmental degradation, and security concerns (Zuboff, 2019; Intergovernmental Panel on Climate Change, 2023). For instance, the rise of automation has improved productivity while simultaneously displacing jobs, leading to complex questions about equity and the future of work (Acemoglu & Restrepo, 2020).

The adaptability of technology also extends to its role in addressing global challenges. Innovations in healthcare, for instance, have enhanced human longevity and quality of life, while breakthroughs in communication technology have fostered unprecedented global connectivity (World Health Organization, 2023; ITU, 2023). Yet, these advancements also necessitate ethical frameworks to ensure their responsible use (UNESCO, 2021; Stilgoe et al., 2013).

Understanding technology as a dynamic and multifaceted concept allows us to appreciate its potential while recognizing its limitations (Arthur, 2009; Feenberg, 2010). It is both a tool for human advancement and a mirror reflecting our values, aspirations, and challenges (Bijker et al., 1987). This dual role makes it an indispensable yet complex element of human existence (Jasanoff, 2016).

As society continues to grapple with the implications of rapid technological change, the importance of fostering a balanced approach becomes evident (Stilgoe et al., 2013). This involves not only leveraging technology to address pressing needs but also considering its broader impacts on humanity and the planet (UNESCO, 2021; Intergovernmental Panel on Climate Change, 2023).

Ultimately, the story of technology is the story of human innovation, resilience, and ambition (Mokyr, 1990). By studying its evolution, we gain insights into our past, understand our present, and equip ourselves to navigate the future (Castells, 2010; Arthur, 2009). Technology, in all its forms, remains a testament to humanity's enduring quest to shape and improve the world (Feenberg, 2010).

THE BIDIRECTIONAL INFLUENCE OF TECHNOLOGY AND SOCIETY

The relationship between technology and society is inherently bidirectional (Misa, 2021; OECD, 2023; UNESCO, 2021). Societal needs and challenges often serve as catalysts for technological innovation. For example, the invention of the printing press in the 15th century arose from the growing demand for the dissemination of knowledge and literacy, revolutionizing education, religion, and communication (Raven, 2020).

Similarly, the industrial machinery of the 18th and 19th centuries emerged in response to the need for greater production efficiency during the Industrial Revolution (Misa, 2021). These innovations were not merely technological advancements; they were solutions to the specific economic and social demands of their time (Misa, 2021; OECD, 2023).

In the modern era, societal needs continue to drive technological development in areas such as renewable energy, healthcare, and artificial intelligence (IEA, 2023; Intergovernmental Panel on Climate Change, 2023; OECD, 2023). For instance, the global push toward sustainability has spurred innovations in solar panels, wind turbines, and electric vehicles, showcasing the alignment of technological progress with societal priorities (IEA, 2023; Intergovernmental Panel on Climate Change, 2023).

Conversely, technology exerts a profound influence on society, often reshaping norms, structures, and relationships (UNESCO, 2021). The internet, for instance, has transformed how people communicate, access information, and engage in commerce (ITU, 2023; Pew Research Center, 2023). Its impact extends to global communication, fostering a more interconnected and interdependent world (ITU, 2023).

The transformative power of technology can also be seen in the way it redefines social norms. Social media platforms have introduced new forms of self-expression and community-building, while also challenging traditional notions of privacy and identity (Pew Research Center, 2023;

Zuboff, 2019). These changes highlight the complex and sometimes unintended consequences of technological innovation (UNESCO, 2021).

The reciprocal nature of this relationship means that technology not only adapts to society but also compels society to adapt (OECD, 2023; UNESCO, 2021). As new technologies emerge, individuals, institutions, and governments are often required to modify their practices, policies, and expectations to align with these advancements (OECD, 2023; WEF, 2023).

An example of this adaptation is the shift in workforce dynamics brought about by automation and artificial intelligence (OECD, 2023; WEF, 2023). While these technologies address efficiency and productivity needs, they also prompt societal debates about employment, skills training, and economic equity (Acemoglu & Restrepo, 2020; OECD, 2023).

The bidirectional influence of technology and society underscores the importance of viewing technological innovation within its broader context (UNESCO, 2021). It is not enough to consider the functionality or efficiency of a technology; we must also examine its societal implications, ethical considerations, and long-term consequences (UNESCO, 2021; OECD, 2022).

Historical case studies provide valuable insights into this dynamic interplay. For instance, the introduction of railroads in the 19th century not only revolutionized transportation but also spurred urbanization, altered economic systems, and changed how people perceived time and distance (Misa, 2021). Similarly, the development of the telephone in the late 19th and early 20th centuries reshaped personal and professional communication, bridging physical distances and fostering a sense of immediacy that continues to influence modern communication technologies (Misa, 2021).

Understanding the bidirectional relationship between technology and society is essential for navigating the challenges and opportunities presented by innovation (UNESCO, 2021). By recognizing how societal needs drive technological development and how technology, in turn, reshapes societal structures, we can adopt a more holistic approach to addressing the complexities of the modern world (OECD, 2023).

This exploration sets the stage for the broader themes of this chapter and the book as a whole, encouraging students to critically engage with the dynamic relationship between technology and society. By applying this understanding to contemporary issues, students will be better equipped to navigate the ethical, social, and cultural dimensions of a tech-driven future (UNESCO, 2021; OECD, 2023).

KEY TECHNOLOGICAL ERAS AND THEIR IMPACTS

Prehistoric and Ancient Eras

The relationship between technology and human society can be traced back to the earliest moments of human history. The development of basic tools during the prehistoric era marks one of the first significant technological advances in human evolution. Stone tools, for instance, enabled early humans to hunt more efficiently, build shelter, and eventually engage in artistic expression (Smithsonian National Museum of Natural History, 2024). The mastery of fire, which allowed humans to cook food and stay warm, marked a significant technological leap that greatly improved survival rates (Smithsonian National Museum of Natural History, 2024). These early innovations were fundamental not just for physical survival but for the formation of early social structures.

As communities grew, early human societies began developing more complex technologies that were essential for group cohesion and growth. One such technological advance was the creation of irrigation systems, which revolutionized agriculture and allowed civilizations to establish themselves in previously inhospitable environments (Food and Agriculture Organization [FAO], 2021). By harnessing water to irrigate crops, societies were able to produce surplus food, which led to the rise of permanent settlements and the establishment of early cities. This allowed for the growth of specialized labor, as individuals could focus on professions other than food production, setting the stage for the development of trade, governance, and cultural development.

Agricultural innovations such as plowing and crop rotation furthered the development of ancient civilizations, enabling populations to grow and thrive. With a steady food supply, societies could focus on expanding their knowledge and capabilities, creating a cycle of innovation and progress. The ancient Egyptians, Mesopotamians, and Indus Valley civilizations, among others, developed advanced systems of writing, construction, and governance that were directly linked to their technological advancements (Roser & Ritchie, 2023). These early technologies laid the foundation for the complex societies we recognize today.

As societies evolved, so too did the sophistication of their technological achievements. The Iron Age brought advancements in metallurgy, allowing for stronger tools, weapons, and infrastructure. These innovations supported the expansion of empires, improved agricultural yields, and strengthened defense mechanisms. Technological progress during this era often dictated the balance of power between civilizations, as those with superior tools and techniques gained a strategic advantage over their neighbors (British Museum, 2022).

The development of transportation technologies was another pivotal moment in human history. From the invention of the wheel to the creation of sea vessels, these innovations enabled humans to travel farther, trade more extensively, and connect disparate societies. The establishment of trade networks not only facilitated the exchange of goods but also ideas, furthering technological and cultural diffusion. For example, the Silk Road served as a conduit for the exchange of technological knowledge between East and West (UNESCO Silk Roads Programme, 2023).

The role of technology in shaping societies became even more pronounced during the Industrial Revolution. This period saw the invention of machines that could perform tasks previously reliant on human or animal labor. The steam engine, for example, revolutionized industries such as manufacturing and transportation, leading to the growth of urban centers and altering the structure of the workforce (Misa, 2021). With industrialization came significant social and economic changes, including the rise of capitalism and the emergence of a distinct working class (Misa, 2021).

While the Industrial Revolution brought unprecedented progress, it also introduced new challenges. The mechanization of labor led to widespread unemployment for artisans, while factory work often involved harsh conditions and exploitation. These ethical dilemmas highlighted the complex interplay between technological advancement and societal well-being, a theme that continues to resonate in contemporary discussions about technology's role in society (International Labour Organization, 2021).

The advent of electricity marked another transformative era in human history. Electricity not only revolutionized industries but also reshaped daily life, making modern conveniences such as lighting, refrigeration, and telecommunications possible. It democratized access to knowledge and services, laying the groundwork for the digital age. With these advancements, societies entered an era of rapid progress and interconnectedness, accelerating the pace of technological and cultural change (International Energy Agency, 2023).

In the 20th century, the rise of digital technology fundamentally altered the way societies function. The development of computers, the internet, and mobile devices created a global digital infrastructure that continues to shape communication, commerce, and governance. These technologies have made the world more interconnected, but they have also introduced complex ethical issues, such as data privacy, cybersecurity, and digital inequality (International Telecommunication Union, 2023; UNESCO, 2021).

The evolution of technology has consistently been a double-edged sword, offering solutions to pressing problems while introducing new challenges. The digital revolution, for example, has bridged geographical divides and enabled global collaboration, but it has also exacerbated disparities in access to resources and information. These tensions underscore the need for thoughtful and ethical approaches to technological development and integration (UNESCO, 2021; OECD, 2023).

Throughout history, technology has also played a crucial role in shaping cultural identity and expression. From the creation of musical instruments to the advent of film and digital media, technological innovations have expanded the ways in which humans can create, share, and preserve culture. These developments have fostered greater understanding and appreciation of diversity, even as they raise questions about cultural homogenization in an increasingly globalized world (Pew Research Center, 2023).

Ultimately, the history of technology is a testament to humanity's ingenuity and resilience. By examining how technological advancements have shaped societies, we gain valuable insights into the forces that drive progress and the challenges that accompany it. This understanding is crucial as we navigate the complexities of the modern technological landscape, ensuring that future innovations align with ethical principles and contribute to the betterment of society (UNESCO, 2021; OECD, 2023).

The Industrial Revolution

The dawn of the Industrial Revolution in the 18th and 19th centuries marked another monumental shift in the relationship between technology and society. The invention of the steam engine, driven by pioneers like James Watt, played a central role in mechanizing labor and transforming industries. The steam engine facilitated the mass production of goods, opening up new possibilities for transportation, energy production, and manufacturing. It allowed factories to operate more efficiently, which in turn led to an explosion of industrialization across Europe and North America (Misa, 2021).

As factories grew in size and complexity, they began to alter the very fabric of society. The mechanization of labor shifted the workforce from agricultural and artisan work to factory-based production. Urbanization accelerated as people moved from rural areas to cities in search of factory work, giving rise to new social dynamics. The growth of cities brought with it both opportunities and challenges, as people sought to adapt to rapidly changing living conditions. The concentration of workers in urban areas also led to the development of labor unions and early social movements aimed at improving working conditions, wages, and workers' rights (ILO, 2021).

The Industrial Revolution also transformed economic systems. It fueled the rise of capitalism, as factory owners and entrepreneurs sought to maximize production and profit. New markets were created for goods, and international trade flourished as transportation methods, such as trains and steamships, facilitated the exchange of products across vast distances. The burgeoning industrial economy encouraged innovation, with new machines, tools, and processes emerging to meet the needs of an expanding global market (Misa, 2021).

However, the Industrial Revolution also gave rise to significant ethical and social debates. Factory labor, often grueling and dangerous, sparked widespread discussions about workers' rights, child labor, and workplace safety. Reform movements began to take shape in response to the exploitation of workers, raising important questions about the ethical implications of labor practices in the face of technological advancement. The development of technologies that enabled industrial efficiency did not automatically ensure the fair and humane treatment of workers, illustrating the need for ethical considerations alongside innovation (ILO, 2021).

The environmental impact of the Industrial Revolution also became evident as societies began to grapple with the unintended consequences of industrialization. The increased use of coal and other fossil fuels led to pollution and environmental degradation, issues that remain relevant today (Intergovernmental Panel on Climate Change, 2023). While the Industrial Revolution heralded unprecedented economic growth and societal progress, it also initiated patterns of resource exploitation and environmental harm that persist in modern industrial practices (IEA, 2023).

Advancements in transportation during the Industrial Revolution revolutionized the movement of people and goods. Railways, steamships, and later the advent of automobiles and airplanes connected distant regions, facilitating trade, migration, and cultural exchange. This newfound mobility transformed the global economy and allowed for the rapid dissemination of ideas, further fueling innovation. However, these advancements also brought about geopolitical tensions, as nations competed for control over transportation networks and the resources necessary to sustain them (Misa, 2021).

One of the most profound impacts of the Industrial Revolution was its influence on communication technologies. The invention of the telegraph, followed by the telephone, radically transformed how information was shared across distances. For the first time, messages could be transmitted almost instantaneously, reshaping commerce, governance, and personal communication. This leap in connectivity laid the groundwork for the modern communication infrastructure we rely on today (Misa, 2021).

The rise of industrialized societies also led to shifts in cultural and social norms. The mechanization of labor and the proliferation of consumer goods created new standards of living and consumption patterns. The concept of leisure time emerged as work schedules became more regimented, giving rise to cultural phenomena such as organized sports, entertainment industries, and tourism. These shifts not only redefined individual lifestyles but also reshaped collective societal values (OECD, 2023).

The educational landscape was similarly transformed by the Industrial Revolution. As societies became more industrialized, the demand for an educated workforce grew. This need gave rise to public education systems aimed at equipping individuals with the skills necessary for factory work and other industrial occupations. The Industrial Revolution thus played a pivotal role in shaping modern education, emphasizing technical and vocational training alongside traditional academic subjects (OECD, 2023).

The Industrial Revolution's technological and societal changes also influenced the arts and literature of the time. Writers and artists grappled with the implications of industrialization, often critiquing its dehumanizing effects while celebrating its transformative power. Works such as Charles Dickens's *Hard Times* and J. M. W. Turner's industrial landscapes captured the complexities of this era, reflecting both its promises and its perils (Misa, 2021).

The legacy of the Industrial Revolution extends to contemporary technological revolutions, such as the digital and green revolutions. The lessons of industrialization—its benefits, challenges, and ethical dilemmas—offer valuable insights for navigating today's technological advancements. By understanding the Industrial Revolution's transformative impact on society, we can better anticipate the societal shifts accompanying current and future innovations (IEA, 2023; Intergovernmental Panel on Climate Change, 2023).

In sum, the Industrial Revolution was a pivotal period that redefined the relationship between technology and society. It brought about unparalleled advancements in production, transportation, and communication, while also raising profound ethical and environmental questions. This era serves as a powerful reminder of the dual nature of technological progress: its ability to drive societal growth and its potential to create new challenges that demand thoughtful and equitable solutions (UNESCO, 2021; OECD, 2023).

The Information Age

The late 20th and early 21st centuries ushered in the Information Age, a period characterized by the rapid growth of digital technologies, including computers, the internet, and digital communication systems. The invention of the personal computer revolutionized how individuals interacted with information, making it possible to store, process, and transmit vast amounts of data at unprecedented speeds. Computers enabled businesses, governments, and individuals to operate more efficiently, setting the stage for global connectivity and digital networks (ITU, 2023).

The rise of the internet further expanded the possibilities of digital communication, allowing individuals to connect with one another in real time, regardless of geographic location. The internet democratized access to information, providing a platform for ideas, creativity, and collaboration. Online education, social media, and e-commerce became integral parts of daily life, reshaping how people communicate, learn, and conduct business. The internet also accelerated globalization by facilitating the rapid exchange of ideas, goods, and services, breaking down traditional barriers of time and space (Pew Research Center, 2023; ITU, 2023).

With the democratization of information came both opportunities and challenges. The ease of accessing information has empowered individuals and communities, but it has also led to concerns about information overload and the spread of misinformation. The rapid growth of digital technologies has raised important ethical questions about data privacy, digital rights, and cybersecurity. As vast amounts of personal data are collected, stored, and analyzed, individuals and organizations are faced with the responsibility of safeguarding that information and protecting against misuse (UNESCO, 2021; World Economic Forum, 2023, 2024).

One of the most pressing ethical issues of the Information Age is the digital divide—the gap between those who have access to technology and those who do not. While many parts of the world have seen rapid technological development, large portions of the global population still lack reliable access to the internet, computers, and other digital tools. This disparity creates a significant social divide, limiting opportunities for education, employment, and social participation for millions of people worldwide (ITU, 2023).

Alongside these concerns, the Information Age has also brought about new debates regarding cybersecurity. With the increasing amount of sensitive personal and financial information stored online, the risk of cyberattacks has become a major concern for individuals, corporations, and governments. Data breaches, identity theft, and the growing prevalence of cybercrime have highlighted the need for robust security systems and policies to protect individuals' rights and maintain the integrity of digital infrastructures (WEF, 2023).

The rise of digital technologies has also raised important ethical questions about the power and responsibility of corporations and governments in shaping the digital landscape. Issues such as surveillance, censorship, and the concentration of data in the hands of a few large tech companies have sparked debates about the ethical implications of technological control and the balance between security and freedom. The increasing influence of technology companies on political and social life has raised concerns about corporate responsibility and the potential for exploitation (UNESCO, 2021; Pew Research Center, 2023).

As the Information Age continues to evolve, the ethical implications of emerging technologies, such as artificial intelligence and biotechnology, will likely take center stage. These innovations hold great promise for addressing global challenges, such as climate change, healthcare, and resource scarcity, but they also come with significant risks and uncertainties. The responsibility of technology creators and users to navigate these challenges ethically will be crucial in shaping the future of society and technology (OECD, 2023; UNESCO, 2021).

In sum, the historical evolution of technology from the prehistoric and ancient eras to the Industrial Revolution and the Information Age has had profound impacts on human society. Each technological advancement has brought about new opportunities for progress, as well as new ethical, social, and economic challenges. As technology continues to advance, understanding its historical context and the reciprocal relationship between technology and society is essential for navigating the complexities of the future (UNESCO, 2021; OECD, 2023).

TRANSFORMATIVE IMPACTS ON HUMAN DEVELOPMENT

Social Structures and Cultural Norms

Technological advancements have continually reshaped the ways individuals and societies communicate, interact, and perceive the world. The invention of the telephone allowed for real-time, long-distance communication, breaking physical barriers that previously limited relationships and business operations. This revolutionized how people maintained personal connections and conducted business, creating a new era of accessibility and immediacy. With the advent of the radio, shared cultural experiences became possible, as communities tuned in to the same broadcasts for news, music, and entertainment. Radio brought the world closer, creating a sense of shared identity and belonging among listeners (Misa, 2021).

Television further expanded this influence by visually connecting audiences to events and stories, solidifying its role as a medium that both reflected and shaped cultural identities. As televisions entered homes worldwide, they became a powerful force for cultural dissemination, allowing audiences to witness historic events, participate in collective moments of joy, and engage with a wide array of narratives. From the moon landing to global sporting events, television created a shared cultural consciousness that bridged national and regional divides (Pew Research Center, 2023).

These earlier technologies not only influenced personal interactions but also drove significant cultural exchanges. Through radio and television, ideas, traditions, and narratives traversed geographical boundaries, fostering a shared understanding of global events. However, they also centralized the dissemination of information, giving unprecedented power to media corporations and governments to shape public opinion. While this centralization brought societies together, it also posed risks of censorship and propaganda, as the control over information became concentrated in the hands of a few (UNESCO, 2021).

In the digital age, social media platforms have amplified these cultural dynamics. Platforms such as Instagram, X (formerly Twitter), and TikTok allow individuals to share personal narratives and cultural artifacts with global audiences, fostering a democratization of expression. Social media has enabled individuals from diverse backgrounds to share their stories, connect across borders, and influence cultural trends. However, this democratization comes with complexities, including the blurring of boundaries between authenticity and performance. Users curate their online personas to align with societal ideals or personal aspirations, influencing how identity is constructed and perceived (Pew Research Center, 2023).

Relationships have also evolved in this digital landscape, as connections once rooted in physical proximity now thrive in virtual spaces. Social media allows individuals to maintain friendships, build professional networks, and even find romantic partners across vast distances. Yet, this convenience is tempered by concerns about superficial interactions, reduced face-to-face communication, and the potential for online relationships to lack depth and authenticity. Despite these challenges, virtual communities have become vital for those who might otherwise feel isolated (OECD, 2023).

Beyond individual identity, social media has played a pivotal role in social movements. It has given marginalized communities tools to voice their concerns, raise awareness, and mobilize for change. Movements like #MeToo and #BlackLivesMatter exemplify how technology can amplify societal issues, fostering collective action and empowering individuals to demand justice and equality. These platforms have demonstrated the power of digital activism to challenge entrenched systems and create lasting societal change (UNESCO, 2021; Pew Research Center, 2023).

However, these platforms are double-edged swords, as they also spread misinformation and perpetuate echo chambers. The algorithms driving social media often prioritize engagement over accuracy, enabling the rapid spread of false information. These echo chambers polarize users by reinforcing preexisting beliefs and limiting exposure to diverse perspectives. This dynamic challenges societal cohesion and the authenticity of cultural exchange, as digital communities risk becoming insular and divisive (Pew Research Center, 2023; OECD, 2023).

In addition to social and cultural implications, advancements in communication technology have transformed industries such as marketing, journalism, and education. Companies leverage social media to reach global audiences, tailoring advertisements to individual preferences with unprecedented precision. Meanwhile, journalism has adapted to the immediacy of digital platforms, delivering breaking news in real time but often struggling to maintain depth and context. Education has also embraced these technologies, with virtual classrooms and online resources expanding access to learning while reshaping traditional pedagogical models (OECD, 2023; UNESCO, 2021).

Despite their challenges, modern communication technologies hold immense potential for fostering global understanding. They provide platforms for cross-cultural dialogue, enabling individuals to learn about diverse perspectives and traditions. For example, online language exchange programs, cultural forums, and virtual collaborations bring people together, fostering empathy and cooperation. These interactions highlight the positive aspects of globalization, where technology serves as a bridge rather than a barrier (UNESCO, 2021).

Looking forward, the future of communication technology will likely involve even greater integration of virtual and augmented reality, artificial intelligence, and immersive experiences. These advancements promise to deepen global connectivity but also raise ethical questions about privacy, accessibility, and the digital divide. Ensuring equitable access to these technologies will be critical to their potential to foster inclusivity and understanding (OECD, 2023; ITU, 2023).

In sum, communication technologies have profoundly influenced human society, from the invention of the telephone to the rise of social media. They have redefined how individuals connect, share, and engage with the world, offering opportunities for progress while presenting new challenges. As these technologies continue to evolve, their impact will depend on how societies navigate the balance between innovation, ethics, and inclusivity (UNESCO, 2021; OECD, 2023).

Economic Shifts

The relationship between technology and economic structures is profound, as innovations have historically driven significant transitions. The shift from agrarian to industrial economies, catalyzed by the Industrial Revolution, introduced machinery that replaced manual labor, increasing productivity and altering the workforce structure. The invention of the steam engine, textile machinery, and other mechanical innovations marked a turning point in manufacturing, allowing industries to scale production at an unprecedented rate. This mechanization not only boosted the economy but also reshaped the role of workers in society. Cities grew as people migrated from rural areas to urban centers for industrial jobs, fundamentally changing societal organization and economic interdependence. Urbanization became synonymous with industrialization, as people sought new opportunities in the growing factories and businesses that characterized this new economic model (Misa, 2021).

This industrial transformation also highlighted inequalities. While machinery replaced manual labor in agriculture and craft industries, many traditional workers found themselves displaced and struggling to find new employment opportunities. The transition to factory work created a class of wage laborers who were often subject to long hours, low pay, and poor working conditions. Simultaneously, the rise of factory owners and entrepreneurs highlighted a growing divide between the wealthy elite and the working poor. The rapid pace of technological and economic change left many workers and policymakers grappling with the implications of this progress. This sparked early debates about labor rights, fair wages, and working conditions, leading to the formation of labor unions and the development of worker protection laws that aimed to address the social consequences of industrialization (ILO, 2021).

In the contemporary era, the transition to knowledge-based economies marks another significant shift. Information technologies, from computers to the internet, have made intellectual capital a primary driver of economic growth. Unlike the industrial economy, which was built on the production of tangible goods, the modern economy thrives on the creation and exchange of information. Industries such as software development, biotechnology, and digital marketing thrive on innovation and adaptability, relying heavily on highly skilled workers who can navigate rapidly changing technological landscapes. The digital revolution has made it possible for businesses to operate on a global scale, with new sectors emerging almost overnight. However, these changes have intensified economic disparities, as those with access to education, advanced skills, and technological resources benefit disproportionately. The so-called "digital divide" has left many behind, particularly in developing countries or areas where access to technology and education is limited (OECD, 2023; ITU, 2023).

Automation has been a disruptive force in the modern economy, particularly in manufacturing and other sectors reliant on repetitive tasks. The rise of robotic systems and artificial intelligence has streamlined operations, reduced costs, and increased efficiency in many industries. However, it has also displaced jobs that once provided stable employment for millions of workers. Many industries, from automotive manufacturing to logistics, have embraced automation to improve productivity and profitability, leaving workers in those fields with fewer opportunities. The displacement of labor through automation has underscored the need for workers to acquire new skills to remain competitive. As a result, the need for reskilling and upskilling has become a central theme in contemporary labor markets. Workers who were once employed in tasks that could be automated now face the challenge of adapting to new roles in an increasingly digital and tech-driven economy (Acemoglu & Restrepo, 2020; WEF, 2023).

At the same time, new technological roles have emerged, such as app developers, data analysts, cybersecurity experts, and artificial intelligence specialists, illustrating how innovation continues to create economic opportunities even as it disrupts established systems. While automation has displaced certain job categories, it has simultaneously generated demand for workers with the expertise to develop, manage, and maintain the technologies driving economic progress. This shift highlights the paradox of technological advancement: while it displaces some workers, it creates new roles that often require specialized knowledge and training. As a result, the workforce has become more segmented, with some individuals reaping the benefits of the digital economy while others struggle to adapt to the changing demands of the labor market (OECD, 2023; WEF, 2023).

The gig economy is another consequence of technological advancement. Platforms like Uber, TaskRabbit, and Etsy have enabled individuals to monetize their skills and assets, offering flexibility and autonomy. These platforms have empowered workers to become entrepreneurs, offering services and products directly to consumers without the need for traditional employers. The gig economy has transformed the way people approach work, enabling them to earn income on their own terms, whether by driving for rideshare services or selling handmade goods online. However, this new economy also raises questions about job security, benefits, and the classification of workers. Many gig workers lack access to health insurance, retirement benefits, and other protections typically associated with full-time employment. The classification of gig workers as independent contractors rather than employees has led to debates about workers' rights and the need for new labor laws to address the realities of this evolving economy (ILO, 2021; OECD, 2023).

The rise of the gig economy has also sparked a broader conversation about the future of work. As technology enables greater flexibility and autonomy, questions arise about the long-term sustainability of gig work as a primary source of income for large segments of the population. While some individuals thrive in the gig economy, others face income instability, lack of job security, and challenges related to work–life balance. The challenge for policymakers is to find ways to ensure that technological innovations benefit workers while also addressing the potential drawbacks of this new economic structure. This may involve creating policies that offer gig workers greater protections, such as access to healthcare, paid leave, and labor rights that are more consistent with traditional employment (OECD, 2023; ILO, 2021).

Moreover, the gig economy has profound implications for the broader economic structure. As more people engage in gig work, traditional industries and business models may face disruption. Large companies that once employed thousands of workers may shift toward more flexible, project-based models, relying on freelance workers rather than permanent staff. This transformation could lead to a more fragmented labor market, where workers have less job security and fewer benefits but more control over their work schedules. The rise of platform-based work may also alter the dynamics of consumer behavior, as individuals increasingly turn to digital platforms to access services and products. As a result, businesses must adapt to this new economic reality by rethinking their employment models and embracing new forms of collaboration and work structures (OECD, 2023).

The impact of technology on economic structures extends beyond just the workforce. Innovations in financial technology (fintech), for example, have transformed the way money is managed, transferred, and invested. The rise of digital payment systems, blockchain, and cryptocurrencies has disrupted traditional banking systems, enabling faster, more efficient financial transactions across borders. These technological advancements have democratized access to financial services, allowing individuals and businesses to participate in global markets in ways that were once impossible. However, they also raise concerns about security, regulation, and the potential for financial instability in an increasingly digital world (Bank for International Settlements [BIS], 2023; International Monetary Fund, 2023).

In conclusion, the relationship between technology and economic structures is complex and multifaceted. Technological innovations have historically driven economic transitions, from the Industrial Revolution to the rise of knowledge-based economies. While technology continues to create new opportunities, it also presents challenges, including job displacement, inequality, and the need for reskilling. The gig economy and digital platforms have redefined work and labor markets, raising important questions about job security and workers' rights. As technology continues to shape economic structures, it will be essential for policymakers to address the evolving needs of workers and businesses, ensuring that technological progress benefits all members of society (OECD, 2023; ILO, 2021).

Governance and Political Systems

Technology has long influenced governance, shaping how governments operate and interact with citizens (UN DESA, 2022; OECD, 2023). Early postal systems enabled rulers to maintain control over vast territories by facilitating communication. Later, innovations such as the printing press revolutionized governance by democratizing access to information (Raven, 2020). For example, during the Protestant Reformation, printed texts allowed reformers to challenge the Catholic Church's authority, catalyzing social and political upheaval. Similarly, revolutionary pamphlets played a critical role in mobilizing the masses during the American and French Revolutions (Raven, 2020).

In the modern era, digital technologies have transformed governance further. E-governance platforms have improved efficiency and transparency, enabling citizens to access services, file taxes, and vote online (UN DESA, 2022; OECD, 2023). Countries like Estonia have pioneered digital governance, offering citizens secure identification systems and comprehensive online access to government services (OECD, 2022). This approach has streamlined bureaucracy, reduced corruption, and increased civic participation (UN DESA, 2022).

However, digital governance also faces significant challenges. Cybersecurity threats, such as hacking and data breaches, pose risks to sensitive information and critical infrastructure (EUAC, 2023; WEF, 2024). Moreover, the digital divide prevents equal access to e-governance benefits, particularly for marginalized communities lacking internet access or digital literacy (ITU, 2023). These issues highlight the importance of equitable and secure technological integration in governance systems (UNESCO, 2021).

Social media's role in modern political movements further exemplifies the complex interplay between technology and governance. Platforms like Twitter and Facebook have been instrumental in organizing protests, spreading revolutionary ideas, and holding governments accountable (Pew Research Center, 2023). The Arab Spring is a notable example, where social media enabled activists to coordinate efforts and share their struggles with the world. Yet, these same platforms have been weaponized for misinformation campaigns and election interference, illustrating their potential to undermine democratic processes (Freedom House, 2023; WEF, 2024).

Historical case studies underscore the transformative power of technology in governance. The printing press demonstrated how access to information could empower citizens and challenge authoritarian regimes (Raven, 2020). Similarly, social media highlights both the opportunities and risks of technological advancements in political systems (Pew Research Center, 2023; Freedom House, 2023). As technology continues to evolve, it is essential to navigate its implications thoughtfully, ensuring that it strengthens governance structures while safeguarding democratic values (UNESCO, 2021).

The interconnectedness of technology with social, economic, and political systems is a defining characteristic of human history. From the cultural shifts brought about by early communication technologies to the economic transformations driven by industrial and digital innovations, technology has continually shaped the trajectory of societies (Misa, 2021; OECD, 2023). Governance, too, has been deeply influenced, with technology acting as both a tool for empowerment and a source of disruption (UN DESA, 2022).

Understanding these historical and contemporary dynamics is crucial for navigating the challenges and opportunities posed by technological advancements. As societies become increasingly reliant on technology, addressing issues such as economic inequality, cultural cohesion, and the ethical implications of innovation will require a collaborative and proactive approach (OECD, 2023; UNESCO, 2021). By critically examining the relationship between technology and human systems, we can strive to create a future that balances progress with equity, ethics, and sustainability.

ETHICAL CONSIDERATIONS IN HISTORICAL CONTEXT

Technological Progress vs. Ethical Concerns

The advancement of technology has consistently brought with it both opportunities and ethical dilemmas (UNESCO, 2021). Throughout history, some of the most transformative innovations have been misused, leading to unintended consequences for society. For example, the development of nuclear technology initially promised a cleaner and more efficient energy source. However, its weaponization during World War II demonstrated the catastrophic potential of technological misuse. The bombings of Hiroshima and Nagasaki highlighted the devastating consequences of prioritizing military advantage over ethical considerations, sparking debates about the morality of technological progress (ICAN, 2023; United Nations Office for Disarmament Affairs, 2022).

Similarly, the Industrial Revolution, while heralded for its advancements in production and efficiency, exploited labor on a massive scale. Factory conditions were often inhumane, with child labor, low wages, and unsafe environments becoming the norm (ILO, 2021). These practices reflected a troubling prioritization of profit over human dignity, underscoring the dangers of unchecked technological adoption. The lack of early regulation allowed these issues to proliferate, offering a cautionary tale about the balance between innovation and social responsibility (ILO, 2021).

Ethical concerns remain central to discussions about technology in the contemporary era. Emerging fields such as artificial intelligence and biotechnology raise profound questions about the boundaries of human intervention (OECD, 2023; UNESCO, 2021). For instance, genetic editing technologies like CRISPR promise breakthroughs in curing diseases but also pose ethical dilemmas regarding "designer babies" and unintended genetic consequences (World Health Organization, 2021; Nature Editorial Board, 2023). The societal values we uphold today will shape how these technologies are adopted, regulated, and integrated into our lives (UNESCO, 2021).

Societal values play a crucial role in guiding the ethical development of technology. Regulations, informed by public discourse and expert input, ensure that advancements align with shared ethical principles (UNESCO, 2021). For example, international agreements on nuclear nonproliferation aim to prevent the misuse of nuclear technology, demonstrating the importance of collective action in addressing global challenges (UNITED NATIONS OFFICE FOR DISARMAMENT AFFAIRS, 2022; ICAN, 2023). Similarly, labor laws established in the aftermath of industrial exploitation have improved working conditions and promoted corporate accountability, offering a framework for navigating the complexities of technological progress (ILO, 2021).

Lessons from History

History offers valuable lessons on the impact of technology and the importance of thoughtful integration into society. By examining past transformations, we gain insights into the dynamics of technological adoption, resistance, and consequences (Misa, 2021). Understanding these patterns can inform decision-making in the present, helping societies mitigate risks while maximizing benefits (UNESCO, 2021).

One of the key takeaways from history is the need for proactive governance. During the Industrial Revolution, the absence of robust labor protections allowed exploitative practices to flourish. It was only through social movements and legislative action that these issues were addressed (ILO, 2021). Similarly, the rise of the internet has revealed gaps in contemporary regulation, such as data privacy and cybersecurity concerns. Learning from past oversights can guide the creation of policies that anticipate challenges rather than reacting to crises (WEF, 2024; OECD, 2023).

Another lesson is the importance of inclusivity in technological adoption. Innovations that serve only a privileged few often exacerbate existing inequalities. For instance, the digital divide illustrates how access to technology is unevenly distributed, with marginalized communities facing barriers to participation in the digital economy (ITU, 2023). Historical examples, such as the uneven benefits of industrialization across regions and classes, underscore the necessity of equitable access to technological advancements (OECD, 2023).

History also highlights the resilience of societies in adapting to change. The transition from agrarian to industrial economies, while disruptive, ultimately led to new opportunities and advancements. Similarly, the Information Age has required individuals and institutions to develop new skills, adapt to changing labor markets, and navigate complex digital environments (OECD, 2023).

By understanding these patterns, we can better prepare for the challenges and opportunities posed by emerging technologies (UNESCO, 2021).

Reflecting on past transformations also fosters a sense of responsibility. Technological progress is not inherently positive or negative; its impact depends on how societies choose to wield it (UNESCO, 2021). As we stand on the cusp of innovations such as artificial intelligence, biotechnology, and renewable energy, history reminds us that thoughtful decision-making, guided by ethical principles and a commitment to the common good, is essential for shaping a future that benefits all (Intergovernmental Panel on Climate Change, 2023; IEA, 2023).

REFLECTION AND ANALYSIS

Case Study Analysis

Case studies provide a compelling way to understand the profound impacts of technological advancements on society. The printing press, developed by Johannes Gutenberg in the 15th century, serves as a transformative example of how a single invention reshaped communication, knowledge dissemination, and cultural development (Raven, 2020). Before the printing press, knowledge was primarily shared through handwritten manuscripts, making information accessible only to a privileged few. With the advent of movable type printing, books became more affordable and widely available, democratizing knowledge and enabling the spread of ideas (Raven, 2020).

The societal impacts of the printing press were multifaceted. On the one hand, it empowered the common people by giving them access to information previously monopolized by elites. This democratization of knowledge fueled the Renaissance, the Protestant Reformation, and the Scientific Revolution, fostering intellectual growth and societal progress. On the other hand, it also created challenges, such as the rapid spread of propaganda and misinformation, which influenced political and religious conflicts (Raven, 2020). The printing press exemplifies how technological innovations can both solve problems and introduce new complexities.

The development of the internet offers another powerful case study. Initially conceived as a tool for academic and military communication, the internet evolved into a global platform that transformed every aspect of life, from commerce to education. The rise of the World Wide Web in the 1990s connected billions of people, creating unprecedented opportunities for collaboration, knowledge sharing, and economic growth. Yet, the internet also introduced challenges such as data breaches, cybercrime, and the proliferation of misinformation, raising questions about security and regulation (Freedom House, 2023; WEF, 2024).

A closer examination of these case studies reveals recurring patterns in the relationship between technology and society. Both the printing press and the internet illustrate the bidirectional influence of societal needs and technological innovation. In the case of the printing press, the growing demand for accessible knowledge drove its development, while the internet was born out of a need for efficient communication in a globalizing world. Once introduced, these technologies reshaped societal structures, creating ripple effects across education, politics, and culture (UNESCO, 2021; OECD, 2023).

By analyzing these historical examples, students gain insights into the complexities of technological progress. These case studies highlight the importance of context, showing how societal values, economic conditions, and cultural dynamics shape and are shaped by technological advancements. Understanding these patterns equips students with a deeper appreciation of the ethical considerations and responsibilities involved in adopting new technologies (UNESCO, 2021).

Discussion Questions

Discussion questions encourage students to critically engage with the themes and concepts presented in this chapter. By reflecting on historical and contemporary examples, students can explore the nuanced interplay between technology, society, and ethics.

How do societal values influence technological innovation? This question prompts students to consider the role of cultural priorities, ethical frameworks, and economic incentives in driving technological progress. For instance, societies that value efficiency and connectivity may prioritize advancements in communication technologies, while those with a strong environmental ethic may invest in renewable energy (IEA, 2023; Intergovernmental Panel on Climate Change, 2023). Exploring this question allows students to examine how societal goals shape technological trajectories.

In what ways have historical technologies both solved problems and created new challenges? This question encourages students to think critically about the dual nature of technological progress. For example, the automobile revolutionized transportation, enabling greater mobility and economic growth. However, it also introduced challenges such as traffic congestion, environmental pollution, and reliance on fossil fuels. By discussing such examples, students can appreciate the complexity of technological innovation and the importance of addressing unintended consequences (UNESCO, 2021; OECD, 2023).

Discussion questions also foster dialogue about the ethical dimensions of technology. For instance, should societies regulate technological advancements to minimize potential harms, or should innovation be allowed to evolve freely? These debates encourage students to consider the balance between fostering innovation and ensuring responsible use, preparing them to navigate the ethical dilemmas of emerging technologies (OECD, 2023; UNESCO, 2021).

Another line of inquiry involves examining the role of technology in perpetuating or alleviating inequalities. Students can explore how historical technologies, such as the industrial machinery of the 19th century, concentrated wealth and power in industrialized nations while marginalizing agrarian economies. Similarly, they can analyze contemporary issues such as the digital divide, which limits access to the benefits of the internet for underserved communities (ITU, 2023; OECD, 2023).

The case study analysis and discussion questions presented in this chapter are designed to deepen students' understanding of the complex relationship between technology and society. By examining historical examples such as the printing press and the internet, students gain insights into the bidirectional influence of societal needs and technological advancements. These case studies illustrate how technologies can solve problems, create opportunities, and introduce new challenges, offering valuable lessons for navigating the complexities of the modern technological landscape (UNESCO, 2021).

The discussion questions encourage students to engage critically with the themes of this chapter, fostering dialogue about the ethical, cultural, and economic dimensions of technology. By reflecting on the influence of societal values, the dual nature of technological progress, and the ethical responsibilities of innovation, students are better equipped to understand the impact of technology on their own lives and communities (OECD, 2023; UNESCO, 2021).

Through these exercises, students develop a nuanced appreciation of the interconnectedness of technology and society. They are encouraged to think critically about how technological advancements shape and are shaped by cultural, economic, and political forces. By exploring these dynamics, students gain a deeper understanding of the ethical considerations involved in technological adoption, preparing them to contribute thoughtfully to discussions about the future of innovation (UNESCO, 2021).

As we continue to grapple with the implications of emerging technologies, the lessons of history and the insights gained from critical inquiry are more relevant than ever. This chapter lays the foundation for a thoughtful and informed exploration of technology's role in shaping society, empowering students to navigate the challenges and opportunities of the 21st century with wisdom and responsibility (OECD, 2023; UNESCO, 2021).

Conclusion

This chapter establishes a foundational understanding of technology as a dynamic and reciprocal force within society. By examining its evolution from rudimentary tools in prehistoric times to the sophisticated digital systems of today, it becomes clear that technology is more than a collection of innovations; it is a driving force that influences how societies grow, function, and interact (Misa, 2021; UN DESA, 2022). This historical lens underscores that technology is not merely a product of human ingenuity but also a reflection of societal needs, priorities, and values. It serves as a mirror of humanity's aspirations, ingenuity, and, at times, its ethical dilemmas (UNESCO, 2021).

Understanding the evolution of technology allows students to see its transformative power, both in solving complex problems and in creating new challenges. From the printing press enabling mass communication to the internet revolutionizing connectivity, these advancements have consistently reshaped economic systems, cultural practices, and governance structures (Raven, 2020; OECD, 2023). However, this transformation has not been without ethical concerns, such as the misuse of technology, the exacerbation of inequalities, and the unforeseen consequences of rapid innovation. By critically analyzing these patterns, students can recognize the importance of balancing technological progress with ethical stewardship (UNESCO, 2021).

This chapter also highlights the bidirectional relationship between technology and society. While technological advancements often emerge to address specific societal needs—such as the steam engine during the Industrial Revolution or the rise of renewable energy in response to climate change—they, in turn, redefine societal structures, cultural norms, and individual behavior (IEA, 2023; Intergovernmental Panel on Climate Change, 2023). This interplay emphasizes that technology does not exist in a vacuum; it is intrinsically linked to the historical, cultural, and economic contexts from which it arises. Such a perspective equips students to better understand how today's innovations are shaped by and will shape the global society they inhabit (OECD, 2023).

The historical perspective provided in this chapter also underscores the iterative nature of technological progress. Each innovation builds upon the successes and failures of its predecessors, creating a continuous cycle of learning and adaptation. This awareness is crucial for fostering a mindset of critical engagement, enabling students to approach new and emerging technologies with both curiosity and caution (UNESCO, 2021). It encourages them to ask not only what a technology can achieve but also at what cost and for whom it will deliver its benefits.

Moreover, the ethical considerations discussed in this chapter set the stage for students to explore their roles as informed citizens, professionals, and leaders in a tech-driven world. The historical case studies, such as the printing press and the internet, illustrate that technological innovation often raises questions about equity, access, and power (Raven, 2020; Freedom House, 2023). By reflecting on these examples, students are challenged to consider their responsibilities in ensuring that technological advancements contribute positively to society and do not perpetuate harm or inequality (UNESCO, 2021).

This chapter also lays the groundwork for a deeper examination of current and emerging technologies in subsequent chapters. By providing a comprehensive historical context, it enables students to connect past lessons to present and future challenges. Topics such as artificial intelligence, automation, and digital governance will be more accessible and meaningful when viewed through the lens of history (OECD, 2023; UN DESA, 2022). This perspective helps students identify recurring themes, such as the tension between innovation and regulation, the interplay of ethics and progress, and the enduring impact of technology on human development.

As we move forward, the knowledge gained from this chapter becomes a critical tool for understanding the complexities of the modern technological landscape. It invites students to engage with questions about how technology can be harnessed to address global challenges, from climate change to social inequality, while also recognizing the unintended consequences that may arise (Intergovernmental Panel on Climate Change, 2023; IEA, 2023). This balanced approach fosters a nuanced understanding of technology as a force that can both unite and divide, empower and exploit, create opportunities and pose risks.

In essence, this chapter equips students with a dual lens: one that looks back to learn from history and one that looks forward to anticipate and shape the future. It encourages a reflective and proactive approach to technology, urging students to think not only about what is possible but also about what is desirable and sustainable. This mindset is essential for navigating the rapidly evolving technological landscape of the 21st century (UNESCO, 2021; OECD, 2023).

The insights gained here serve as a call to action, inspiring students to become thoughtful participants in the ongoing dialogue about technology's role in society. Whether as innovators, policymakers, educators, or consumers, they are reminded that they hold the power to influence how technology is developed and used. By understanding the historical interplay between technology and society, students are better prepared to make informed decisions that prioritize ethical considerations, equity, and the common good (UN DESA, 2022; UNESCO, 2021).

In conclusion, this chapter provides a vital foundation for exploring the broader implications of technology in the modern world. By tracing its historical trajectory and examining its societal impacts, students are empowered to critically evaluate the technologies that shape their lives. This understanding is not only an academic exercise but a practical necessity in a world increasingly defined by technological change. It invites students to be both historians of the past and architects of the future, using the lessons of history to guide the ethical and innovative use of technology in the years to come (OECD, 2023).

CHAPTER 1 ESSAY PROMPTS

Technology and Society Through a Historical Lens

1. **What is Technology?**
   - Discuss the evolution of the concept of technology, from basic tools to complex systems, and explain how technology functions as an adaptive system that addresses human needs and solves problems.

2. **The Interplay Between Technology and Society**
   - Analyze how societal needs have driven technological innovation throughout history. Provide examples of how technological advancements, in turn, have reshaped societal norms and structures.

3. **Prehistoric and Ancient Technological Milestones**
   - Explore how early tools and agricultural innovations contributed to the survival and development of human communities. What lessons can we draw from these early advancements about the relationship between technology and human progress?

4. **The Industrial Revolution and Its Transformative Effects**
   - Examine the role of the Industrial Revolution in shaping urbanization, labor systems, and economic structures. How did this era highlight both the benefits and challenges of technological progress?

5. **Ethical Considerations in the Information Age**
   - Evaluate the ethical dilemmas that have emerged in the digital era, such as data privacy, cybersecurity, and the digital divide. How can these challenges inform responsible technological development moving forward?

6. **Technology's Impact on Social Structures and Cultural Norms**
   - Discuss how inventions like the telephone, radio, and television transformed interpersonal communication and cultural exchange. How have social media platforms amplified or altered these dynamics in contemporary society?

7. **Economic Transformations Through Technology**
   - Trace the transition from agrarian to industrial and knowledge-based economies. How has technological innovation simultaneously created and displaced jobs, and what implications does this have for workforce development today?

8. **Technology's Role in Governance and Political Systems**
   - Analyze how technological advancements have shaped governance structures, using case studies like the printing press and its role in revolutionary movements or social media's influence on modern political campaigns.

9. **Technological Progress vs. Ethical Concerns**
   - Discuss historical examples where technology was misused (e.g., exploitation of labor during industrialization or the development of nuclear weapons). How can societal values guide ethical adoption and regulation of technology?

10. **Lessons from History for the Future**
    - Reflect on how understanding historical technological transformations can help us make informed decisions about emerging technologies. What strategies should society adopt to ensure technology is used for the common good?

REFERENCES

Acemoglu, D., & Restrepo, P. (2020). Robots and jobs: Evidence from US labor markets. *Journal of Political Economy, 128*(6), 2188–2244.

Allen, R. C. (2009). *The British industrial revolution in global perspective.* Oxford University Press.

Arthur, W. B. (2009). *The nature of technology: What it is and how it evolves.* Free Press.

Bank for International Settlements. (2023). *Annual economic report 2023.* BIS.

Bijker, W. E., Hughes, T. P., & Pinch, T. (Eds.). (1987). *The social construction of technological systems.* MIT Press.

boyd, d. (2014). *It's complicated: The social lives of networked teens.* Yale University Press.

British Museum. (2022). Iron Age technologies and society. *British Museum.* https://www.britishmuseum.org/

Castells, M. (2010). *The rise of the network society* (2nd ed.). Wiley-Blackwell.

Diamond, J. (1997). *Guns, germs, and steel: The fates of human societies.* W. W. Norton.

Eisenstein, E. L. (1980). *The printing press as an agent of change.* Cambridge University Press.

European Union Agency for Cybersecurity. (2023). *ENISA threat landscape 2023.* ENISA.

Food and Agriculture Organization. (2021). *The state of the world's land and water resources for food and agriculture 2021.* FAO.

Feenberg, A. (2010). *Between reason and experience: Essays in technology and modernity.* MIT Press.

Floridi, L. (2014). *The ethics of information.* Oxford University Press.

Freedom House. (2023). *Freedom on the Net 2023: The repressive power of artificial intelligence.* Freedom House.

Gordon, R. J. (2016). *The rise and fall of American growth: The U.S. standard of living since the Civil War.* Princeton University Press.

International Campaign to Abolish Nuclear Weapons. (2023). *Nuclear weapons: Risks and humanitarian impacts.* ICAN.

International Energy Agency. (2023). *World energy outlook 2023.* IEA.

International Labour Organization. (2021). *World employment and social outlook: The role of digital labour platforms in transforming the world of work.* ILO.

International Monetary Fund. (2023). *Fintech and the digital economy.* IMF.

International Telecommunication Union. (2023). *Measuring digital development: Facts and figures 2023.* ITU.

Intergovernmental Panel on Climate Change. (2023). *Climate change 2023: Synthesis report.* INTERGOVERNMENTAL PANEL ON CLIMATE CHANGE.

Jasanoff, S. (2016). *The ethics of invention: Technology and the human future.* W. W. Norton.

Jasanoff, S., & Kim, S.-H. (Eds.). (2015). *Dreamscapes of modernity: Sociotechnical imaginaries and the fabrication of power.* University of Chicago Press.

Kranzberg, M. (1986). Technology and history: Kranzberg's laws. *Technology and Culture, 27*(3), 544–560. https://doi.org/10.2307/3105385

Misa, T. J. (2021). *Leonardo to the internet: Technology and culture from the Renaissance to the present* (3rd ed.). Johns Hopkins University Press.

Mitcham, C. (1994). *Thinking through technology: The path between engineering and philosophy.* University of Chicago Press.

Mokyr, J. (1990). *The lever of riches: Technological creativity and economic progress.* Oxford University Press.

Nature Editorial Board. (2023, April 25). The gene-therapy revolution risks stalling if we don't talk about drug pricing [Editorial]. *Nature, 616*, 629–630. https://doi.org/10.1038/d41586-023-01389-z

OECD. (2022). *OECD digital government review of Estonia: Human-centric digital government in practice.* OECD Publishing.

OECD. (2023). *Government at a glance 2023.* OECD Publishing.

OECD. (2023). *OECD employment outlook 2023: Artificial intelligence and the labour market.* OECD Publishing.

Pacey, A. (1983). *The culture of technology.* MIT Press.

Perez, C. (2002). *Technological revolutions and financial capital: The dynamics of bubbles and golden ages.* Edward Elgar.

Pew Research Center. (2023). *Teens, social media and technology 2023.* Pew Research Center.

Raven, J. (Ed.). (2020). *The Oxford illustrated history of the book.* Oxford University Press.

Rogers, E. M. (2003). *Diffusion of innovations* (5th ed.). Free Press.

Roser, M., & Ritchie, H. (2023). Urbanization. *Our World in Data.* https://ourworldindata.org/

Scott, J. C. (2017). *Against the grain: A deep history of the earliest states.* Yale University Press.

Smithsonian National Museum of Natural History. (2024). What does it mean to be human? *Smithsonian Institution.* https://humanorigins.si.edu/

Stilgoe, J., Owen, R., & Macnaghten, P. (2013). Developing a framework for responsible innovation. *Research Policy, 42*(9), 1568–1580. https://doi.org/10.1016/j.respol.2013.05.008

UN DESA. (2022). *United Nations e-government survey 2022: The future of digital government.* United Nations Department of Economic and Social Affairs.

UNESCO. (2021). *Recommendation on the ethics of artificial intelligence.* UNESCO.

UNESCO Silk Roads Programme. (2023). *The Silk Roads: A route of dialogue.* UNESCO.

United Nations Office for Disarmament Affairs. (2022). *Treaty on the Non-Proliferation of Nuclear Weapons (NPT): Review and developments.* UNITED NATIONS OFFICE FOR DISARMAMENT AFFAIRS.

World Economic Forum. (2023). *The future of jobs report 2023.* World Economic Forum.

World Economic Forum. (2024). *Global cybersecurity outlook 2024.* World Economic Forum.

World Health Organization. (2021). *Human genome editing: A WHO governance framework.* WHO.

World Health Organization. (2023). *World health statistics 2023.* WHO.

Zuboff, S. (2019). *The age of surveillance capitalism: The fight for a human future at the new frontier of power.* PublicAffairs.

Technological Innovation and Cultural Shifts

2

ABSTRACT

Chapter 2, *Technological Innovation and Cultural Shifts*, explores the dynamic relationship between technological advancements and cultural transformation. It examines how innovations, from the printing press to the internet, have reshaped societal norms, communication practices, and social structures. The chapter discusses the profound impacts of technology on family dynamics, identity formation, and global connectivity, while also addressing the challenges of inequality and cultural homogenization that arise from technological progress. Through a critical lens, it highlights both the opportunities and complexities technology presents in influencing cultural evolution, and how these shifts shape contemporary life and societal organization.

INTRODUCTION

Technological innovations are powerful agents of change that shape not only the way we interact with the world but also how we perceive and organize society. Over the course of history, advances in technology have fundamentally altered cultural practices, societal norms, and the everyday lives of individuals. The link between technological progress and cultural transformation is often complex, with innovations both influencing and being influenced by the cultural contexts in which they emerge. This chapter explores how technological advancements have contributed to cultural shifts, focusing on how innovations reshape societal behaviors, beliefs, and practices (Castells, 2010).

The earliest examples of technological innovations, such as the invention of the wheel and the development of simple tools, had profound effects on the way early human societies organized themselves. These basic tools allowed humans to interact with their environment in new ways, from hunting and gathering to building shelters and creating art. As civilizations developed, more sophisticated technologies emerged, such as the plow, which revolutionized agriculture and allowed societies to grow in size and complexity. The cultural shift brought about by these innovations was significant, as humans moved from nomadic lifestyles to settled agricultural communities, leading to the formation of villages, cities, and the rise of social hierarchies (Encyclopaedia Britannica, 2025a, 2025b).

In the ancient world, technologies like writing and the printing press played crucial roles in shaping cultural and intellectual developments. The advent of writing systems allowed societies to record information, communicate across distances, and develop legal and administrative systems. Writing also facilitated the growth of literature and philosophy, which had lasting impacts on culture. The printing press, invented by Johannes Gutenberg in the 15th century, was a technological breakthrough that revolutionized the dissemination of knowledge. It made books more accessible, led to the spread of new ideas, and played a key role in the Protestant Reformation and the Renaissance. These shifts in the accessibility and spread of knowledge fundamentally altered the way individuals and societies understood the world (Eisenstein, 1979; Encyclopaedia Britannica, 2025c).

The Industrial Revolution, which began in the late 18th century, marked another turning point in the relationship between technology and culture. Innovations such as the steam engine, mechanized looms, and the development of factory systems transformed not only economies but also the social fabric of society. As industrialization spread, rural communities began to shrink, and urban centers grew rapidly as people migrated to cities in search of factory work. This shift from agrarian to industrial economies fundamentally altered social dynamics, as traditional rural lifestyles gave way to urbanization, and class structures were redefined. The rise of industrialization also brought about a new emphasis on productivity, efficiency, and technological progress, leading to cultural shifts that prioritized economic growth over individual well-being and family life (Encyclopaedia Britannica, 2025d).

The development of transportation technologies, such as the steamship, train, and automobile, further accelerated these changes, making it easier for people, goods, and ideas to move across vast distances. This created a more interconnected world, where cultural exchanges occurred more rapidly, and new ideas were more likely to spread and influence societies. As people moved to cities for work, traditional family structures began to shift as well. Women, for example, began entering the workforce in larger numbers, and the concept of the "nuclear family" started to take shape. This shift in familial and gender roles reflects a broader cultural transformation influenced by technological advancements (Encyclopaedia Britannica, 2025d).

In the 20th century, the rise of mass media technologies such as radio, television, and film further accelerated the cultural transformations brought about by earlier technological innovations. These technologies brought entertainment, news, and advertisements into homes around the world, shaping cultural norms and expectations. For instance, television programs and advertisements influenced consumer behavior, and new forms of mass entertainment helped to shape public opinions about social issues. The media also played a significant role in the formation of national identities, as people from different regions and backgrounds began to share similar cultural experiences through popular media (Encyclopaedia Britannica, 2025e, 2025f).

The digital revolution of the late 20th and early 21st centuries has perhaps brought about the most profound cultural shifts of all. The internet, mobile phones, and social media have radically transformed the way people communicate, access information, and interact with one another. Social media platforms like Facebook, Instagram, Twitter, and TikTok have reshaped cultural norms around identity, self-presentation, and interpersonal relationships. The ability to connect with others instantaneously across the globe has created a more interconnected world, but it has also raised questions about privacy, authenticity, and the impact of digital culture on mental health (Baym, 2015; boyd, 2014; Pew Research Center, 2025).

One of the most significant cultural shifts brought about by digital technology is the changing nature of identity. In the past, individuals' identities were largely shaped by their immediate social circles and local communities. Today, digital technologies allow people to curate and project their identities to a global audience. Social media platforms provide individuals with the ability to present idealized versions of themselves, often blurring the line between reality and online personas. This has led to new cultural norms around self-presentation, social validation, and the constant pressure to maintain an idealized image. The rise of influencer culture, where individuals gain fame and financial success through their online presence, has also influenced how people view success and status in contemporary society (Abidin, 2018; Pew Research Center, 2025).

Cultural shifts have also been influenced by the ways in which technology has changed how people consume and interact with media. The shift from traditional forms of media consumption, such as reading newspapers and watching scheduled television programs, to on-demand streaming and social media content has transformed how people access information and entertainment. This has led to changes in attention spans, with shorter, more fragmented forms of content gaining popularity. The rise of "clickbait" and sensationalized headlines is an example of how the demands of digital media have influenced journalistic practices and shaped public perception of news (Newman et al., 2025).

Moreover, the accessibility of information in the digital age has raised important questions about the authenticity of cultural narratives. In the past, information was often filtered through traditional institutions such as newspapers, universities, and government bodies. Today, anyone with an internet connection can access and share information, which has democratized the creation and dissemination of cultural narratives. However, this democratization of information has also led to the proliferation of misinformation and "fake news," challenging the authenticity and trustworthiness of cultural discourses. The impact of this shift on public trust, political polarization, and social cohesion remains a critical area of study in the context of technological advancements (Vosoughi et al., 2018; Newman et al., 2025).

The relationship between technology and culture is not one-sided, however. Cultural values and societal norms also influence the development and adoption of new technologies. In some cases, technologies emerge in response to cultural needs or challenges. For example, the development of renewable energy technologies has been driven in part by cultural concerns about environmental sustainability and the desire to mitigate the effects of climate change. Similarly, cultural attitudes toward privacy, security, and personal freedom shape the design and regulation of digital technologies. The ongoing debates around data privacy, government surveillance, and the ethical implications of artificial intelligence reflect the ways in which societal values inform technological innovation (International Energy Agency, 2024; European Union, 2024).

As technology continues to advance, it will undoubtedly continue to shape and be shaped by cultural shifts. The rise of artificial intelligence, biotechnology, and advanced robotics presents both opportunities and challenges for societies worldwide. These technologies have the potential to revolutionize industries, improve quality of life, and address some of humanity's most pressing challenges. However, they also raise complex ethical questions about the role of technology in human life, the distribution of power, and the potential risks associated with new innovations (European Union, 2024).

In conclusion, technological innovations have played a fundamental role in shaping cultural shifts throughout history. From the early tools of prehistoric humans to the digital technologies of the 21st century, each advancement has influenced societal norms, behaviors, and identities. As technology continues to evolve, its impact on culture will only become more profound, raising new questions about ethics, privacy, and the future of human society. Understanding the relationship between technology and cultural shifts is essential for navigating the complexities of the modern world and preparing for the challenges and opportunities that lie ahead (Castells, 2010).

From printing press to platforms: how innovation reshapes culture, identity, and everyday life

Read through a culture-and-identity lens, the timeline shows how each communication leap—print, broadcast, internet, platforms—widens who can speak while subtly steering what gets heard. It highlights a double bind: more connection and creative self-expression alongside algorithmic gatekeeping, data surveillance, and uneven access. The practical prompt beneath the image is this: how do we widen digital opportunity and protect community voice while steering platforms toward public value?

THE ROLE OF TECHNOLOGY IN CULTURAL EVOLUTION

Technology does not exist in a vacuum; it is intertwined with the cultural environment in which it is developed. Each technological advancement—whether in communication, transportation, or manufacturing—reflects the values, priorities, and needs of the society that created it. At the same time, these technologies influence how cultures evolve, changing everything from interpersonal relationships to global interactions. Technology and culture are inseparable, each shaping and transforming the other in a dynamic, reciprocal relationship (Castells, 2010).

For example, the printing press, invented by Johannes Gutenberg in the 15th century, revolutionized the way knowledge was disseminated. Before its advent, books were rare and expensive, typically hand-copied by monks. The printing press made books more accessible, leading to an explosion of knowledge that fueled the Renaissance, the Enlightenment, and the scientific revolution. This shift in access to information altered social norms and hierarchies, empowering the middle class and allowing for more widespread education. The printed word created a new cultural landscape where ideas could be spread quickly and widely, fostering an environment for debate, collaboration, and cultural transformation (Eisenstein, 1979; Encyclopaedia Britannica, 2025c).

This democratization of knowledge also played a pivotal role in the rise of new political and religious movements, such as the Protestant Reformation. The ability to distribute pamphlets, books, and translations of the Bible allowed ideas to circulate outside of religious and political elites, challenging established authorities. The printing press facilitated the growth of literacy, shifting cultural practices around reading, education, and intellectual engagement. In many ways, the printing press was the technological catalyst for the modern world, sparking a cultural transformation that reverberated across centuries (Eisenstein, 1979).

In the modern era, the advent of the internet has had a similarly profound effect on culture, democratizing information on an unprecedented scale. Social media platforms, blogs, and websites have replaced traditional forms of communication and have transformed the way people access news, share opinions, and participate in public discourse. This shift has reshaped cultural practices, from how individuals form their identities to how communities interact and organize around shared interests and causes. Social media platforms like Facebook, Twitter, and Instagram allow individuals to curate their lives, broadcast their thoughts, and engage in virtual communities, all of which have contributed to the evolution of cultural norms, especially around communication and self-expression (Baym, 2015; boyd, 2014).

The rise of the internet has also transformed how individuals form and express their identities. The ability to engage in virtual environments allows people to create online personas that may differ from their offline selves. This shift has given rise to new cultural norms around self-presentation, often blurring the boundaries between private and public life. The pressure to maintain a curated, idealized version of oneself online has contributed to changing ideas around authenticity, social validation, and personal image. Social media influencers, who have gained large followings by sharing aspects of their personal lives and opinions, have become central figures in shaping modern cultural trends, further altering the nature of celebrity and influence (Abidin, 2018; Pew Research Center, 2025).

On a larger scale, the internet has reshaped cultural interactions between communities and nations. It has enabled people to connect across borders, fostering global conversations about politics, culture, and social issues. Movements such as #MeToo, Black Lives Matter, and environmental activism have been propelled by social media, demonstrating how technology can give marginalized communities a platform to voice their concerns and mobilize for change. Technology has played an essential role in reshaping cultural understandings of power, identity, and social justice, offering new ways for individuals and groups to advocate for their rights (Newman et al., 2025).

The internet also influences the globalization of culture, allowing for the rapid spread of cultural products such as music, film, fashion, and cuisine. Platforms like YouTube, Spotify, and Netflix have made cultural content from around the world easily accessible to a global audience. This has led to a blending of cultural influences, where individuals are exposed to a wide array of global perspectives and practices. While this can foster understanding and appreciation of diverse cultures, it can also lead to the homogenization of culture, where dominant global trends overshadow local traditions and identities (Nielsen, 2025; Ofcom, 2025).

As technology continues to advance, new forms of communication and cultural exchange are emerging. Virtual reality (VR) and augmented reality (AR), for instance, are beginning to transform the way people experience the world around them. VR allows for immersive experiences in virtual environments, from entertainment and gaming to education and therapy. AR, on the other hand, overlays digital information onto the physical world, enhancing how people interact with their surroundings. These technologies are still in their early stages but are already starting to influence cultural practices, such as how people learn, shop, and socialize (IDC, 2025).

The evolution of transportation technologies has also had significant cultural impacts. The development of the steam engine in the 18th century revolutionized travel, shrinking distances and making it easier for people to explore new places. The advent of automobiles and airplanes further accelerated these changes, fostering mobility and reshaping ideas around space, time, and personal freedom. As travel became more accessible, cultures began to mix in ways that had not been possible before. Globalization of trade and tourism allowed for new cultural exchanges, leading to the diffusion of ideas, goods, and traditions across borders (Encyclopaedia Britannica, 2025d).

The widespread adoption of smartphones has similarly influenced cultural behaviors, particularly around communication. Smartphones have transformed how people stay connected, from texting and calling to instant messaging and video chats. The constant availability of communication through mobile devices has created a cultural expectation of immediate responses, altering norms around time, attention, and personal boundaries. Smartphones have also given rise to new forms of entertainment, such as mobile gaming and social media, which have created new cultural industries and impacted how people spend their leisure time (American Psychological Association, 2023; U.S. Department of Health and Human Services, 2023).

The growth of artificial intelligence (AI) and machine learning technologies is beginning to shape culture in profound ways as well. AI is increasingly being integrated into everyday life, from personal assistants like Siri and Alexa to AI-driven recommendations on streaming platforms. These technologies are changing how people interact with information, entertainment, and services, and may even reshape cultural norms around work, creativity, and privacy. AI's potential to automate tasks and make decisions has raised important questions about its ethical implications, particularly regarding labor displacement, privacy concerns, and bias in algorithms. As these technologies continue to develop, they will likely have far-reaching impacts on cultural practices and societal norms (European Union, 2024; Zuboff, 2019).

In addition to the positive impacts of technology, there are significant challenges that accompany its rapid advancement. One major issue is the digital divide, which refers to the gap between those who have access to technology and those who do not. This divide can exacerbate existing social inequalities, limiting opportunities for education, employment, and civic participation for marginalized groups. The unequal distribution of technological resources has cultural implications, as access to technology increasingly shapes individuals' ability to participate in the digital economy and engage in global cultural exchanges (International Telecommunication Union, 2024).

Furthermore, the rise of surveillance technologies has introduced new challenges to privacy and personal freedoms. As more aspects of life move online, individuals are increasingly subject to surveillance, both by governments and private companies. This has raised important cultural questions about the balance between security and privacy, freedom and control. The widespread collection of personal data through online activities has transformed how people view their relationship to technology, with increasing concerns about the implications of data breaches, identity theft, and surveillance capitalism (Zuboff, 2019; European Union, 2024).

The relationship between technology and culture is also evident in the ways new technologies shape creative practices. Digital technologies have transformed art, music, and literature, providing new tools for creation and distribution. Artists can now create music using digital instruments and software, photographers can edit their images on computers, and writers can publish their works online without needing traditional publishers. These technological innovations have made it easier for people to express themselves and share their creativity with a global audience, fostering a more inclusive and diverse cultural landscape. However, the ease of digital reproduction also raises questions about intellectual property and the value of creative work in a digital age (Baym, 2015).

As technology continues to evolve, its impact on culture will only intensify. Innovations in robotics, biotechnology, and space exploration have the potential to transform society in ways that are difficult to predict. However, as these technologies develop, they will continue to shape the values, priorities, and social structures of the societies that create and adopt them. The interplay between technology and culture is complex, as each influences the other in a dynamic process that continuously reshapes the world we live in (Castells, 2010).

In conclusion, technological innovations are not just tools that serve practical purposes; they are powerful forces that shape and are shaped by cultural practices. From the printing press to the internet, from smartphones to artificial intelligence, each technological advancement has left its mark on society, transforming how people communicate, work, and interact with the world. These changes are not always linear or predictable, but they are fundamental to understanding the cultural shifts that define the modern era. As technology continues to advance, its relationship with culture will remain a central area of exploration for scholars, policymakers, and individuals alike, as it offers both opportunities and challenges in shaping the future of human society (Eisenstein, 1979; Castells, 2010).

TECHNOLOGICAL INNOVATION AND CHANGING SOCIAL NORMS

As technology evolves, it has a profound impact on societal norms, particularly with respect to communication, work, and leisure. The way people relate to each other, work, and live is constantly reshaped by new innovations. For instance, the rise of the smartphone has radically changed communication norms. Where once people communicated through face-to-face interactions or letters, the smartphone has ushered in an era of instantaneous digital communication. Social norms surrounding availability and response time have shifted dramatically, with expectations for immediate replies via text, email, or social media messages becoming the standard. The immediacy of digital communication has altered how people manage their time, expectations of personal availability, and the speed at which decisions are made, affecting both personal and professional relationships (American Psychological Association, 2023; U.S. Department of Health and Human Services, 2023).

Smartphones have not only accelerated the pace of communication but also transformed how people maintain social connections. Social media platforms have become central to social interaction, allowing individuals to maintain relationships, engage with friends and family, and build communities online. These platforms have reshaped norms around privacy, self-presentation, and connection, as individuals carefully curate their online personas and share aspects of their lives with a global audience. Additionally, the rise of messaging apps like WhatsApp, Telegram, and WeChat has facilitated communication across borders, allowing for instant exchanges of information and ideas, which has contributed to the globalization of communication (Baym, 2015; boyd, 2014).

The workplace has also undergone significant changes due to technological innovations. The industrial revolution's mechanization of labor shifted workers from home-based or agricultural work to factory-based labor. This marked a shift in societal norms around the nature of work, as labor became more regimented and time-bound. Workers were required to adapt to rigid schedules and standardized processes, which in turn influenced the way societies viewed time and productivity. The factory-based model became the dominant structure for work in the industrial era, leading to a division between personal and professional life that emphasized punctuality, uniformity, and adherence to schedules (Encyclopaedia Britannica, 2025d).

In the present day, the advent of telecommuting and the gig economy, fueled by platforms like Zoom, Uber, and Etsy, has shifted the norm toward more flexible and self-directed forms of work. What was once seen as the standard of work—fixed hours in a physical location—has been replaced with new expectations for work-life balance, autonomy, and online collaboration. Technology has enabled workers to engage in remote work, allowing for more control over their schedules and locations. This shift has also led to the rise of freelance and gig work, where individuals are hired for specific tasks or short-term projects rather than traditional full-time employment. The gig economy has empowered workers to seek alternative ways of earning income and to build careers based on their own skills, preferences, and interests (International Labour Organization, 2024).

Technology's impact on leisure and entertainment has also been profound. The rise of video games, streaming services, and social media has shifted cultural practices around entertainment. In the past, people often engaged in community-based recreational activities, such as attending live theater performances or participating in local sports leagues. Today, much of this has been replaced by digital entertainment options that transcend physical spaces. Streaming platforms like Netflix, YouTube, and Spotify have changed how people consume content, allowing for on-demand access to music, television, movies, and podcasts. This shift has led to new social practices, such as binge-watching TV shows, and has fostered global communities around shared media consumption (Nielsen, 2025; Ofcom, 2025).

The accessibility of entertainment has also led to significant cultural shifts in how people spend their leisure time. With streaming services offering an endless catalog of films, documentaries, and series, the traditional rhythm of TV programming—tied to specific airing times—has been replaced by the ability to watch content at any time. This shift has not only transformed individual viewing habits but has also changed family and group dynamics around entertainment, as shared experiences of watching a particular show can create new kinds of social bonding and discussion. Additionally, video games have become a dominant form of entertainment, contributing to the growth of e-sports, streaming platforms like Twitch, and online multiplayer games that enable global interaction (Newman et al., 2025).

In contrast to the once-static nature of leisure activities, technology now allows for dynamic, personalized, and interactive experiences. For example, the growth of virtual reality (VR) and augmented reality (AR) technologies has introduced new forms of entertainment and recreation that blend the digital and physical worlds. VR games provide immersive environments where players can engage in highly interactive experiences, while AR enhances the real world with digital overlays, allowing users to interact with both physical and virtual elements simultaneously. These technologies are changing the way people engage with entertainment, creating new possibilities for personal involvement and transforming cultural perceptions of what constitutes leisure (IDC, 2025).

Television and film have also been deeply impacted by digital technologies. The shift from traditional cable and satellite TV to online streaming has changed both the production and consumption of media. Content creators now have more freedom to produce niche content for specific audiences, and viewers can access a wider array of global media than ever before. This democratization of content creation has opened the door for independent filmmakers and smaller creators to gain visibility, which has led to a diversification of the media landscape. The cultural norm around watching TV shows has evolved from scheduled broadcasts to on-demand viewing, where individuals can watch what they want, when they want, and often on any device (Ofcom, 2025).

In the workplace, the impact of technology extends beyond communication and flexibility to include automation and artificial intelligence (AI). The development of AI and machine learning has started to reshape entire industries, automating tasks that were once performed by humans. This has had a profound effect on labor markets, as many traditional jobs are being replaced by machines or software programs. While this has led to greater efficiency and reduced labor costs in certain sectors, it has also raised questions about job displacement, the future of work, and the skills required for future employees (European Union, 2024; International Labour Organization, 2024).

The rise of AI has sparked debates about the ethical implications of technology and its role in shaping cultural norms around work. Automation may alleviate human workers from repetitive or dangerous tasks, but it also has the potential to widen the income gap and create new forms of inequality. The cultural shifts surrounding work are no longer just about job types but also about how people define their relationship to work. As AI and automation redefine traditional job structures, new norms are emerging about what it means to be employed, productive, or successful (European Union, 2024).

Technology is also reshaping education, offering new opportunities for learning and cultural exchange. Online education platforms, digital textbooks, and collaborative tools like Google Docs have made learning more accessible and flexible. This shift has allowed for a greater democratization of education, enabling students from different parts of the world to learn from the same resources and educators. Online education has not only changed how and where people learn but has also transformed cultural attitudes about the value of education, making it possible for individuals to pursue lifelong learning and adapt to changing job markets (Baym, 2015).

As education becomes more digitized, it raises questions about the digital divide and the challenges of ensuring equitable access to technology. The global spread of educational tools has the potential to equalize educational opportunities, but disparities in access to technology—whether due to geography, income, or infrastructure—can exacerbate existing social inequalities. The cultural implications of these disparities are profound, as they limit opportunities for certain populations to participate fully in the digital age, perpetuating cycles of poverty and marginalization (International Telecommunication Union, 2024).

While technology continues to advance and shape societal norms, its impact on leisure, work, and communication also presents new challenges. One of the most pressing concerns is the erosion of privacy. As individuals engage more deeply with digital platforms, they leave behind trails of personal data, which can be used by corporations and governments to monitor behaviors, predict preferences, and even manipulate decisions. This shift has led to cultural shifts regarding personal autonomy and security, with growing concerns about surveillance and data collection (Zuboff, 2019; European Union, 2024).

Social media platforms and digital technologies have also contributed to changing societal norms around identity and self-representation. Online platforms have become spaces where individuals craft and perform their identities, often curated to present an idealized version of their lives. The pressure to present a perfect image can lead to mental health challenges, as people struggle to live up to the expectations set by their online personas. This new form of self-presentation has also influenced how individuals interact with others, as social media platforms can both connect people across vast distances and create feelings of isolation and alienation when individuals feel excluded from the virtual communities they engage with (American Psychological Association, 2023; U.S. Department of Health and Human Services, 2023).

Moreover, the nature of work itself is evolving in the digital age, where traditional career paths and job stability are being replaced by the fluidity of gig work, freelancing, and remote jobs. These changes reflect the growing influence of technology on labor practices, but they also challenge cultural norms regarding career success and job security. For many individuals, the rise of the gig economy has led to an increase in flexibility and control over their work schedules, but it has also created a culture of instability and precarity, where job benefits, healthcare, and retirement plans are no longer guaranteed (International Labour Organization, 2024).

As technology continues to advance, its impact on leisure, work, and communication will continue to unfold in complex and interconnected ways. These changes present both opportunities and challenges for individuals and societies, forcing cultural norms to adapt to the ever-evolving technological landscape. The profound shifts in societal behavior and practices, influenced by ongoing technological innovations, will continue to shape the future of culture, presenting new ways of connecting, working, and living (Newman et al., 2025).

THE IMPACT OF TECHNOLOGICAL INNOVATIONS ON FAMILY AND SOCIAL STRUCTURES

One of the most significant areas affected by technological advancement is the structure and dynamics of family and social life. Historically, technological innovations like the telephone, television, and later the internet, have influenced how families interact, how individuals form relationships, and how communities are structured. These technologies have acted as catalysts for change in how people connect, communicate, and relate to one another, transforming the very fabric of social life (Encyclopaedia Britannica, 2025e, 2025f).

The advent of the telephone, for example, altered family communication dynamics. Prior to its invention, families maintained close, localized relationships with their neighbors and extended families. Social interactions were typically face-to-face, and long-distance relationships were limited to written correspondence. With the introduction of the telephone, families could now communicate instantly over vast distances, breaking down geographical barriers. The telephone allowed for immediate communication, enabling family members and friends to stay in touch regardless of where they were located. This change not only helped maintain personal relationships but also allowed families to stay connected during times of crisis, illness, or joy, providing a sense of continuity and emotional support (Encyclopaedia Britannica, 2025f).

The telephone also paved the way for greater mobility, as people could move away from their hometowns or families without the fear of losing touch. As people moved to different cities or countries for work or education, the ability to call home whenever they wished kept them connected to their roots. This shift played a major role in the rise of urbanization, where families became more dispersed, and individuals created new social circles in different regions, sometimes entirely detached from their original family units (Encyclopaedia Britannica, 2025d).

The internet has since taken this to a new level. Online communication platforms such as email, instant messaging, and social media have reshaped family structures, especially in the context of globalization. Families that were once separated by vast distances can now maintain constant contact through video calls, group chats, and social media updates. This constant connectivity allows families to share life moments in real-time, from birthdays to holidays to everyday happenings, fostering a sense of closeness despite physical distance. Social media platforms, like Facebook and Instagram, have become digital spaces where family members share photos, updates, and milestones, which can be instantly viewed by relatives across the world (Baym, 2015; boyd, 2014).

This transformation of family communication, however, comes with its own set of challenges. While technology has made it easier to stay connected, it has also raised questions about the quality of relationships in a digital age. The rise of instant communication, though convenient, often leads to shallow, less meaningful interactions compared to in-person conversations. Text-based messages, for example, lack the nuances of face-to-face conversations, such as tone of voice and body language, which are essential for building deep emotional connections. The constant stream of digital interactions can lead to misunderstandings, misinterpretations, and a diminished sense of intimacy in relationships (American Psychological Association, 2023).

Moreover, the increase in virtual relationships over physical ones has led to concerns about social isolation. People, especially younger generations, are spending more time online and less time engaging in face-to-face activities. In some cases, virtual relationships may replace in-person interactions, creating an illusion of connection without the emotional depth that comes from real-world socializing. This shift has sparked debates about the long-term effects of digital communication on mental health and emotional well-being, with some experts warning about the risks of loneliness and disconnection in a world that is increasingly reliant on technology for social interaction (U.S. Department of Health and Human Services, 2023).

Social media platforms, while allowing for enhanced connectivity, have also altered the nature of family dynamics. Social networks often prioritize curated, idealized versions of family life, which can lead to unrealistic expectations and comparisons. Families might feel pressure to present a picture-perfect version of their lives, resulting in increased stress and anxiety. The constant comparison to others on social media can undermine feelings of contentment and self-worth, particularly for young people who are more susceptible to external validation. This phenomenon, known as social comparison, has raised concerns about its negative effects on self-esteem and mental health (American Psychological Association, 2023).

Furthermore, technology has influenced broader social structures, including how communities are formed and sustained. Online communities, forums, and social media groups have replaced traditional community-based organizations like local clubs or neighborhood associations. These digital communities offer a space for individuals to connect with others who share similar interests, beliefs, or experiences, no matter where they are located geographically. The ability to join digital communities around specific hobbies, causes, or professional interests has expanded the notion of community, allowing people to find support, guidance, and a sense of belonging online (Newman et al., 2025).

While online communities can foster a sense of belonging, they also bring new challenges related to privacy, misinformation, and the commercialization of personal data. Unlike traditional communities, where members interact face-to-face and can hold each other accountable, digital spaces are often less regulated and can become breeding grounds for harmful behavior, such as cyberbullying, trolling, and harassment. In some cases, the anonymity provided by the internet enables individuals to behave in ways they would not in real life, leading to negative consequences for community members (Vosoughi et al., 2018).

Moreover, the ease with which information is shared online has led to the rapid spread of misinformation and disinformation. Social media platforms, news websites, and blogs often circulate inaccurate or misleading information, which can shape public opinions and influence political outcomes. The rise of "fake news" has prompted discussions about the need for digital literacy and critical thinking skills to help individuals navigate the complexities of online information and distinguish between fact and fiction. Misinformation can also contribute to the polarization of society, as people gravitate toward echo chambers where their existing beliefs are reinforced, further dividing communities along ideological lines (Newman et al., 2025; Vosoughi et al., 2018).

Another significant concern with digital communities is the commercialization of personal data. Social media platforms and online services collect vast amounts of data on their users, including personal preferences, browsing habits, and interactions. This data is then used for targeted advertising, influencing consumer behavior and shaping individuals' online experiences. The commodification of personal data raises ethical questions about privacy and consent, as individuals often provide personal information without fully understanding how it will be used. The increasing monetization of online interactions has led to a growing sense of distrust toward digital platforms, with many users becoming more aware of how their data is being exploited for profit (Zuboff, 2019).

Despite these challenges, technology has also enabled the creation of new types of community-based organizations, such as crowdfunding platforms, social enterprises, and activist groups. These digital initiatives have empowered individuals to come together for causes they care about, whether it's raising money for a charitable project, advocating for social justice, or mobilizing for political change. Technology has allowed for the democratization of community-building, enabling people to collaborate across borders and take collective action on a global scale. This new form of community-building has expanded the possibilities for social activism and solidarity in ways that were not previously possible (Newman et al., 2025).

In addition to transforming communication and community, technology has also had a significant impact on work-life balance. The ability to work remotely, thanks to advances in communication tools like email, video conferencing, and collaborative software, has led to a more flexible approach to work. While this has allowed people to better manage their time and integrate work with personal life, it has also blurred the boundaries between the two. The expectation that workers will be available 24/7 due to the connectivity enabled by technology has led to concerns about burnout and the erosion of personal time (International Labour Organization, 2024).

The rise of the gig economy, facilitated by platforms like Uber, Lyft, and Etsy, has further complicated the concept of work-life balance. Gig workers have greater autonomy over their schedules, but they also face insecurity in terms of income and benefits. The lack of traditional employment protections, such as healthcare and retirement plans, leaves gig workers vulnerable to financial instability. This shift has raised questions about the future of work and the need for policies that support workers in non-traditional employment arrangements (International Labour Organization, 2024).

The impact of technology on social and family dynamics also extends to how people engage in leisure activities. The rise of online gaming, streaming services, and social media has redefined how people spend their free time. While digital entertainment options have created new ways for people to relax and have fun, they have also contributed to a decline in physical activity and outdoor recreation. The sedentary nature of screen-based entertainment has led to concerns about the physical and mental health implications of prolonged screen time, particularly for children and adolescents (American Psychological Association, 2023; U.S. Department of Health and Human Services, 2023).

While technology has transformed family and social life in many positive ways, it has also introduced new challenges. The constant connectivity facilitated by digital tools has created new pressures and expectations, while also contributing to the erosion of privacy and face-to-face interactions. The increasing reliance on technology for communication and socialization has reshaped the way people build relationships, engage with communities, and balance their personal and professional lives. As society continues to adapt to these changes, it will be important to consider the implications of technology on family life, social structures, and individual well-being. Balancing the benefits of technology with the need for genuine human connection will be key in shaping the future of social and family life (Baym, 2015; boyd, 2014).

TECHNOLOGY AND CULTURAL IDENTITY

Technological innovations also play a crucial role in shaping cultural identity. In a world increasingly connected by technology, individuals often draw upon digital spaces to explore and express their cultural identities. The internet, social media, and other digital platforms have created spaces where people can connect with others who share similar backgrounds, traditions, and experiences. These platforms allow individuals to present their identities to a global audience, redefining what it means to belong to a particular cultural group. Through digital tools, individuals can connect with distant family members, fellow countrymen, or people with shared interests, reinforcing a sense of community that transcends geographic and social boundaries (Baym, 2015; Pew Research Center, 2025a).

For instance, social media platforms like Instagram, TikTok, and Twitter have given rise to the influencer culture, where individuals are able to create and share content that reflects their personal identities. These platforms provide a means for individuals to craft their own narratives and challenge traditional notions of identity, culture, and self-representation. By posting videos, photos, and blogs, individuals can highlight their cultural practices, values, and beliefs, often bypassing traditional media gatekeepers. In this way, these platforms empower users to shape their own public image and to explore identities in new, dynamic ways. This autonomy over self-representation has led to more diverse voices being heard in mainstream discourse, particularly among younger generations who are keen to express individuality and challenge societal norms (Abidin, 2018; Pew Research Center, 2025b).

In addition to providing individuals with platforms for self-expression, social media has also enabled marginalized groups to gain visibility. Communities that were once relegated to the periphery of public consciousness now have access to a global stage where they can share their experiences and advocate for social justice causes. Movements such as Black Lives Matter, #MeToo, and LGBTQ+ rights have found powerful expressions in digital spaces, with social media serving as a vehicle for organizing, spreading awareness, and mobilizing people. The ability to unite over shared experiences and struggles has given rise to solidarity that is not bound by physical borders, allowing individuals to connect with others who share similar challenges or aspirations (Reuters Institute, 2025; Reuters, 2024).

Moreover, the digital space allows for the celebration of cultural diversity. Platforms like YouTube and Instagram provide an opportunity for people to share their traditional music, dance, art, and cuisine with the world. For example, a young person in an urban center can discover traditional dances from a rural part of the world or access tutorials on how to cook dishes from different countries. This cultural exchange enriches global understanding, creating opportunities for people to explore cultures outside their own, broadening their horizons in ways that were not previously possible. In this sense, technology has become a bridge for cross-cultural dialogue, fostering appreciation and respect for cultural diversity (UNESCO, 2024b; UNESCO, 2025a).

However, the increasing dominance of global digital platforms has also raised concerns about cultural homogenization. As technology connects the world in new ways, it can sometimes diminish the distinctiveness of local cultures. The global reach of platforms like Netflix, YouTube, and Spotify often results in the widespread dissemination of certain cultural norms and preferences, which can overshadow indigenous practices or regional traditions. Popular culture, driven by global media corporations, tends to favor mainstream, westernized ideas and representations, which can sometimes marginalize or dilute indigenous voices (Ofcom, 2025; Reuters Institute, 2025).

This phenomenon is evident in the way Western popular culture, particularly American movies, television shows, and music, dominates global media consumption. As streaming services make Hollywood films and TV shows available worldwide, many local traditions and forms of expression are overshadowed by the appeal of high-budget entertainment. In some cases, the global popularity of certain media content can lead to the erosion of regional traditions, as younger generations may increasingly adopt globalized cultural norms, such as fashion trends, music genres, and entertainment preferences, at the expense of their own heritage (Ofcom, 2025).

The challenge, then, becomes how to preserve and celebrate diverse cultural identities while embracing the benefits of global connectivity. In an era where globalization allows for the sharing of cultural products across borders, the idea of a single, dominant global culture has become more apparent. However, the homogenization of culture does not mean that local cultures are necessarily doomed to disappear. Many cultural practitioners and activists are working to preserve and revitalize traditional arts, languages, and practices in response to the increasing dominance of global culture. Social media and other digital platforms can serve as tools for cultural preservation as much as for cultural dissemination (UNESCO, 2025b; WIPO, 2024a).

For example, indigenous communities around the world have used social media to document and promote their traditional knowledge, language, and rituals. In regions where indigenous cultures have been threatened by colonialism or modernity, technology offers a means of safeguarding cultural heritage. Online platforms allow indigenous people to pass on knowledge, share stories, and teach their languages to younger generations, even in the face of growing global influences. This digital cultural preservation ensures that traditional practices and values are not lost, while simultaneously allowing these cultures to engage with the global community on their own terms (WIPO, 2024a, 2024b).

In addition to preserving cultural heritage, technology can also promote cultural resilience. Digital spaces allow for the celebration of cultural identity, even in the face of external pressures to conform. Cultural festivals, traditional ceremonies, and rituals can be shared with a wider audience, offering an alternative narrative to mainstream portrayals of cultural identity. Digital platforms, such as YouTube (2025) and Instagram, can host content that showcases traditional forms of art, music, and dance, serving as a counterpoint to the homogenizing forces of globalization (UNESCO, 2024b).

Moreover, technology enables people to explore cultural identities in complex and fluid ways. While some individuals may define their cultural identity through a specific ethnic or national lens, others are increasingly exploring hybrid identities that incorporate elements from multiple cultural influences. In a multicultural world, where migration, travel, and intermarriage are more common, many people now embrace a cosmopolitan identity that blends elements from various cultures. Technology facilitates this exploration by providing access to a vast array of cultural materials, from foreign films and music to culinary traditions and language tutorials. As a result, individuals can develop multifaceted identities that reflect the diverse cultural influences they have experienced (Roudometof, 2023; Pew Research Center, 2025a).

This fluidity in cultural identity is particularly relevant for younger generations who are growing up in a globalized world. For many, their sense of identity is shaped by the cross-pollination of various cultural influences rather than by a singular, fixed tradition. Technology has given rise to the idea of "glocalization," where global trends and local customs blend together to create unique cultural expressions. Social media allows for individuals to engage with global movements, trends, and ideas while simultaneously expressing their own cultural values and traditions. This dynamic process leads to new forms of cultural hybridity that challenge rigid notions of identity and belonging (Roudometof, 2023).

However, the rise of digital platforms has not been without its drawbacks in terms of cultural preservation. As global digital platforms push local content into the background, there is a risk that cultural traditions might become commodified or misrepresented. The commercialization of culture—where traditional music, fashion, or art forms are adapted for mass consumption—can lead to the loss of cultural meaning and context. The process of "cultural appropriation," where elements of one culture are taken out of context and used for profit by another group, has become a significant issue in discussions of cultural identity in the digital age (WIPO, 2024a).

The commercialization of cultural products can also lead to exploitation. In some cases, traditional knowledge and practices are appropriated by larger entities without proper acknowledgment or compensation to the communities that created them. This has raised ethical concerns about intellectual property rights and the need for greater protection of cultural resources in the digital era. Indigenous and local communities are increasingly calling for stronger regulations around the use of their cultural heritage, particularly in the digital space, where content is easily reproduced and shared without consent (WIPO, 2024a, 2024b).

Furthermore, the way cultural products are represented in the media can have lasting effects on how cultural groups are perceived globally. The digital age has amplified the potential for misrepresentation, as biased portrayals of certain cultures or groups can quickly go viral, shaping public perceptions. For example, stereotypes of certain ethnic groups, religions, or nationalities can spread rapidly through social media, reinforcing harmful biases and prejudices. As global platforms dominate cultural production and consumption, there is a growing need for greater diversity and accuracy in the way cultures are represented in digital spaces (Pew Research Center, 2025c; Reuters Institute, 2025).

In response to these concerns, efforts are underway to create more inclusive and representative digital platforms that empower cultural expression in a way that is both authentic and respectful. Platforms like YouTube have begun to implement guidelines that encourage diversity and prohibit harmful stereotypes. Additionally, content creators from marginalized groups are advocating for more opportunities to present their cultural narratives and challenge mainstream portrayals. This includes a push for greater representation in the film industry, advertising, and news media, where cultural diversity is often underrepresented or misrepresented (European Commission, 2024; YouTube, 2025).

As digital technology continues to reshape cultural identity, it will be important to strike a balance between global connectivity and the preservation of cultural diversity. While technology provides unprecedented opportunities for cross-cultural exchange and self-expression, it also raises questions about the impact of global media dominance on local traditions. Ultimately, the challenge lies in harnessing technology to create a world where cultural identities are celebrated, respected, and protected, while embracing the benefits of a more interconnected and digitally savvy global society (UNESCO, 2025b).

As the future unfolds, it will be crucial to continue fostering a global dialogue on cultural representation, identity, and preservation in the digital age. Digital platforms have the potential to enrich our understanding of cultural diversity and to provide new spaces for marginalized voices. At the same time, they must be used responsibly to ensure that cultural heritage is respected and protected from exploitation or misrepresentation. In the end, the goal should be to celebrate the multiplicity of cultural identities and experiences, ensuring that technology serves as a tool for empowerment, inclusion, and cultural exchange in an increasingly connected world (UNESCO, 2024a, 2024b).

THE PARADOX OF TECHNOLOGICAL PROGRESS AND SOCIAL INEQUALITY

While technological advancements have often led to positive cultural shifts, they have also contributed to deepening inequalities. Access to technology is not evenly distributed, and those without access to the latest innovations may find themselves excluded from the cultural shifts that technology drives. The digital divide, which refers to the gap between those with and without access to digital technologies, remains a significant barrier in many societies. In an era where digital tools are central to participating in the global economy and cultural exchange, this divide is not only a matter of technological access but also a social and economic issue with far-reaching consequences (ITU, 2024; Broadband Commission, 2024).

In many developing countries, lack of access to the internet, smartphones, and computers limits the ability of individuals to participate in the global cultural exchange facilitated by technology. Without reliable internet access, many individuals are unable to access educational resources, engage in online communities, or participate in the global economy through e-commerce or remote work. The internet has become a central part of global life, shaping everything from social interaction to employment opportunities, and those without access are often left behind, unable to tap into these transformative experiences. This technological gap deepens existing inequalities, making it harder for people in these regions to access critical opportunities for advancement (World Bank, 2025a; ITU, 2024).

In these regions, the digital divide often intersects with other forms of inequality, including economic disparities, lack of infrastructure, and education. For example, rural areas and low-income households are more likely to lack the necessary resources to access and benefit from technological advances. Even in regions where smartphones are widely available, the lack of affordable data plans or high-speed internet further exacerbates the gap. These barriers create a vicious cycle, where individuals without access to technology are excluded from opportunities for skill-building, employment, and participation in society's evolving cultural conversations (GSMA, 2025; World Bank, 2025b).

This exclusion has profound implications not only for individuals but also for entire societies. When large segments of the population are left behind in the technological race, it impedes social and economic development. In addition to missing out on educational and economic opportunities, those without access to technology are also less likely to have a voice in political or social issues that are increasingly debated in online spaces. The absence of technology hinders people's ability to advocate for themselves, participate in social movements, or access vital information on health, safety, and governance (Reuters Institute, 2025; Pew Research Center, 2025c).

On the other hand, in more developed nations, the rapid pace of technological change has exacerbated inequalities in terms of education, income, and employment opportunities. In countries with advanced technological infrastructure, technology is a powerful driver of economic growth, but it also increases the gap between those who have the skills to thrive in the digital economy and those who do not. As industries evolve, workers who have not had access to quality education or upskilling programs may struggle to keep up with technological changes. This creates a bifurcated labor market, where those with advanced technical skills and education benefit from high-paying, flexible, and often remote work, while others are left with lower-wage, lower-skill jobs (OECD, 2024a; IMF, 2024).

This divide is particularly stark in areas like automation and artificial intelligence. The increasing reliance on AI and robotics has resulted in the displacement of workers in low-skilled or repetitive jobs. Tasks that were once performed by humans—such as data entry, customer service, or manual labor—are increasingly being handled by machines or automated systems. While automation improves efficiency and reduces costs for businesses, it creates a job market where low-skill workers are at a significant disadvantage. The result is a growing inequality between those who have the skills to work with new technologies and those who are replaced by them (ILO, 2024; OECD, 2023, 2024b).

Meanwhile, the digital economy rewards individuals with advanced technical skills, such as coding, data analysis, and software development. These skills are in high demand, and those who possess them have the potential to thrive in the digital economy, earning high salaries and enjoying flexible working conditions. However, the skills required to participate in this economy are often inaccessible to individuals in marginalized communities or rural areas, who may not have access to high-quality education or the internet resources necessary to acquire these skills. This exacerbates the gap between the highly skilled and the under-skilled, making it more difficult for disadvantaged individuals to catch up (World Bank, 2025b; OECD, 2024a).

In some cases, technology has exacerbated existing social divisions, reinforcing the power structures that determine access to resources. The concentration of technological wealth and power in the hands of a few large corporations has led to a situation where the benefits of technological advancements are disproportionately distributed. Major tech companies such as Google, Amazon, and Facebook control vast amounts of data, shaping the way people interact with technology and how information is distributed. These companies have significant influence over the digital economy, controlling access to information, entertainment, and communication platforms. While they reap the rewards of innovation, those who rely on their services for employment or access to information are often left with little power to shape the systems that govern their lives (IMF, 2024).

In this environment, individuals who are already at a disadvantage may find themselves even more marginalized. The concentration of power in the hands of a few major players means that smaller, local businesses and individuals are less likely to benefit from technological progress. Smaller companies often lack the resources to invest in new technologies or to compete with the dominant corporations in the digital marketplace. As a result, the digital economy can favor large corporations at the expense of smaller businesses, further entrenching existing economic inequalities (OECD, 2024a; IMF, 2024).

Furthermore, as technology becomes more integrated into everyday life, the ability to access and control personal data has become an increasingly important form of power. Those who control data—whether it be governments, tech companies, or other organizations—hold the keys to economic and social influence. The digital divide extends into data access and privacy, with marginalized communities often facing greater risks in terms of data exploitation, surveillance, and lack of control over personal information. The uneven distribution of data rights and digital privacy can have long-lasting implications for social equity and personal autonomy (Zuboff, 2019).

The social and economic inequalities created by the digital divide also have implications for education. As education systems increasingly integrate technology into classrooms, students who do not have access to the necessary tools and resources are at a disadvantage. In many places, schools and universities have moved to online learning environments, which requires reliable internet access and personal devices. Students without these resources may struggle to complete coursework, participate in class discussions, or access educational materials, further widening the achievement gap between privileged and underprivileged students (UNESCO, 2025a; OECD, 2023a).

Moreover, the shift to digital learning has raised questions about the future of education in an increasingly technology-driven world. The rapid pace of digital change means that educators must continuously update their skills and knowledge to keep up with new tools and methods. This creates a pressure on teachers, especially those in underfunded or rural areas, who may not have the support they need to integrate technology into their teaching effectively. The disparity between well-resourced schools and those with fewer resources exacerbates existing educational inequalities, hindering the ability of students in disadvantaged schools to compete in the digital economy (OECD, 2024c; UNESCO, 2025b).

Despite the many challenges, some governments, organizations, and initiatives are working to bridge the digital divide and mitigate the inequalities that arise from unequal access to technology. For example, programs that provide affordable internet access and subsidized devices to low-income families can help ensure that more people have the opportunity to participate in the digital world. Governments are also increasingly focused on policies to ensure that all citizens, regardless of their socio-economic background, can benefit from technological advancements (World Bank, 2025a; GSMA, 2024a).

At the same time, there is growing awareness of the importance of digital literacy as a critical skill for the 21st century. Digital literacy programs aim to equip individuals with the skills they need to navigate the digital landscape, from understanding online privacy to mastering essential tech tools. Such programs are essential for closing the knowledge gap and providing marginalized individuals with the tools they need to succeed in the digital economy (UNESCO, 2024a; OECD, 2024c).

Ultimately, addressing the deepening inequalities driven by technological advancements will require concerted efforts at multiple levels—individual, institutional, and governmental. While technology has the potential to bring people together and create new opportunities, it also has the power to deepen existing divisions. By ensuring equitable access to technology, promoting digital literacy, and advocating for policies that prioritize inclusivity, we can create a more just and fair technological future for all (Broadband Commission, 2024; ITU, 2024).

As the digital divide persists, it is crucial to keep in mind the role of technology in fostering equality and opportunity. Access to technology is no longer just a luxury or a convenience; it is a fundamental aspect of participation in modern society. Without access to digital tools, individuals are excluded from the cultural, educational, and economic benefits that technology offers. The digital divide is therefore not just about technological access, but about social and economic justice. Bridging the gap requires a commitment to ensuring that all individuals, regardless of their background or location, can benefit from the opportunities that technology provides (GSMA, 2025; World Bank, 2025a).

The impact of technological inequality is not just confined to the digital realm—it extends to broader societal issues, including access to healthcare, political participation, and social mobility. Those without access to technology are less likely to be able to access telemedicine, participate in online civic engagement, or engage in the new forms of work that the digital economy enables. As technology continues to evolve, it is imperative that we address these disparities and ensure that the benefits of technological progress are shared by all. This requires a concerted effort to bridge the digital divide and ensure that no one is left behind in the age of technology (ITU, 2024; World Bank, 2025b).

CONCLUSION

Technological innovations have always been intertwined with cultural shifts, reshaping societal norms, practices, and identities. From the printing press to the internet, advancements in technology have opened new channels for communication, transformed the nature of work and leisure, and redefined family and social structures. Throughout history, technological progress has acted as a catalyst for cultural evolution, enabling the dissemination of ideas, facilitating social movements, and broadening the scope of human experience. These innovations have allowed people to connect across vast distances, share knowledge, and collaborate in ways that were previously unimaginable. Yet, as transformative as these advancements have been, they also come with unintended consequences—both positive and negative—that affect the fabric of society (Reuters Institute, 2025; UNESCO, 2024a).

On one hand, technology has provided significant benefits in terms of connectivity, education, and access to information. It has democratized knowledge, making it more accessible to individuals worldwide, regardless of geographical location or socioeconomic background. The internet, for example, has enabled people to access online learning resources, participate in virtual communities, and gain exposure to global perspectives. These advancements have empowered individuals to pursue self-directed learning, enhance their skills, and engage with the world in new and meaningful ways. Social media platforms have given rise to diverse voices, allowing marginalized groups to advocate for their rights, challenge oppressive systems, and build communities that transcend traditional boundaries (UNESCO, 2025a; Pew Research Center, 2025a).

Moreover, technology has revolutionized the way we work and engage with leisure. The rise of remote work, facilitated by digital tools and platforms, has enabled greater flexibility in employment, allowing individuals to balance personal and professional lives more effectively. The digital economy has opened up new forms of entrepreneurship and innovation, fostering an environment where creativity can flourish across borders. At the same time, technological advances in entertainment, such as streaming services and gaming platforms, have changed the way people consume media and interact with one another. This shift has fostered new forms of cultural expression, collaboration, and creativity that were previously impossible or limited by physical constraints (Ofcom, 2025).

However, while technology has undeniably brought about positive cultural shifts, it has also created new challenges. One of the most pressing concerns is the widening gap between those with access to technology and those without, often referred to as the digital divide. This divide is not just about access to devices or internet connectivity, but about the skills, education, and infrastructure required to fully participate in a digitally driven world. As technology continues to advance, those without access to the latest innovations find themselves excluded from educational opportunities, job markets, and social networks, perpetuating cycles of inequality. In both developed and developing nations, the digital divide remains a significant barrier to achieving social and economic equity, and addressing it requires concerted efforts to ensure that technology is accessible, affordable, and inclusive (ITU, 2024; World Bank, 2025a).

Another challenge that technology has introduced is the potential for cultural homogenization. While digital platforms have allowed for the widespread sharing of cultural expressions, they have also led to the dominance of certain global cultures at the expense of others. The algorithms that power platforms like YouTube, Instagram, and Netflix prioritize content that appeals to large, global audiences, often at the cost of local or indigenous cultures. This can result in the erosion of unique cultural identities as people around the world adopt similar values, tastes, and behaviors shaped by the digital content they consume. The global reach of technology, while fostering greater connectivity, can sometimes dilute the diversity that exists in local traditions, languages, and practices, leading to a cultural conformity that threatens the rich tapestry of human expression (Ofcom, 2025; UNESCO, 2024b).

Furthermore, technology has disrupted traditional social structures and norms, particularly in relation to family dynamics, social relationships, and community life. The rise of social media has transformed the way people interact with one another, often prioritizing virtual connections over face-to-face interactions. This shift has led to concerns about the decline of social capital, as individuals increasingly engage in online communities that may not offer the same depth of personal connection as traditional social networks. While digital platforms allow for greater freedom of expression and connection, they also raise questions about the authenticity and quality of relationships formed in the digital realm. The erosion of traditional social structures, such as extended families or close-knit local communities, may have long-term implications for mental health, emotional well-being, and societal cohesion (Pew Research Center, 2025c; Reuters Institute, 2025).

As we continue to navigate the ever-evolving landscape of technological progress, it will be essential to critically examine the cultural shifts that accompany these innovations. While technology offers immense potential for positive change, it is important to ensure that these changes are equitable and inclusive. The benefits of technology should not be confined to a select few but should be accessible to all members of society, regardless of their background, location, or economic status. This requires addressing the digital divide and creating policies that ensure equal access to technology and digital literacy for all individuals. Moreover, as we embrace the benefits of technological advancements, we must also be mindful of the potential consequences they bring, particularly in terms of cultural homogenization and the erosion of traditional social structures (ITU, 2024; GSMA, 2025).

In addition to ensuring equitable access to technology, it is crucial to foster a culture of digital literacy and critical engagement. As digital spaces become increasingly central to our social, political, and economic lives, it is essential that individuals possess the skills and knowledge to navigate these spaces responsibly. Digital literacy extends beyond technical proficiency—it involves understanding the ethical implications of online behavior, being able to critically evaluate information, and engaging in digital spaces in ways that promote inclusivity and social good. By cultivating a society that values digital literacy, we can empower individuals to use technology in ways that enhance their lives and contribute positively to their communities (UNESCO, 2025a; OECD, 2024c).

Moreover, there is a need to preserve and celebrate diverse cultural identities in the digital age. While technology has the potential to connect people across the globe, it is essential to ensure that local cultures, languages, and traditions are not marginalized or erased in the process. This can be achieved by promoting the creation and sharing of content that reflects the diversity of human experience. Additionally, supporting policies that protect cultural heritage and encourage the preservation of indigenous languages and traditions can help to maintain cultural diversity in a rapidly globalizing world. Technology should be viewed as a tool for amplifying, not diminishing, cultural diversity (UNESCO, 2024b; WIPO, 2024a).

The future of technological innovation and cultural shifts will be shaped by our ability to balance progress with inclusivity, equity, and respect for diversity. As we move forward, it will be crucial to engage in ongoing dialogue about the ethical, social, and cultural implications of technology. By fostering a collective understanding of the role technology plays in shaping our world, we can ensure that its benefits are harnessed in ways that uplift society as a whole. This will require collaboration across industries, governments, and communities to create frameworks that promote responsible innovation, protect cultural heritage, and address inequalities that technology may exacerbate (Broadband Commission, 2024; OECD, 2024a).

Ultimately, technology has the potential to drive profound cultural transformation, but this transformation must be guided by a commitment to equity, justice, and inclusivity. As we continue to develop and integrate new technologies into our lives, we must remain vigilant in our efforts to ensure that these innovations serve the collective good. By embracing technology as a tool for positive change, while being mindful of its challenges and risks, we can create a future where technology enhances human life, preserves cultural diversity, and fosters a more just and connected world (UNESCO, 2024a; ITU, 2024).

CHAPTER 2 ESSAY PROMPTS

Technological Innovation and Cultural Shifts

1. How have technological innovations historically influenced cultural shifts in society? Provide examples of specific technologies and their impacts on social norms, practices, and identities.
2. Discuss the role of technology in democratizing knowledge and education. What are the potential benefits and challenges of online learning in fostering global access to education?
3. Examine the relationship between technology and the changing nature of work and leisure. How have digital tools transformed professional lives and personal recreation?
4. In what ways has the rise of remote work affected social structures and relationships? Analyze both the positive and negative impacts on individuals and families.
5. What is the digital divide, and why is it a critical issue in today's global society? Discuss the factors contributing to the divide and the potential consequences for economic and social equity.
6. How has technology contributed to cultural homogenization? Explore the potential risks of global digital platforms on the preservation of local cultures, traditions, and languages.
7. To what extent has technology disrupted traditional social structures such as families and communities? Discuss the implications for mental health, social cohesion, and interpersonal relationships.
8. What role should digital literacy play in addressing the challenges posed by technological progress? How can society ensure that all individuals possess the necessary skills to navigate digital spaces responsibly?
9. How can technology be used to preserve and celebrate diverse cultural identities in the digital age? Provide examples of initiatives or strategies that support cultural diversity online.
10. In your opinion, what is the future of technological innovation in relation to cultural shifts? How can society balance the benefits of progress with inclusivity, equity, and the protection of cultural heritage?

REFERENCES

Abidin, C. (2018). *Internet celebrity: Understanding fame online*. Emerald.

American Psychological Association. (2023). *Health advisory on social media use in adolescence.* https://www.apa.org/topics/social-media-internet/health-advisory-adolescents

Baym, N. K. (2015). *Personal connections in the digital age* (2nd ed.). Polity.

boyd, d. (2014). *It's complicated: The social lives of networked teens*. Yale University Press.

Castells, M. (2010). *The rise of the network society* (2nd ed.). Wiley-Blackwell.B

Eisenstein, E. L. (1979). *The printing press as an agent of change*. Cambridge University Press.

Encyclopaedia Britannica. (2025a). Wheel. https://www.britannica.com/technology/wheel (Retrieved November 9, 2025).

Encyclopaedia Britannica. (2025b). Plow. https://www.britannica.com/technology/plow (Retrieved November 9, 2025).

Encyclopaedia Britannica. (2025c). Printing press. https://www.britannica.com/technology/printing-press

Encyclopaedia Britannica. (2025d). Industrial Revolution. https://www.britannica.com/event/Industrial-Revolution

Encyclopaedia Britannica. (2025e). Television. https://www.britannica.com/technology/television-technology

Encyclopaedia Britannica. (2025f). Radio. https://www.britannica.com/topic/radio

European Commission. (2025, February 13). *Commission endorses the integration of the voluntary Code of Practice on Disinformation into the Digital Services Act (DSA)* [Press release]. https://ec.europa.eu/commission/presscorner/detail/en/ip_25_505

European Union. (2024). *Regulation (EU) 2024/1689 … laying down harmonised rules on artificial intelligence (AI Act). Official Journal of the European Union.* https://eur-lex.europa.eu/eli/reg/2024/1689/oj

IDC. (2025, June 18). *AR/VR market rebounds with 18.1% growth in latest quarter* [Press release]. https://my.idc.com/getdoc.jsp?containerId=prAP53608225

International Energy Agency. (2024). *Renewables 2024: Analysis and forecasts to 2030*. https://www.iea.org/reports/renewables-2024

International Labour Organization. (2024). *Mind the AI divide: Shaping a global perspective on the future of work.* https://www.ilo.org/global/research/publications/issue-briefs/WCMS_935637/lang--en/index.htm

International Labour Organization. (2024). *Realizing decent work in the platform economy (ILC.113/Report V(1)).* https://www.ilo.org/resource/conference-paper/ilc/113/realizing-decent-work-platform-economy

International Telecommunication Union. (2024). *Measuring digital development: Facts and figures 2024.* https://www.itu.int/itu-d/reports/statistics/facts-figures-2024/

Nielsen. (2025, May 30). *Time spent streaming reaches all-time high in April (The Gauge).* https://www.nielsen.com/insights/2025/time-spent-streaming-reaches-all-time-high-in-april/

Newman, N., Fletcher, R., Robertson, C. T., Ross Arguedas, A., & Nielsen, R. K. (2025). *Digital news report 2025.* Reuters Institute for the Study of Journalism. https://www.digitalnewsreport.org/

OECD. (2023). *OECD employment outlook 2023.* https://www.oecd.org/employment-outlook/

OECD. (2024a). *Job creation and local economic development 2024: The geography of generative AI.* OECD Publishing. https://doi.org/10.1787/83325127-en

OECD. (2024b). *The impact of AI on the workplace: Main findings from the OECD AI surveys of employers and workers* (OECD Social, Employment and Migration Working Paper No. 288; M. Lane, M. Williams, & S. Broecke, Authors). OECD Publishing. https://doi.org/10.1787/ea0a0fe1-en

OECD. (2024c). *Equity in education and on the labour market (Education at a Glance 2024 Spotlight).* https://www.oecd.org/en/publications/equity-in-education-and-on-the-labour-market_b502b9a6-en.html

Ofcom. (2025, August). *Media Nations: UK 2025.* https://www.ofcom.org.uk/research-and-data/tv-radio-and-on-demand/media-nations/2025/uk-report

Pew Research Center. (2024, August 21). *Social media and news fact sheet.* https://www.pewresearch.org/journalism/fact-sheet/social-media-and-news-fact-sheet/ International

Pew Research Center. (2025, May 7). *Teens, social media and mental health.* https://www.pewresearch.org/internet/2025/05/07/teens-social-media-and-mental-health/

Pew Research Center. (2025, May). *1 in 5 Americans now regularly get news on TikTok.* https://www.pewresearch.org/short-reads/ (article series)

Pew Research Center. (2025, July 2). *News influencers: Fact sheet.* https://www.pewresearch.org/journalism/fact-sheet/news-influencers/

Pew Research Center. (2025). *Many Americans say they often come across inaccurate news—and have a hard time knowing what's true.* https://www.pewresearch.org/

Reuters Institute for the Study of Journalism. (2025). *Digital news report 2025* (see Newman et al., 2025).

Roudometof, V. (2023). Digital glocalization: Theorizing the twenty-first-century ICT revolution. *Frontiers in Communication, 8*, 1244614. https://doi.org/10.3389/fcomm.2023.1244614

U.S. Department of Health and Human Services. (2023). *Surgeon General's advisory: Social media and youth mental health.* https://www.hhs.gov/surgeongeneral/priorities/youth-mental-health/social-media/index.html

UNESCO. (2024a). *Digital education—Policies and transformation.* https://www.unesco.org/en/digital-education/policies

UNESCO. (2024b). *Digital learning and transformation of education.* https://www.unesco.org/en/digital-education

UNESCO. (2025a). *Digital learning can promote equity in low-resource contexts* (International Day for Digital Learning). https://www.unesco.org/en/articles/unesco-spotlights-how-digital-learning-can-promote-equity-low-resource-contexts

UNESCO. (2025b). *Gateways to Public Digital Learning Initiative: Progress report.* https://unesdoc.unesco.org/ark:/48223/pf0000388597

WIPO. (2024a). *Traditional knowledge (TK) and traditional cultural expressions (TCEs).* https://www.wipo.int/tk/en/

WIPO. (2024b, May 24). *Member States adopt historic new treaty on intellectual property, genetic resources and associated traditional knowledge* [Press release]. https://www.wipo.int/pressroom/en/articles/2024/article_0007.html (see also the treaty resource page: https://www.wipo.int/en/web/treaties/ip/gratk/index)

YouTube. (2025). *Community Guidelines: Hate speech and harassment policies (updated).* https://support.google.com/youtube/answer/2801939 merlin.obs.coe.int

World Bank. (2025a). *Global Findex Database 2025: Digital Connectivity Tracker.* https://globalfindex.worldbank.org/

World Bank. (2025b). *Digital progress and trends report.* https://www.worldbank.org/en/publication/digital-progress-and-trends-report IDC

Vosoughi, S., Roy, D., & Aral, S. (2018). The spread of true and false news online. *Science, 359*(6380), 1146–1151. https://doi.org/10.1126/science.aap9559

Zuboff, S. (2019). *The age of surveillance capitalism.* PublicAffairs.

The Economic Impacts of Technology: Growth, Inequality, and Disruption

3

ABSTRACT

Chapter 3 explores the profound economic impacts of technology, focusing on its role in driving growth, exacerbating inequality, and disrupting traditional industries and labor markets. While technological advancements have fueled economic expansion by increasing productivity, creating new markets, and enhancing business efficiency, they have also contributed to rising economic disparities. Automation and AI have displaced low- and middle-skilled jobs, while rewarding high-skill workers with greater opportunities, leading to labor market polarization. Additionally, the concentration of wealth in tech giants and geographic disparities in access to technology further deepen inequality. The chapter examines how technology is reshaping industries and labor, highlighting both the challenges of job displacement and the emergence of new employment models, such as the gig economy and remote work, and emphasizes the need for policies that ensure equitable growth and worker support in an ever-evolving economy.

INTRODUCTION

Technology has dramatically transformed economies worldwide, reshaping industries, labor markets, and global commerce in ways that were once unimaginable. From the rise of automation and artificial intelligence (AI) to the expansion of the digital economy, the economic implications of technological advancements are profound, multifaceted, and, at times, paradoxical. While technology has fueled unprecedented growth and innovation, it has also contributed to growing economic inequality and disruption in traditional labor markets. This chapter analyzes the broad economic impacts of technology, focusing on growth, inequality, and the disruption of industries and labor markets (World Economic Forum, 2025; Cazzaniga et al., 2024).

The relationship between technology and economic growth is not new; historically, technological advancements have served as catalysts for economic expansion. The Industrial Revolution, for instance, brought about mass production, urbanization, and the rise of new industries, fundamentally altering economies across the globe. In the modern era, technology continues to drive growth through innovation, increasing productivity, and the creation of entirely new markets and business models. The growth of industries like information technology, telecommunications, and digital media demonstrates the transformative power of technology in expanding the global economy (OECD, 2024a).

At the same time, the rise of automation, artificial intelligence (AI), and robotics has significantly changed the labor market. While these technologies have led to increases in efficiency and productivity, they have also displaced many traditional jobs, particularly in industries like manufacturing and retail. Automation has taken over repetitive tasks that were once performed by humans, leading to job losses and shifts in employment patterns. These disruptions are felt most acutely by workers in low-skill, labor-intensive sectors, contributing to a widening gap between high- and low-wage earners (Huang, 2024; OECD, 2024a).

Furthermore, technology has fueled the rise of the gig economy, where individuals engage in short-term, freelance, or contract work rather than permanent employment. Digital platforms like Uber, Lyft, and Airbnb have enabled millions of people worldwide to earn income through flexible, on-demand jobs. While this has created new opportunities for many workers, it has also raised concerns about job security, income instability, and the erosion of traditional labor protections. The gig economy represents a shift away from the standard employer-employee relationship and presents challenges in terms of fair wages, benefits, and working conditions (International Labour Organization, 2024).

Alongside the disruption of labor markets, the economic impacts of technology are also evident in the growing concentration of wealth within certain sectors. The rise of tech giants like Amazon, Apple, Google, and Microsoft has created a concentration of power and wealth in the hands of a few companies. These companies benefit from economies of scale, network effects, and intellectual property rights, enabling them to dominate global markets and amass enormous wealth. This concentration of economic power has raised concerns about monopolistic practices and the unequal distribution of economic benefits, especially as these companies continue to shape the future of industries such as retail, finance, and media (European Commission, 2024; OECD, 2024b).

While technological advancements have created vast wealth and opportunities for some, they have also led to increased economic inequality. The benefits of technology have not been evenly distributed across society, with many individuals and regions left behind in the digital revolution. The divide between those who have access to technology and those who do not, both within countries and across the globe, has grown significantly. Developed nations and urban areas have been quick to adopt new technologies, while rural areas and developing countries often lag behind, exacerbating existing inequalities (International Telecommunication Union, 2024; UNCTAD, 2024).

The digital divide is not just a matter of access to technology but also the ability to leverage it for economic gain. In many low-income and developing areas, people may have access to the internet or smartphones, but they lack the skills, resources, and infrastructure needed to fully participate in the digital economy. This technological exclusion limits their ability to access online education, e-commerce, and remote work opportunities, further entrenching economic disparities. As technology continues to evolve, it is crucial to address these inequalities to ensure that the benefits of innovation are shared more equitably (OECD, 2024b; OECD, 2025).

The rise of technology has also had a profound impact on global markets. Digital platforms have enabled businesses to reach global audiences, disrupting traditional trade patterns and creating new forms of international commerce. The ability to sell products and services online has lowered barriers to entry for small businesses and entrepreneurs, creating a more competitive global marketplace. At the same time, global supply chains have become more interconnected and complex, with technology facilitating the movement of goods and services across borders in ways that were previously impossible (WTO, 2024; OECD, 2024b).

However, the globalization of markets has also led to significant economic disruption, particularly for workers in industries that have been outsourced to countries with lower labor costs. As companies seek to reduce production costs, many have moved manufacturing and customer service jobs to developing countries. While this has created jobs in those regions, it has also led to the decline of certain industries in developed countries, leaving workers in these sectors vulnerable to unemployment and wage stagnation (WTO, 2024).

The rapid pace of technological change has also introduced significant challenges in terms of workforce preparedness and job displacement. Many workers, particularly in low-skilled and routine jobs, find themselves ill-equipped to adapt to the changing demands of the labor market. As automation and AI continue to evolve, there is a growing need for policies that promote education, training, and reskilling to help workers transition to new roles. Without these efforts, there is a risk that large segments of the workforce will be left behind as technology continues to reshape industries and employment opportunities (OECD, 2024a; OECD, 2025).

Technology's impact on labor markets has raised important questions about the future of work and the role of the human workforce. As machines take over more tasks, there are concerns about the potential for widespread job losses and the societal implications of large-scale unemployment. Some argue that technology could lead to a future where human workers are no longer needed for many tasks, while others believe that new technologies will create new forms of work that we cannot yet imagine. What is clear, however, is that the relationship between technology and labor is evolving rapidly, and the future of work will likely look very different from the past (World Economic Forum, 2025; Cazzaniga et al., 2024).

In response to these changes, governments and businesses are increasingly recognizing the need to create policies that address the economic challenges posed by technological advancement. Universal basic income (UBI), for example, has emerged as a potential solution to provide financial security for individuals whose jobs are displaced by automation. Additionally, there is growing support for education reforms that focus on developing skills that are adaptable to an ever-changing job market, such as digital literacy, critical thinking, and problem-solving (OECD, 2024a; Executive Office of the President of the United States, 2024).

The economic impacts of technology also extend to the broader social and cultural landscape. As technology continues to shape the global economy (United Nations Conference on Trade and Development, 2024), it influences not only how people work but also how they live, interact, and consume goods and services. Digital technologies have transformed social interactions, with platforms like Facebook, Twitter, and Instagram changing the way people connect, communicate, and share information. The rise of e-commerce has revolutionized consumer behavior, with more people shopping online than ever before (OECD, 2024b).

At the same time, technology has raised questions about privacy, security, and the ethical implications of its use. As personal data becomes a commodity, concerns about surveillance, data breaches, and the exploitation of consumer information have become more prevalent. These issues have sparked debates about how to regulate the digital economy and ensure that technology is used responsibly and ethically (OECD, 2024b).

In conclusion, while technology has undoubtedly brought about significant economic growth and innovation, its impacts are complex and multifaceted. The growth of the digital economy, the rise of automation and AI, and the disruption of traditional labor markets have all contributed to a rapidly changing economic landscape. However, this transformation has also led to increased inequality, job displacement, and economic concentration. As technology continues to reshape the global economy, it is essential to consider its broader societal implications and develop policies that ensure that its benefits are distributed more equitably across society (World Economic Forum, 2025; OECD, 2024a).

Growth, inequality, and disruption: how automation and platforms raise productivity, polarize work, and concentrate wealth

Read through a political-economy lens, the timeline suggests a dual reality: automation and platforms lift aggregate productivity even as they hollow out middle-skill jobs, push more workers into contingent roles, and concentrate gains in a few firms and regions. What looks like unambiguous "growth" masks redistribution—toward capital, data owners, and scarce skills—while offloading risk onto households. The prompt beneath the image is pragmatic: pair innovation with counterweights (skills pathways, wage floors, portable benefits, and pro-competition rules) so the next productivity surge broadens security and mobility instead of deepening polarization.

TECHNOLOGICAL GROWTH AND ECONOMIC EXPANSION

The economic impacts of technology are undeniably multifaceted and deeply intertwined with broader societal issues. While technology undeniably fuels economic growth, drives innovation, and opens doors to new opportunities, it also introduces a range of challenges that must be addressed. As technology continues to evolve, it is reshaping the global economy in ways that fundamentally alter how work is structured, who benefits from technological advancements, and how workers are supported. The shift toward automation, digital platforms, and AI technologies is creating new efficiencies but also presenting risks of widespread job displacement, widening economic disparities, and challenging traditional models of employment. These challenges require a thoughtful and proactive approach to ensure that the benefits of technology are shared equitably (World Economic Forum, 2025; OECD, 2024a).

At the heart of these challenges is the increasing digital divide, which exacerbates inequalities within society. Those with access to technology and the necessary skills to navigate the digital economy are reaping the rewards of a growing tech-driven labor market. In contrast, those without access to advanced technology or the ability to adapt to new skills are at risk of being left behind. As industries such as manufacturing, transportation, and healthcare increasingly adopt automation and AI systems, low-skilled workers in these sectors face significant job displacement, while others—often in high-tech fields—see their wages and opportunities expand. This has created a stark divide between those who can adapt to the digital economy and those who struggle to keep pace, intensifying existing economic inequalities. Furthermore, the rise of the gig economy, characterized by precarious work arrangements and lack of worker protections, adds another layer of insecurity, leaving many workers without the stability or benefits that traditionally accompanied full-time employment (International Telecommunication Union, 2024; International Labour Organization, 2024).

Addressing these inequalities requires more than just technological innovation; it necessitates a concerted effort to create policies that foster inclusivity, equity, and fairness. Governments, businesses, and educational institutions must collaborate to ensure that technological progress does not exacerbate existing disparities. This includes ensuring equitable access to technology, providing resources for reskilling and upskilling, and rethinking labor protections for gig workers. Additionally, there is a pressing need to reframe education and workforce development in ways that prepare workers for a rapidly changing labor market. By emphasizing lifelong learning, vocational training, and investment in STEM education, we can equip workers with the skills needed to thrive in emerging industries. Policymakers should also consider innovative solutions like universal basic income (UBI) and other social safety nets to help mitigate the displacement effects of automation, providing a safety cushion for workers who may struggle to transition into new roles. By taking these steps, we can foster a more equitable distribution of the economic benefits generated by technological advancements (OECD, 2024a; OECD, 2025).

Ultimately, the future of work depends on our ability to navigate the complexities of technological disruption while ensuring that its benefits are shared broadly. The transformation of work, whether through automation, gig work, or remote models, presents both challenges and opportunities. The key will be creating a balance where flexibility, autonomy, and creativity can thrive, while workers are protected by strong social safety nets and fair labor practices. New models of work that prioritize well-being, inclusivity, and collaboration will be essential to building a future where technology enhances lives rather than deepens inequality. By fostering a cooperative, forward-thinking approach that includes diverse stakeholders, we can ensure that technology is a force for good, helping to build a future of work that is not only prosperous but also just, equitable, and sustainable for all. This will require continuous effort, innovation, and commitment, but it is a goal worth striving for to ensure that the digital revolution benefits everyone (Cazzaniga et al., 2024; World Economic Forum, 2025).

INCREASED PRODUCTIVITY AND EFFICIENCY

One of the most immediate and noticeable impacts of technology is its ability to enhance productivity. In industries ranging from manufacturing to services, automation and digital tools have allowed businesses to produce more goods and services with fewer resources. For example, in manufacturing, robotics and AI-driven systems can perform tasks such as assembly and quality control at speeds and levels of precision unattainable by human workers. These advancements allow for the production of goods at a much faster rate, increasing the overall output of factories and reducing production costs. Similarly, in the service sector, technologies like cloud computing, customer relationship management (CRM) systems, and AI-driven analytics have streamlined operations, allowing businesses to serve customers more efficiently and scale operations rapidly. These innovations reduce the time and effort spent on manual tasks, enabling employees to focus on higher-value work (World Economic Forum, 2025; OECD, 2024a).

This increased productivity contributes to economic growth by enabling firms to reduce costs, expand their output, and increase profitability. As businesses generate higher profits, they can reinvest those funds into further innovation, expansion, and workforce development. In addition, enhanced productivity often leads to lower prices for consumers, improving living standards and making products more accessible to a wider population. As businesses continue to leverage technology to streamline operations and improve performance, the overall economy benefits from the cumulative effects of increased efficiency and productivity (OECD, 2024a).

CREATION OF NEW MARKETS AND OPPORTUNITIES

Technological advancements have also given rise to entirely new markets and business models. The emergence of the internet economy has created global platforms for e-commerce, digital marketing, and content creation. Companies like Amazon, Google, and Facebook have not only disrupted traditional retail and media industries but have also spawned new sectors, including the gig economy, fintech, and data analytics. These companies have reshaped the way we consume products, access services, and communicate with one another, contributing to the rapid expansion of the digital economy (OECD, 2024b).

In the gig economy, for instance, digital platforms have facilitated the rise of freelancing and on-demand work. Platforms like Uber, Lyft, and Upwork connect workers with consumers directly, bypassing traditional employer-employee relationships. This has led to the creation of millions of flexible, short-term job opportunities around the world. At the same time, the fintech sector has revolutionized the financial industry by offering alternative banking solutions, mobile payment systems, and blockchain technology, creating new opportunities for financial inclusion and innovation. The rise of digital currencies and peer-to-peer payment systems has further disrupted traditional financial systems, presenting new challenges and opportunities for investors, regulators, and consumers alike (International Labour Organization, 2024).

Moreover, new industries, such as electric vehicles, renewable energy, and biotechnology, have flourished as a result of advances in clean energy technologies, biotechnology, and AI. The growing demand for sustainable energy solutions has led to innovations in solar power, wind energy, and electric transportation, helping to reduce dependence on fossil fuels and combat climate change. Similarly, breakthroughs in biotechnology and medicine have resulted in new treatments, therapies, and diagnostic tools, improving healthcare outcomes and creating new business opportunities in the healthcare sector. These emerging markets not only offer exciting growth prospects but also contribute to addressing some of the world's most pressing challenges, such as environmental sustainability and public health (OECD, 2024a).

These new markets provide businesses with the opportunity to diversify their revenue streams and tap into growing consumer demand for innovative products and services. They also stimulate job creation in emerging fields that require specialized skills, from app developers to data scientists. The rapid pace of technological innovation means that companies must continually adapt to stay competitive, often requiring new skill sets and business strategies. As a result, education and training have become increasingly important, with an emphasis on equipping workers with the skills needed to succeed in these evolving industries (OECD, 2025).

THE ROLE OF TECHNOLOGICAL INNOVATION IN GLOBAL TRADE

Global trade has undergone a significant transformation due to technological innovations, particularly in communication and transportation. The development of the internet has been a key catalyst in facilitating the growth of international commerce, allowing businesses to reach customers, suppliers, and partners from all corners of the world. E-commerce platforms have democratized access to global markets, enabling small and medium-sized enterprises (SMEs) to compete on a more level playing field with larger corporations. These advancements have opened up new opportunities for businesses of all sizes to grow and expand beyond their local markets. The ability to seamlessly connect with a global customer base, coupled with tools that allow for efficient marketing and transactions, has revolutionized the way businesses operate (WTO, 2024; OECD, 2024b).

Additionally, technological advancements in logistics, such as real-time tracking systems and automated warehousing, have streamlined the movement of goods across borders, reducing costs and improving the speed of deliveries. Artificial intelligence (AI), machine learning, and data analytics are being increasingly utilized to optimize supply chains by predicting demand, identifying inefficiencies, and improving inventory management. These innovations allow businesses to respond more rapidly to changes in consumer preferences, global supply chain disruptions, or shifts in market conditions, ultimately enhancing the overall efficiency of international trade. AI-powered algorithms also enable businesses to personalize customer experiences, leading to increased customer satisfaction and loyalty, which is particularly valuable in competitive global markets (OECD, 2024b; World Economic Forum, 2025).

The rise of global trade and interconnected economies has also fostered the exchange of knowledge, culture, and ideas between nations, contributing to economic growth and shared global prosperity. As businesses and consumers increasingly turn to digital platforms for purchasing goods and services, the global economy becomes more integrated. This interconnectedness is evident in the growing interdependence between countries, where a disruption in one part of the world can have ripple effects across global markets. This interconnected nature of trade has led to greater innovation and the sharing of technological advancements, providing opportunities for developing countries to access cutting-edge tools and systems that improve productivity and foster economic development (WTO, 2024).

However, this increasing interconnectedness presents several challenges, particularly in the areas of regulation and risk management. As global trade expands, there is a growing need for international regulatory frameworks that can address issues such as data privacy, cybersecurity, and fair trade practices. Without standardized regulations, businesses may face obstacles in navigating the complexities of cross-border trade, and consumers may be vulnerable to fraud or exploitation. Furthermore, the reliance on global supply chains makes economies more susceptible to disruptions caused by unforeseen events such as pandemics, natural disasters, or geopolitical tensions. These risks highlight the need for businesses and governments to invest in resilient systems and policies that can withstand external shocks, ensuring that global trade remains robust and adaptable in the face of uncertainty (European Parliament, 2024; UNCTAD, 2024).

TECHNOLOGICAL DISRUPTIONS IN TRADITIONAL INDUSTRIES

While technology has undoubtedly created new markets and opportunities, it has also disrupted traditional industries in profound ways. One of the most notable examples of this disruption is the rise of automation and artificial intelligence (AI), which have significantly reshaped sectors such as manufacturing, transportation, and retail. Automation, particularly in manufacturing, has enabled companies to reduce reliance on human labor by utilizing machines to perform repetitive and labor-intensive tasks. The automotive industry, for instance, has seen a dramatic reduction in demand for human workers in areas like assembly line production, as robots and automated systems increasingly take over these roles. In transportation, autonomous vehicles and drones are poised to revolutionize delivery systems, displacing truck drivers, couriers, and other workers in logistics. Similarly, the growth of e-commerce has brought about a seismic shift in retail, with online shopping platforms taking market share from traditional brick-and-mortar stores, leading to closures and job losses in the physical retail sector (OECD, 2024a; World Economic Forum, 2025).

This disruption, however, is not confined to low-wage or manual labor. As AI and automation continue to evolve, even highly skilled professions are at risk of being displaced or transformed. Fields such as data entry, customer service, and legal work are increasingly being automated through machine learning algorithms and AI-powered tools, which are capable of handling tasks that once required human input. In data entry, for example, AI systems can now process and organize vast amounts of information far more efficiently than a human worker could. In customer service, chatbots and virtual assistants are taking over routine inquiries, reducing the need for human customer service representatives. Legal work, too, is being impacted, as AI is used to conduct legal research, draft contracts, and even review documents, displacing workers who once performed these tasks. These advancements challenge the traditional understanding of work and create a growing need for individuals to adapt to new technologies and acquire different skill sets to remain relevant in their fields (Huang, 2024; OECD, 2024a).

As these technological changes continue to unfold, they present significant challenges for workers who must constantly adapt to new realities in the labor market. The pace of technological change is accelerating, and workers are finding it increasingly difficult to keep up. This presents a particular challenge for those in industries that are being heavily impacted by automation, as many lack the skills or resources to transition into new, technology-driven roles. The need for reskilling has never been more urgent, as workers must develop competencies in emerging fields such as data analysis, software development, and digital marketing. While technological advancements often create new job opportunities, these opportunities require a different set of skills and a mindset that embraces continuous learning. Workers must not only keep pace with technological changes but also anticipate them, in order to stay competitive in an evolving job market (OECD, 2025).

For businesses and governments, the rapid pace of technological disruption calls for proactive measures to ensure that workers are not left behind. Companies must invest in workforce development programs that offer training and reskilling opportunities, particularly for employees whose roles are at risk of automation. Governments also have a role to play by fostering educational systems that prepare individuals for the jobs of tomorrow, which may require new curricula and a focus on STEM (Science, Technology, Engineering, and Math) education. Additionally, safety nets and policies to support workers during transitions, such as unemployment benefits, universal basic income, or retraining subsidies, can help alleviate the impacts of job displacement. The challenge, then, is to create a workforce that is not only prepared for the jobs of the present but is also equipped to navigate the uncertainties of an increasingly automated and AI-driven future. In this way, while technology disrupts traditional industries, it also opens up pathways for a more adaptive and resilient workforce (OECD, 2024a; Cazzaniga et al., 2024).

THE UNEQUAL DISTRIBUTION OF TECHNOLOGICAL BENEFITS

As technological advancements continue to reshape the global economy, they have exacerbated income inequality, creating a widening divide between those who have access to the benefits of innovation and those who do not. High-skilled workers in industries such as technology, finance, and healthcare have seen significant increases in wages and job opportunities as a result of digital transformation. These sectors have been at the forefront of adopting new technologies like artificial intelligence, data analytics, and automation, which have driven growth and enhanced productivity. However, this has created a stark contrast for low-skilled workers, particularly those in traditional sectors such as manufacturing, retail, and agriculture. Automation and outsourcing have led to the loss of millions of manufacturing jobs, especially in developed countries, while the rise of the gig economy has introduced precarious work arrangements that offer limited benefits and job security. As a result, wage stagnation and job displacement have become prevalent challenges for workers in lower-income brackets, deepening the economic divide (Huang, 2024; International Labour Organization, 2024; OECD, 2024a; World Economic Forum, 2025).

Moreover, the economic benefits of technology are not distributed evenly across regions, with urban areas and developed nations reaping the most rewards. Metropolitan areas, with their higher concentrations of skilled labor, access to investment, and advanced infrastructure, have seen substantial economic growth driven by the adoption of cutting-edge technologies. In contrast, rural areas and developing countries have often been left behind in this digital revolution. Many rural regions lack the necessary infrastructure to support digital connectivity, such as high-speed internet, which hampers their ability to engage in digital economies. In developing countries, limited access to technology and education further exacerbates the disparity, preventing these regions from fully participating in the global economy. This growing divide between urban and rural, developed and developing regions reflects how technological progress can benefit some while marginalizing others, reinforcing global inequalities (International Telecommunication Union, 2024; OECD, 2024b; UNCTAD, 2024).

The digital divide, which refers to the gap between those who have access to digital technologies and those who do not, has been a significant factor in widening income inequality. While access to digital tools is a key component, the ability to effectively leverage technology for economic gain is equally important. In many low-income and rural communities, access to technology is hindered by a lack of digital literacy and technical support. Even when the infrastructure is available, individuals may not have the necessary skills to use technology in ways that enhance their economic opportunities. This lack of education and training means that marginalized communities are less likely to secure high-paying jobs in technology-driven sectors or benefit from the digital economy. Consequently, these communities remain excluded from many of the opportunities that technology offers, perpetuating cycles of poverty and inequality (International Telecommunication Union, 2024; OECD, 2025; UNCTAD, 2024).

Addressing this widening gap requires a multi-faceted approach that includes improving access to technology, expanding digital literacy, and providing targeted support to underserved populations. Governments and businesses must invest in infrastructure to ensure that rural areas and developing countries are not left behind in the digital age. Additionally, educational systems must adapt to equip individuals with the skills necessary to succeed in a technology-driven world. This includes focusing on STEM education, digital literacy, and reskilling programs that can help workers transition into emerging sectors. By closing the digital divide and providing equitable access to technology, societies can ensure that the benefits of innovation are more evenly distributed, creating a more inclusive and just global economy. In doing so, we can help prevent technology from exacerbating inequality and instead use it as a tool for broader economic advancement (OECD, 2025; OECD, 2024a; World Economic Forum, 2025).

TECHNOLOGICAL INNOVATION AND WEALTH CONCENTRATION

Another critical concern related to the economic impact of technology is the concentration of wealth in the hands of a few powerful companies and individuals. The rapid rise of technology giants such as Amazon, Apple, Google, and Facebook has created enormous wealth for their founders and investors, but this wealth is often not shared equitably across society. These companies benefit from economies of scale, network effects, and intellectual property rights, allowing them to dominate their respective markets and accumulate vast fortunes. While these companies contribute significantly to economic growth and technological innovation, their market dominance raises concerns about monopolistic practices and the unequal distribution of economic power. As these companies grow larger and more influential, they often enjoy significant advantages over smaller competitors, making it increasingly difficult for new entrants to compete, thus stifling innovation and limiting market diversity (European Commission, 2024; European Commission, 2025; OECD, 2024b).

As technology companies continue to grow and expand their influence, there is increasing concern about the social and economic consequences of wealth concentration. Large tech firms wield immense power, not only over the markets they operate in but also over the political and social landscape. These companies have the ability to shape consumer behavior through targeted advertising, manipulate public discourse via social media algorithms, and even influence election outcomes. The ability of these firms to control vast amounts of data further amplifies their power, allowing them to dictate the terms of engagement for both consumers and businesses. This concentration of power limits competition, reduces consumer choice, and contributes to a growing sense of social and economic inequality, as wealth continues to flow into the hands of a few at the expense of the broader public (OECD, 2024b; OECD, 2024d; UNCTAD, 2024).

Furthermore, the wealth generated by these tech companies is often concentrated among a small group of investors and executives, leaving the majority of workers in lower-wage positions with limited opportunities for upward mobility. While high-ranking employees and founders of these companies often see enormous financial gains, those working in entry-level positions, customer service, or logistics may experience stagnant wages, poor working conditions, and little to no opportunity for advancement. This wealth disparity is particularly evident in the gig economy, where many workers, such as delivery drivers and freelance contractors, are classified as independent contractors rather than full-time employees, depriving them of benefits like health insurance, paid leave, and retirement savings. As a result, the benefits of technological growth are not evenly distributed, further entrenching income inequality and undermining the promise of upward mobility (International Labour Organization, 2024; International Labour Organization, 2025; World Economic Forum, 2025).

Addressing the issue of wealth concentration in the tech industry requires a combination of regulatory oversight, labor reforms, and wealth redistribution policies. Governments must ensure that large tech companies are held accountable for monopolistic practices and that antitrust laws are effectively enforced to prevent market concentration. Additionally, there must be a focus on improving labor rights and working conditions for those in lower-wage positions, ensuring that workers in the tech sector are compensated fairly and have opportunities for advancement. Finally, policies aimed at wealth redistribution, such as progressive taxation and universal basic income, could help mitigate the effects of wealth concentration and provide a safety net for those who are left behind by the technological revolution. By addressing these issues, we can create a more equitable economic system where the benefits of technological advancement are shared more widely, rather than concentrated in the hands of a few powerful entities (European Commission, 2025; OECD, 2024c; OECD, 2024a).

THE FUTURE OF WORK AND JOB CREATION

As technology continues to evolve, it is crucial to consider its impact on the future of work. While automation and AI have led to job displacement in certain sectors, they have also created new opportunities in emerging industries. The rise of technology-driven fields such as software development, data analytics, cybersecurity, and renewable energy has generated millions of new jobs, many of which require specialized skills and higher education. However, these new jobs often require workers to adapt to changing skill requirements, and not all workers are able to make the transition successfully (OECD, 2025; World Economic Forum, 2025; OECD, 2024e).

The future of work will likely be shaped by the continued rise of automation, artificial intelligence, and digital platforms. As machines take over more tasks, there will be a greater need for human workers to focus on tasks that require creativity, problem-solving, and emotional intelligence—skills that are difficult for machines to replicate. Additionally, as the gig economy continues to grow, more people will work in non-traditional, flexible roles, which could lead to changes in job security, benefits, and labor rights. Governments, businesses, and educational institutions will need to work together to ensure that workers are adequately prepared for the jobs of the future (International Labour Organization, 2024; International Labour Organization, 2025; World Economic Forum, 2025).

The role of technology in driving economic growth is undeniable. From increasing productivity and creating new markets to disrupting traditional industries and shaping the future of work, technological advancements have reshaped the global economy in profound ways. As technology continues to evolve, it will continue to drive innovation, create new opportunities, and challenge existing business models. However, it is essential to address the inequalities and disruptions that accompany technological progress, ensuring that the benefits of innovation are shared more equitably across society. By investing in education, training, and policies that support workers, we can harness the full potential of technology to create a more prosperous and inclusive global economy (OECD, 2024a; World Economic Forum, 2025).

THE IMPACT OF TECHNOLOGY ON ECONOMIC INEQUALITY

Technological advancements have been a significant force in driving economic growth, creating new opportunities, and enhancing productivity. However, as technology has evolved, it has also been a major contributor to rising economic inequality. The benefits of technological progress have disproportionately favored certain sectors and groups within society, leading to a widening gap between the rich and poor. This chapter will explore the complex relationship between technology, economic growth, and inequality, highlighting both the opportunities and challenges that arise as technology reshapes the global economy (OECD, 2024a; UNCTAD, 2024).

SKILL DISPLACEMENT AND THE POLARIZATION OF LABOR MARKETS

One of the most significant ways that technology has contributed to inequality is through the displacement of low- and middle-skilled jobs. Automation and artificial intelligence (AI) have replaced many routine, manual, and clerical tasks that were once performed by human workers. In manufacturing, for example, robots and AI systems have taken over jobs on assembly lines, reducing the need for human labor in tasks such as product assembly, inspection, and quality control. Similarly, in services, technologies like chatbots and automated customer service tools have minimized the demand for human workers in roles such as call centers and customer support (Congressional Budget Office, 2024; Huang, 2024; OECD, 2024a).

As routine tasks are increasingly automated, the demand for highly skilled workers has surged. Fields like software engineering, data analysis, cybersecurity, and machine learning now require workers with specialized knowledge and technical expertise. This growing demand for high-skill, high-pay jobs has led to the polarization of the labor market. On one hand, well-paid, highly skilled professionals are in high demand, while on the other, low-skill, low-pay jobs are either automated or outsourced. This has led to a dual labor market, where workers with higher levels of education and technical expertise are benefiting from technological progress, while those with fewer skills face job insecurity, stagnant wages, and limited opportunities for career advancement (OECD, 2024e; OECD, 2025; World Economic Forum, 2025).

WEALTH CONCENTRATION AND THE RISE OF TECH GIANTS

Another major contributor to economic inequality in the age of technology is the concentration of wealth in the hands of a few large technology companies. Global giants such as Apple, Amazon, Microsoft, and Google have amassed significant market power and wealth. These companies have capitalized on economies of scale, network effects, and intellectual property rights to become highly profitable. The massive scale at which these firms operate allows them to dominate entire sectors, squeezing out smaller competitors and creating monopolies (European Commission, 2024; European Commission, 2025; OECD, 2024b).

GEOGRAPHIC AND SOCIAL DISPARITIES

The economic inequality exacerbated by technological advancements is not only an issue of individual skill or wealth but also reflects broader geographic and social disparities. In many rural areas and developing countries, there is limited access to the infrastructure, education, and capital needed to take full advantage of technological progress. In contrast, urban areas and developed nations, where technology companies are concentrated, benefit from better access to tech hubs, talent, and innovation (International Telecommunication Union, 2024; UNCTAD, 2024; WTO, 2024).

EDUCATION AND THE DIGITAL DIVIDE

Education is a key factor in determining how well individuals can navigate the challenges and opportunities brought about by technological change. In developed countries, education systems are increasingly focused on preparing students for the demands of the digital economy. However, in less developed regions, access to quality education and training in technological fields remains limited. Without access to the necessary education and training, workers in these regions are less likely to acquire the skills needed for the high-paying, tech-driven jobs that are becoming more prevalent in the global labor market (International Telecommunication Union, 2024; OECD, 2025).

THE CHANGING NATURE OF WORK AND JOB SECURITY

The shift toward automation, AI, and gig work has also led to changes in the nature of work itself. Traditional, stable jobs with benefits are increasingly being replaced by contract work, part-time employment, and freelance opportunities. While these flexible work arrangements can offer greater autonomy and convenience, they also come with a lack of job security, benefits, and long-term career prospects. Many workers in the gig economy, for example, face income instability, limited access to healthcare and retirement benefits, and the absence of legal protections typically afforded to full-time employees (International Labour Organization, 2024; International Labour Organization, 2025).

THE ROLE OF TECHNOLOGICAL INNOVATION IN GLOBAL TRADE

Technological advancements have also transformed global trade (World Trade Organization, 2024), both by making it easier for businesses to connect with suppliers and customers around the world and by creating new opportunities for global commerce. The internet, digital platforms, and advancements in logistics have made it possible for companies to conduct business on a global scale, reaching customers and partners across borders. E-commerce platforms like Amazon, Alibaba, and eBay have allowed even small businesses to access international markets, creating new opportunities for trade and economic growth (OECD, 2024b; WTO, 2024).

At the same time, technology has streamlined supply chains, improved efficiency, and reduced the cost of doing business across borders. Innovations in transportation, such as autonomous vehicles and drones, promise to further reduce costs and increase the speed of international trade. However, these developments are not without their challenges. The growth of global trade and the interconnectedness of economies also means that disruptions in one region, such as natural disasters or geopolitical tensions, can have far-reaching effects on global markets. As trade becomes more digital and automated, the potential for inequality also increases, as some regions and countries may be better positioned to take advantage of technological innovations than others (OECD, 2024b; WTO, 2024).

IMPACT ON LABOR MARKETS IN DEVELOPED VS. DEVELOPING ECONOMIES

While technological advancements have created vast economic opportunities in developed economies, they have also had a profound impact on labor markets in developing countries. In many developing nations, the automation of jobs traditionally performed by low-wage workers has led to job losses and economic displacement. At the same time, the rise of global digital platforms has increased competition for low-wage jobs, making it harder for workers in developing countries to secure well-paying positions. As a result, the gap between developed and developing economies continues to widen, with the benefits of technological progress concentrated in wealthier nations (UNCTAD, 2024; WTO, 2024).

In some cases, developing economies have been able to leverage technology to create new industries and business models, such as mobile banking and fintech startups. However, these opportunities are often limited to specific sectors and require significant investment in infrastructure, education, and innovation. Without access to the necessary resources, many developing countries face difficulties in capitalizing on technological advancements, leaving their populations vulnerable to economic inequality and job displacement (International Telecommunication Union, 2024; UNCTAD, 2024).

TECHNOLOGICAL ADVANCEMENTS AND ECONOMIC POLICY

As technological advancements continue to reshape the global economy, policymakers must consider how best to address the economic challenges they pose. Governments around the world are increasingly grappling with issues related to wealth concentration, job displacement, and the digital divide. Some governments have implemented policies to support workers displaced by automation, such as retraining programs and unemployment benefits. However, these measures are often insufficient to address the scale of the challenges posed by technological change (OECD, 2024a; OECD, 2025).

In addition, policymakers must consider the role of technology companies in driving economic inequality. Many large tech firms operate in a global environment, taking advantage of tax havens and regulatory loopholes to minimize their tax liabilities. This has led to calls for more robust taxation policies that ensure tech giants contribute their fair share to the economies in which they operate. Moreover, governments must work to ensure that the benefits of technology are more equitably distributed, both within their own countries and across the globe (European Commission, 2025; OECD, 2024c).

THE FUTURE OF WORK IN A TECHNOLOGICALLY ADVANCED ECONOMY

As technology continues to evolve, it will inevitably reshape the future of work. While some jobs will be lost to automation and AI, others will be created in emerging industries such as renewable energy, biotechnology, and AI development. However, the jobs of the future will require specialized skills and education, raising concerns about access to training and education for workers in lower-income sectors (OECD, 2025; OECD, 2024e).

The future of work will likely involve a combination of automation, AI, and human labor, with humans focusing on tasks that require creativity, problem-solving, and emotional intelligence—skills that are difficult for machines to replicate. As the nature of work changes, it is essential for workers to continually adapt and acquire new skills to stay competitive in the labor market (World Economic Forum, 2025; OECD, 2024e).

CONCLUSION

While technology has undoubtedly driven significant economic growth, its benefits have not been equally distributed. The rapid pace of technological change has led to job displacement, wealth concentration, and growing disparities between regions and social groups. To ensure that the benefits of technological progress are more equitably shared, it is essential to invest in education and training, implement policies to address the digital divide, and ensure that workers are prepared for the jobs of the future. By addressing these challenges, we can harness the full potential of technology to create a more inclusive and prosperous global economy (OECD, 2024a; UNCTAD, 2024; World Economic Forum, 2025).

DISRUPTION OF INDUSTRIES AND LABOR MARKETS

Automation and Job Displacement

One of the most significant disruptions technology has caused is the automation of jobs across multiple sectors. In manufacturing, for example, the use of robotics has made it possible to automate repetitive tasks that were once performed by human workers. This trend is now extending to other industries, including transportation, logistics, retail, and even healthcare. Autonomous vehicles, for example, threaten to displace truck drivers, while AI-powered diagnostic tools may reduce the demand for human medical professionals in certain tasks (Congressional Budget Office, 2024; OECD, 2024a; World Economic Forum, 2025).

Automation has the potential to replace jobs that involve routine and repetitive tasks, such as data entry, assembly line work, and customer service. While this can lead to cost savings and increased efficiency for businesses, it also poses a major challenge for workers in these roles, as they face the threat of unemployment or underemployment. In some cases, workers may be able to transition into new roles, but this often requires significant retraining or upskilling (Huang, 2024; OECD, 2025; OECD, 2024e).

The displacement effects of technology are particularly pronounced among low-skilled and middle-skilled workers, who may not have the training or resources to transition into more specialized or high-tech jobs. These workers often lack the education or technical skills needed to fill the emerging roles in fields like AI development, robotics maintenance, and data analysis. Without proper support systems and opportunities for skill development, these workers risk being left behind as the labor market evolves (International Labour Organization, 2024; International Labour Organization, 2025; OECD, 2024a).

The Gig Economy and Flexibility

At the same time, technology has given rise to the gig economy, where workers take on short-term contracts or freelance work, often facilitated by digital platforms. Companies like Uber, Airbnb, and TaskRabbit have redefined the nature of employment (Organisation for Economic Co-operation and Development, 2024a, 2024b, 2024c, 2024d, 2025), allowing individuals to access flexible, on-demand work opportunities. While this model offers certain advantages, such as autonomy and flexibility, it also comes with significant downsides, including job insecurity, lack of benefits, and limited wage growth (ILO, 2024; OECD, 2024a; WEF, 2025).

Gig workers typically do not have the same protections and benefits that full-time employees in traditional industries enjoy. They may not have access to health insurance, retirement plans, or paid leave. Furthermore, the income of gig workers can be unstable and unpredictable, as it is dependent on market demand and competition. While some workers may be able to earn a good income through gig work, others may struggle to make ends meet, especially if the demand for their services fluctuates (ILO, 2024; European Parliament & Council, 2024; U.S. Department of Labor, 2024).

The gig economy has also contributed to the erosion of traditional labor protections and benefits. Many gig workers are classified as independent contractors, which means they are not entitled to the same legal protections as employees. This classification allows companies to avoid providing benefits like health insurance, unemployment insurance, and workers' compensation. As a result, many gig workers face financial insecurity and a lack of support in times of illness, injury, or unemployment (European Parliament & Council, 2024; U.S. Department of Labor, 2024; ILO, 2024).

Structural Shifts and New Employment Models

In response to the disruptions caused by technology, some companies and governments are exploring new models of employment and economic organization. For example, some corporations are embracing remote work and hybrid models, allowing workers to access jobs from anywhere in the world. Remote work has become increasingly popular, particularly in the wake of the COVID-19 pandemic, as companies and workers alike have discovered the benefits of flexibility and reduced commuting time (Aksoy et al., 2025; OECD, 2024a; WEF, 2025).

The rise of remote work has led to significant changes in the way businesses operate. Companies no longer need to maintain large office spaces, and employees can work from anywhere with an internet connection. This shift has also opened up opportunities for workers in underserved or rural areas to access jobs that were previously limited to large urban centers. Remote work can be particularly beneficial for workers with disabilities, caregivers, or others who may face barriers to traditional office-based employment (OECD, 2024a; Aksoy et al., 2025).

In addition to remote work, some countries are experimenting with policies like universal basic income (UBI) to offset the potential loss of jobs due to automation. UBI provides individuals with a guaranteed income, regardless of their employment status, and is intended to ensure that all citizens have a basic standard of living, even in the face of technological disruption. While UBI has been proposed as a way to alleviate poverty and reduce economic inequality, it also presents significant challenges in terms of funding, implementation, and political feasibility (Daruich & Fernández, 2024; Welsh Government, 2025; OECD, 2024b).

While these models may offer solutions to the challenges posed by technological disruption, they also present new challenges related to regulation, taxation, and the need to provide workers with sufficient social safety nets in a rapidly changing economic environment. For example, remote work can blur the lines between employees and contractors, making it difficult for governments to implement appropriate labor laws and tax policies. Similarly, UBI may require significant government spending, which could lead to debates about how to fund such programs and whether they would be effective in addressing inequality (OECD, 2024a; OECD, 2025; World Bank, 2024).

Education and Workforce Development

As technology continues to change the nature of work, there is a growing emphasis on education and workforce development. In order to succeed in a technology-driven economy, workers need to acquire new skills and knowledge. This includes not only technical skills in fields like computer science and engineering but also soft skills such as critical thinking, problem-solving, and communication (OECD, 2025; WEF, 2025).

Educational institutions, from primary schools to universities, are increasingly focusing on preparing students for the demands of the digital economy. Many schools are introducing coding, robotics, and STEM (Science, Technology, Engineering, and Mathematics) programs to help students develop the technical skills needed for the future. However, there is also a need to focus on lifelong learning and the development of skills for individuals who are already in the workforce (OECD, 2024a; UNESCO, 2024).

Lifelong learning is crucial in a rapidly changing job market, where new technologies and industries are constantly emerging. Many workers may need to retrain or upskill throughout their careers in order to remain competitive. Governments, businesses, and educational institutions will need to collaborate to create accessible and affordable opportunities for workers to acquire new skills (OECD, 2025; WEF, 2025).

THE ROLE OF GOVERNMENTS IN ADDRESSING TECHNOLOGICAL DISRUPTION

Governments play a critical role in addressing the economic challenges posed by technological disruption. They can implement policies that support workers who are displaced by automation and other technological changes. This may include providing financial support, retraining programs, and access to education and job placement services (OECD, 2024b; World Bank, 2024).

Governments can also help to mitigate the negative impacts of the gig economy by introducing regulations that protect workers' rights and ensure fair compensation. This may involve introducing minimum wage laws for gig workers, providing benefits such as healthcare and retirement plans, and establishing clear rules around classification and employment status (European Parliament & Council, 2024; U.S. Department of Labor, 2024; ILO, 2024).

In addition, governments can play a role in promoting innovation and entrepreneurship by providing incentives for businesses to invest in research and development. This can help to drive economic growth and create new job opportunities in emerging industries. Governments can also invest in infrastructure, such as high-speed internet and digital literacy programs, to ensure that all citizens have the tools they need to succeed in the digital economy (OECD, 2024a; ITU, 2024; GSMA, 2024).

GLOBALIZATION AND TECHNOLOGICAL CHANGE

Technological advancements have also been a driving force behind globalization, as digital platforms and communication technologies have made it easier for businesses to connect with customers, suppliers, and employees around the world. Globalization has led to the outsourcing of jobs to countries with lower labor costs, as businesses seek to reduce expenses and increase profits (WTO, 2024; UNCTAD, 2024).

While globalization has created new economic opportunities, it has also contributed to economic inequality, particularly in developed countries. As businesses outsource manufacturing and other labor-intensive jobs to developing countries, workers in developed countries may face job losses and wage stagnation. In addition, the competition for jobs in the global economy can lead to downward pressure on wages, as workers in both developed and developing countries compete for the same positions (WTO, 2024; WEF, 2025).

Globalization has also led to the rise of multinational corporations, which often have more power and influence than national governments. These corporations can use their global reach to avoid taxes, exploit labor, and influence policy decisions. As a result, there are growing concerns about the concentration of economic power and the erosion of workers' rights in the global economy (WTO, 2024; UNCTAD, 2024).

THE DIGITAL DIVIDE

As technology continues to advance, there is a growing concern about the digital divide—the gap between those who have access to digital technologies and those who do not. In many parts of the world, particularly in rural and low-income areas, access to the internet, smartphones, and other digital tools is limited. This lack of access can create significant barriers to economic participation, education, and social mobility (ITU, 2024; GSMA, 2024; Broadband Commission, 2024).

The digital divide is not only an issue in developing countries but also within developed nations. In many rural and underserved urban areas, people may not have access to high-speed internet or the necessary devices to participate in the digital economy. This can limit their ability to access job opportunities, education, healthcare, and other essential services (GSMA, 2024; ITU, 2024).

Efforts to bridge the digital divide include expanding internet infrastructure, providing affordable access to digital devices, and increasing digital literacy. However, these efforts must be part of a broader strategy to ensure that all individuals, regardless of their socioeconomic status or geographic location, have the tools they need to succeed in the digital economy (Broadband Commission, 2024; ITU, 2024; OECD, 2024a).

TECHNOLOGY AND ECONOMIC INEQUALITY

While technology has driven significant economic growth, it has also exacerbated economic inequality. The benefits of technological advancements have often been concentrated among a small group of individuals and corporations, leaving many others behind. Wealthy tech entrepreneurs and investors have amassed enormous fortunes, while many workers in traditional industries face job displacement, stagnant wages, and insecure employment (OECD, 2024b; IMF, 2024; WEF, 2025).

The rise of automation and the gig economy has contributed to the polarization of the labor market, with high-skill, high-wage jobs on one side and low-skill, low-wage jobs on the other. Workers in low-wage jobs may struggle to find stable employment, while workers in high-wage jobs may see their income and wealth continue to grow (WEF, 2025; OECD, 2024b).

In addition, the concentration of wealth in the hands of a few large corporations has led to concerns about economic power and inequality. As tech giants like Amazon, Google, and Apple continue to dominate the global economy, there are growing calls for policies that address wealth concentration and ensure that the benefits of technological progress are more equitably distributed (OECD, 2024b; IMF, 2024).

THE FUTURE OF WORK AND TECHNOLOGICAL INNOVATION

The future of work will undoubtedly be shaped by technological innovation. As automation and artificial intelligence continue to evolve, many jobs will be lost, while new ones will emerge. The key challenge will be ensuring that workers have the skills and support they need to thrive in this new environment (WEF, 2025; OECD, 2025).

At the same time, technology has the potential to create new opportunities and improve the quality of work. Advances in fields like robotics, AI, and biotechnology could lead to new industries and job categories that we cannot yet predict. However, this will require investment in education, retraining, and workforce development to ensure that workers are prepared for the jobs of the future (OECD, 2025; WEF, 2025).

The future of work will also require new forms of social and economic organization. As traditional labor models are disrupted, businesses, governments, and workers will need to collaborate to create new employment models, policies, and protections. This may include the development of new social safety nets, such as universal basic income or job guarantee programs, to ensure that all individuals have access to economic security in a rapidly changing world (Welsh Government, 2025; OECD, 2024b).

Ultimately, the challenge will be to harness the potential of technology to create a more inclusive and equitable global economy. By addressing issues like automation, the gig economy, and economic inequality, we can ensure that the benefits of technological progress are shared more broadly, creating a world where all individuals have the opportunity to succeed (WTO, 2024; WEF, 2025).

CONCLUSION

Navigating the Complex Landscape of Technology and Economic Change

The economic impacts of technology are undeniably complex and multifaceted. While it is clear that technology serves as a major driver of growth, innovation, and the creation of new opportunities, it also introduces a host of significant challenges. Among these are concerns about inequality, the disruption of traditional industries, and the future of work itself. These challenges are particularly pressing as technology continues to evolve at a rapid pace, fundamentally altering how work is organized, who benefits from economic growth, and how workers are supported (OECD, 2024b; WEF, 2025).

At the heart of the issue is the growing disparity between those who have access to the tools and skills needed to thrive in a digital economy and those who do not. This digital divide has led to a situation where technology, instead of being an equalizing force, is reinforcing existing inequalities. The displacement of low-skilled workers by automation and the rise of precarious gig work, coupled with a lack of adequate worker protections, has led to growing insecurity for many in the labor force. Meanwhile, those with the right education and skills—often in technology-driven fields—are seeing their wages and opportunities increase. This has created a two-tiered labor market where the gap between the haves and have-nots is widening (ITU, 2024; GSMA, 2024; OECD, 2024b).

For policymakers, businesses, and educators, the challenge lies in ensuring that the benefits of technological advancements are distributed more equitably. It is not enough to simply drive technological progress; the focus must also be on how to ensure that no one is left behind. This requires a proactive approach that includes creating policies aimed at reducing the digital divide and increasing access to technology, especially for underserved communities. By addressing issues of access, equity, and fairness, we can help ensure that technology serves as a tool for inclusive growth, rather than exacerbating existing inequalities (Broadband Commission, 2024; WTO, 2024).

One of the most critical areas of focus in this regard is education. The rapid pace of technological change has made it clear that lifelong learning will be essential for workers to remain competitive in the labor market. As automation and AI continue to replace certain jobs, workers must be empowered with new skills to take on roles that require more complex problem-solving, creativity, and emotional intelligence. This necessitates a greater emphasis on vocational training, STEM education, and reskilling programs that help workers transition into emerging sectors. By investing in education and retraining, we can help workers develop the skills necessary to succeed in the digital economy, enabling them to secure fulfilling, stable employment in new industries (OECD, 2025; WEF, 2025; UNESCO, 2024).

The transformation of work itself is another key area that requires attention. The rise of remote work, flexible hours, and the gig economy has shifted the way many workers engage with their jobs. While these changes can offer benefits, such as greater autonomy and work-life balance, they also introduce new challenges, particularly in terms of job security, benefits, and fair compensation. Traditional labor protections, which were designed for full-time, office-based jobs, must be reimagined to account for the growing prevalence of gig and contract work. This includes revisiting policies related to health insurance, paid leave, and retirement savings, ensuring that all workers—regardless of their employment status—have access to the benefits and protections they need to thrive (European Parliament & Council, 2024; U.S. Department of Labor, 2024).

In addition to these specific areas of policy reform, the future of work will require new models of employment and economic organization. Concepts such as universal basic income (UBI) have gained traction as a potential solution to mitigate the impacts of automation on workers who may struggle to find new jobs. By providing a guaranteed income to all citizens, regardless of employment status, UBI could provide a safety net for those who face job displacement due to technological change. Similarly, government initiatives aimed at supporting entrepreneurship and innovation can create new avenues for economic participation, helping workers navigate the shifting landscape of the job market (Daruich & Fernández, 2024; Welsh Government, 2025).

Furthermore, embracing new models of work that prioritize flexibility, creativity, and collaboration will be crucial for ensuring a prosperous future. For instance, hybrid work models, which combine remote and in-office work, offer a balanced approach to addressing the needs of both businesses and employees. These models can enable companies to tap into a global talent pool while providing workers with greater control over their work schedules and locations. However, these models also require the development of new tools, infrastructure, and cultural shifts within organizations to ensure they are successful and inclusive (Aksoy et al., 2025; OECD, 2024a).

Ultimately, the future of work will be shaped by a collective effort to mitigate the negative effects of technology while maximizing its potential to improve lives. By embracing inclusive policies, investing in education and retraining, and fostering new models of work, we can ensure that technological advancements benefit everyone. This vision of a more inclusive future requires collaboration across all sectors—government, business, education, and labor—to create a framework that fosters innovation, supports workers, and promotes economic opportunity for all (WTO, 2024; WEF, 2025).

CHAPTER 3 ESSAY PROMPTS

The Economic Impacts of Technology—Growth, Inequality, and Disruption

1. How has technology acted as a driver of economic growth, and what specific sectors have benefited the most from these advancements?
2. In what ways has technology contributed to the widening gap between high-skilled and low-skilled workers in the global economy? Provide examples.
3. Explain the role of automation and artificial intelligence in disrupting traditional industries. How can businesses and workers adapt to these changes?
4. Discuss the concept of the "digital divide." How has unequal access to technology exacerbated income inequality in both developed and developing nations?
5. Evaluate the positive and negative economic effects of e-commerce on traditional retail industries. How can brick-and-mortar stores compete in the digital age?
6. How has the rise of the gig economy influenced income inequality? What are the potential long-term consequences of gig work on workers' financial stability?
7. Analyze the impact of technological advancements on global trade. How have digital platforms and logistics technologies reshaped international commerce?
8. What are the implications of wealth concentration in the tech industry? How does the dominance of tech giants like Amazon, Apple, and Google affect market competition and income distribution?
9. How can policymakers address the economic challenges created by technological disruption, such as job displacement and wage stagnation, while promoting innovation?
10. Discuss the potential role of education and reskilling in mitigating the negative economic impacts of technology. How can lifelong learning help workers stay competitive in a rapidly changing labor market?

REFERENCES

Aksoy, C. G., Barrero, J. M., Bloom, N., Davis, S. J., Dolls, M., & Zárate, P. (2025). The global persistence of work from home. *Proceedings of the National Academy of Sciences, 122*(27), e2509892122.

Congressional Budget Office. (2024, December 20). *Artificial intelligence and its potential effects on the economy and the federal budget*. Author.

Cazzaniga, M., Jaumotte, F., Li, L., Melina, G., Panton, A. J., Pizzinelli, C., Rockall, E. J., & Tavares, M. M. (2024). *Gen-AI: Artificial intelligence and the future of work* (IMF Staff Discussion Note No. 2024/001). International Monetary Fund.

Daruich, D., & Fernández, R. (2024). Universal basic income: A dynamic assessment. *American Economic Review, 114*(1), 38–88.

European Commission. (2024, March 7). Designated gatekeepers must now comply with all obligations under the Digital Markets Act [Press release]. Author.

European Commission. (2025, April 25). *Annual report on Regulation (EU) 2022/1925 (Digital Markets Act) 2024* (COM(2025) 166 final). Author.

European Parliament. (2024). *WTO agreement on electronic commerce: At a glance*. Author.

European Parliament & Council. (2024, October 23). *Directive (EU) 2024/2831 on improving working conditions in platform work. Official Journal of the European Union*, L (11 November 2024).

GSMA. (2024). *The state of mobile internet connectivity 2024*. GSMA.

Huang, Y. (2024). *The labor market impact of artificial intelligence: Evidence from US regions* (IMF Working Paper WP/24/199). International Monetary Fund.

International Labour Organization. (2024). *Realizing decent work in the platform economy* (ILC.113/Report V(1)). Author.

International Labour Organization. (2025, April 2). *Decent work in the platform economy: Law and practice update* (Factsheet). Author.

International Telecommunication Union. (2024). *Measuring digital development: The ICT Development Index 2024*. Author.

Organisation for Economic Co-operation and Development. (2024a). *Artificial intelligence, data and competition* (DAF/COMP/WD(2024)49). OECD Publishing.

Organisation for Economic Co-operation and Development. (2024b). *OECD Digital Economy Outlook 2024 (Vol. 2): Strengthening connectivity, innovation and trust*. OECD Publishing.

Organisation for Economic Co-operation and Development. (2024c). *OECD Employment Outlook 2024: The net-zero transition and the labour market*. OECD Publishing.

Organisation for Economic Co-operation and Development. (2024d). *2024 progress report on tax co-operation for the 21st century*. OECD Publishing.

Organisation for Economic Co-operation and Development. (2025). *Bridging the AI skills gap*. Paris: OECD Publishing.

U.S. Department of Labor. (2024). Employee or independent contractor classification under the Fair Labor Standards Act (Final rule). *Federal Register, 89*(7), 1638–1743.

United Nations Conference on Trade and Development. (2024). *Digital Economy Report 2024: Shaping an environmentally sustainable and inclusive digital future (Overview).* United Nations.

Welsh Government. (2025). *Basic income for care leavers in Wales pilot evaluation: Annual report 2024–2025 (Summary).* Author.

World Bank. (2024). *Adaptive social protection: Building resilience to shocks (Key findings).* Author.

World Economic Forum. (2025, January). *The Future of Jobs Report 2025.* Author.

World Trade Organization. (2024). *World Trade Report 2024: Trade and inclusiveness—How to make trade work for all.* Author.

Technology and Political Structures: Power, Policy, and Participation

4

ABSTRACT

Chapter 4 explores the profound impact of technology on political structures, examining its role in governance, policymaking, and civic engagement. As digital tools such as big data, artificial intelligence, and social media reshape political power dynamics, governments are increasingly leveraging these technologies to enhance decision-making, increase transparency, and optimize public services. However, the chapter also highlights challenges, including privacy concerns, algorithmic bias, and the digital divide, which may limit equitable participation in political processes. Furthermore, while technology has empowered citizens to engage in political discourse and activism, it has also contributed to issues like misinformation, political polarization, and the concentration of power in the hands of a few corporations. The chapter ultimately underscores the need for a balanced approach to technological integration in politics, ensuring that innovation serves democratic values while mitigating its potential risks.

INTRODUCTION

In the modern era, technology has become a fundamental force in shaping political structures, influencing governance, policymaking, and civic engagement. The rapid evolution of digital tools and platforms has transformed the way governments interact with citizens, how policies are developed, and how individuals participate in political processes. From social media campaigns to artificial intelligence in policy analysis, technology's integration into politics has both enhanced democratic participation and introduced new challenges related to transparency, accountability, and power dynamics. This chapter explores the role of technology in governance, the formulation of policies, and the ways in which it is reshaping civic engagement and political participation (UN DESA, 2024; World Bank, 2024).

One of the most significant ways that technology has reshaped political structures is by altering how information is disseminated and consumed. The internet has provided unprecedented access to information, enabling citizens to stay informed on political issues, government actions, and global events in real-time. Social media platforms like Twitter, Facebook, and Instagram have become primary tools for political communication, allowing politicians, activists, and citizens to interact in ways that were previously impossible. These platforms have democratized the flow of information, giving individuals the power to influence public discourse and mobilize support for causes (Milli et al., 2024; European Commission, 2024a).

However, while social media has empowered individuals to participate more actively in political discussions, it has also led to the spread of misinformation, echo chambers, and political polarization. The algorithms used by social media platforms often prioritize sensational or emotionally charged content, which can distort public understanding of issues and manipulate political opinions. In addition, the anonymity afforded by these platforms has led to the proliferation of fake news and malicious content, further complicating the political landscape. As a result, governments and institutions are facing the challenge of managing these new digital spaces while ensuring that citizens can engage in informed and responsible political discourse (Milli et al., 2024; Freedom House, 2024).

The role of technology in policymaking has also expanded significantly in recent years. Governments and political leaders are increasingly turning to digital tools such as artificial intelligence (AI), big data, and predictive analytics to inform decision-making. AI algorithms can analyze large volumes of data to identify trends, predict outcomes, and assist in the development of policies that address complex societal issues, from healthcare to climate change. These tools have the potential to improve the efficiency and effectiveness of governance by enabling more data-driven, evidence-based decision-making (OMB, 2024; OECD, 2024a).

At the same time, the use of AI in policymaking raises concerns about fairness, transparency, and accountability. AI algorithms are only as good as the data they are trained on, and if that data is biased or incomplete, the resulting policies may inadvertently reinforce existing inequalities. Moreover, the "black box" nature of many AI systems—where it is difficult to understand how decisions are made—has led to concerns about accountability in the policymaking process. If an algorithmic decision negatively impacts a community or individual, it can be difficult to trace responsibility or correct the error. As such, the integration of AI in politics requires careful consideration of ethical principles and safeguards to ensure that technology serves the public good (Council of Europe, 2024; European Parliament & Council, 2024).

Another area in which technology is reshaping political participation is through the rise of digital democracy. E-governance platforms, online petitions, and digital town halls have made it easier for citizens to engage with policymakers, express their opinions, and participate in decision-making processes. In some countries, governments have implemented electronic voting systems that allow citizens to cast their votes securely and conveniently from anywhere in the world. This has the potential to increase voter turnout, particularly among those who face barriers to in-person voting, such as individuals with disabilities or those living in remote areas (UN DESA, 2024; Estonian National Electoral Committee, n.d.).

However, while digital democracy has the potential to increase participation, it also raises concerns about security and accessibility. The digital divide—the gap between those with access to technology and those without—remains a significant challenge. In many parts of the world, particularly in rural and low-income areas, people lack the infrastructure or skills to engage fully in digital political processes. This can exacerbate existing inequalities in political participation and exclude vulnerable populations from important decisions that affect their lives. Additionally, the rise of online voting has raised concerns about cyberattacks and the integrity of electoral processes, particularly in light of recent incidents of foreign interference in elections (ITU, 2024; CISA/EAC/FBI/NIST, 2024).

The concentration of power in the hands of a few large technology companies is another critical issue in the relationship between technology and politics. Tech giants like Amazon, Google, Facebook, and Apple have enormous influence over public discourse, commerce, and even political processes. These companies control vast amounts of data, which they use to shape consumer behavior and influence political opinions through targeted advertising. With such power, these companies have the potential to disrupt or even undermine democratic processes by controlling the flow of information, censoring content, or manipulating political narratives (OECD, 2024a; European Commission, 2024b).

The increasing reliance on private companies for the provision of public services also raises questions about accountability and transparency. Governments often outsource key functions, such as data collection, security, and communication, to private corporations. This has led to concerns about the privatization of governance and the erosion of public trust in institutions. When private companies hold significant power over the information and services that citizens rely on, it becomes more difficult to hold them accountable for their actions (UN DESA, 2024; GAO, 2024).

The impact (U.S. Digital Service, 2024)of technology on political structures is also evident in the way it has transformed civic activism and social movements. Social media has been a powerful tool for organizing protests, advocating for policy change, and raising awareness about social justice issues. Movements like #MeToo, Black Lives Matter, and the Arab Spring have been amplified and spread across the globe through digital platforms. These movements have shown how technology can be used to empower marginalized groups, amplify their voices, and challenge entrenched power structures. However, technology has also been used to suppress dissent, as governments employ surveillance tools to track and monitor activists and protesters (Freedom House, 2024; Amnesty International, 2024).

In many cases, technology has created new opportunities for activism, but it has also introduced new risks. The use of encryption, anonymous browsing, and other digital tools has become essential for protecting the privacy and security of activists in repressive regimes. Governments have increasingly used technology to surveil and control their populations, from facial recognition software to internet censorship. As a result, activists must be more cautious about how they use technology to organize and communicate, lest they expose themselves to government repression or retaliation (Freedom House, 2024).

In the context of global politics, technology has played a pivotal role in shaping international relations and diplomacy. The use of social media and digital platforms has allowed governments to communicate directly with foreign populations, bypassing traditional diplomatic channels. This has led to more public engagement in international affairs and has given ordinary citizens a voice in global issues. At the same time, it has also created new opportunities for cyberattacks, disinformation campaigns, and foreign influence in domestic politics. The rise of digital diplomacy has forced governments to reconsider how they approach international relations, as they must now navigate the complexities of digital influence, cyber warfare, and the rapid spread of information (NATO, 2024; NATO StratCom COE, 2023).

Finally, the intersection of technology and politics calls for a reevaluation of democratic principles and governance structures. As technology becomes increasingly integrated into political systems, governments must ensure that innovation does not erode fundamental democratic values such as transparency, accountability, and individual rights. The power of technology must be harnessed in ways that enhance democratic governance rather than undermine it. Policymakers, civic leaders, and technologists must work together to create frameworks for responsible technology use in politics that prioritize the well-being of citizens and ensure the continued functioning of democratic institutions (European Parliament & Council, 2024; Council of Europe, 2024).

This chapter aims to provide a comprehensive overview of the relationship between technology and political structures, examining both the opportunities and challenges posed by digital innovation in governance and political participation. As we move further into the digital age, it is clear that technology will continue to shape the future of politics in profound ways. By understanding these dynamics, we can better navigate the complex landscape of modern governance and ensure that technology is used to strengthen, rather than undermine, democratic processes (UN DESA, 2024).

Platforms, data, and AI: how digital tools redistribute political power, reshape policy, and rewire civic participation

Read through a civic lens, the image shows how platforms, data, and AI redistribute political power—lowering barriers to voice and mobilization while creating new chokepoints controlled by platform policies and opaque algorithms. It highlights a double edge: greater reach and real-time coordination alongside surveillance, targeted manipulation, and rapid misinformation, all of which reshape how agendas form and who gets heard. The takeaway is pragmatic: technology doesn't determine democracy; design choices and public rules—transparency, accountability, privacy protections, and independent audits—can steer these tools toward broader participation, equity, and trust.

TECHNOLOGY AND GOVERNANCE: SHAPING POLITICAL POWER

Governance, at its core, involves the systems and processes by which decisions are made and implemented for a society. Historically, power was concentrated within governmental institutions, with leaders and bureaucrats making key decisions behind closed doors. These decisions often lacked transparency and accountability, and the process was dominated by a small group of elites. Over time, this system came under increasing pressure from demands for greater democratic participation and public oversight. The rise of technology, particularly in the 20th and 21st centuries, has profoundly shifted the dynamics of political power, decentralizing access to information and providing citizens with more direct channels to influence political outcomes. The advent of the internet, mobile technologies, and social media platforms has created a more connected and informed electorate, and this shift has the potential to radically reshape governance structures across the globe (UN DESA, 2024; World Bank, 2024).

The development of digital platforms has brought new opportunities for political engagement. No longer do citizens need to rely solely on traditional media or government-sanctioned channels to obtain information; instead, they can now access a multitude of sources, contribute to debates, and express opinions directly to policymakers. Social media platforms such as Twitter, Facebook, and YouTube have democratized the dissemination of political information, enabling movements and ideas to gain traction rapidly, sometimes bypassing traditional political gatekeepers. Additionally, technology has provided political leaders with new methods of engaging with constituents, enabling them to directly communicate and interact with citizens in real-time, at a scale never before possible (Milli et al., 2024).

Governments worldwide have begun to incorporate digital tools into their operations, transforming both the way public services are delivered and how policies are formulated. E-government initiatives are now commonplace in many countries, allowing citizens to access public services online, from renewing identification documents to applying for government benefits. This increased accessibility not only improves efficiency but also enhances citizen satisfaction by reducing wait times and making services more available. Moreover, digital platforms have allowed governments to streamline communication with the public, making it easier to disseminate important information about health, safety, and public welfare, especially during times of crisis (World Bank, 2024; USDS, 2024).

In addition to these practical benefits, the integration of technology into governance has played a crucial role in increasing the transparency of government actions. Digital dashboards and open data portals now allow citizens to track public spending, monitor the progress of public projects, and evaluate the outcomes of policy decisions. These tools enable a level of accountability that was previously difficult to achieve, as they allow the public and civil society organizations to hold governments accountable for their actions. With more accessible data, citizens can make more informed decisions about political candidates, party platforms, and public policy issues, fostering a more engaged and participatory democracy (UN DESA, 2024; Maali & Perrault, 2024).

The use of big data and artificial intelligence (AI) has further enhanced the policymaking process. Governments are increasingly turning to AI and machine learning algorithms to process vast amounts of data and derive insights that can improve decision-making. By analyzing patterns in data, governments can identify emerging issues, predict trends, and tailor policies to address the specific needs of different population groups. For example, AI can be used to identify regions where health services are lacking, helping to direct resources more effectively. Similarly, big data is being used in areas such as crime prevention, climate change adaptation, and transportation planning to create more targeted and efficient policies (OMB, 2024; OECD, 2024a).

These technological advancements are not without their challenges, however. One of the primary concerns is the potential for the misuse of personal data. Governments often collect vast amounts of data on their citizens in order to improve public services, enhance policy outcomes, or address social issues. However, the use of this data can be problematic if it infringes on privacy rights. For example, data collection for the purpose of social welfare programs may inadvertently expose sensitive information about individuals, potentially leading to discrimination or stigmatization. Governments must balance the benefits of data-driven policymaking with the need to protect citizens' privacy and safeguard against misuse of personal information (USCCR, 2024).

The rise of surveillance technologies has further complicated the intersection of technology and governance. Tools such as facial recognition, geolocation tracking, and biometric data collection are increasingly being employed by governments to monitor public spaces and identify potential threats to national security. While these technologies can enhance safety and security, they also raise ethical questions about the balance between national security and individual freedoms. The implementation of mass surveillance programs can lead to overreach, where governments monitor the behavior of ordinary citizens without sufficient justification or oversight. This has raised alarm among privacy advocates and human rights organizations, who warn that such practices could erode civil liberties and lead to a "Big Brother" scenario (Amnesty International, 2024; European Parliament & Council, 2024).

The challenge for governments is to harness the benefits of technology while ensuring that citizens' rights are protected. As more decisions are made based on data-driven insights, the risk of excluding or marginalizing certain populations grows. For instance, if AI systems are not designed to account for social, cultural, or economic differences, they could inadvertently perpetuate biases and reinforce existing inequalities. This is particularly problematic in areas like criminal justice, where algorithmic decision-making is increasingly used to determine sentencing, parole, and bail. The lack of transparency in how these algorithms function further exacerbates concerns, as it becomes difficult to challenge decisions that are made without full understanding of the processes involved (Council of Europe, 2024; OMB, 2024).

Governments must also contend with the issue of digital divides—disparities in access to technology and digital literacy that exist both within countries and between the global north and south. While some citizens have easy access to high-speed internet, smartphones, and digital tools, others are excluded from participating in the digital economy due to geographic, financial, or educational barriers. This divide not only prevents marginalized groups from benefiting from digital governance tools but also hampers their ability to engage meaningfully in political processes. As a result, governments must invest in bridging these gaps by improving infrastructure, expanding access to affordable technology, and providing digital literacy programs to ensure that all citizens can participate in the digital age (ITU, 2024).

The increasing reliance on technology for governance also raises concerns about the centralization of power. While technology can democratize information and participation, it can also concentrate power in the hands of a few individuals or organizations. Major tech companies, such as Google, Amazon, and Facebook, have enormous influence over the digital landscape, shaping everything from the information people see to the products they buy. These companies control vast amounts of personal data, which they use to target political advertisements and influence consumer behavior. As these companies continue to grow, they wield increasing influence over political outcomes, raising questions about whether democratic governance can truly thrive when so much power is concentrated in the private sector (OECD, 2024a; European Commission, 2024b).

Furthermore, the increasing use of technology in governance has introduced new challenges regarding cybersecurity and the integrity of political processes. Cyberattacks, disinformation campaigns, and electoral manipulation are growing concerns, particularly as governments move toward digital voting systems and online political participation. The potential for foreign interference in national elections through hacking and the spread of fake news has prompted many to question the security and legitimacy of digital platforms. As more governments rely on these platforms for communication, policymaking, and public engagement, the risk of cyberattacks grows, requiring greater investment in cybersecurity and the development of robust protocols to ensure the integrity of democratic processes (CISA/EAC/FBI/NIST, 2024; Saeva & Tasheva, 2024).

Technology also plays a pivotal role in shaping public opinion and political mobilization. Online platforms allow individuals to form communities, organize protests, and advocate for policy changes at an unprecedented scale. Movements like #MeToo, Black Lives Matter, and the Arab Spring demonstrate how social media and digital platforms can amplify voices that were previously marginalized or ignored. These movements have not only led to real-world political changes but have also redefined how people engage with politics. Technology has given citizens new ways to hold governments accountable, challenge oppressive systems, and demand justice. However, these movements also face significant challenges, including online harassment, misinformation campaigns, and government crackdowns on digital organizing (Freedom House, 2024).

One of the defining features of modern politics is the increasing role of data analytics in shaping public opinion. Politicians, political parties, and interest groups now rely heavily on data-driven strategies to target voters and shape political discourse. The use of data analytics in campaigns allows for more personalized, persuasive messaging that resonates with specific groups of voters. While this has the potential to make political campaigns more efficient and responsive to the needs of constituents, it also raises concerns about privacy, manipulation, and the ethical implications of using personal data to influence electoral outcomes (OMB, 2024; OECD, 2024a).

The use of technology in governance also raises critical questions about the future of democracy itself. As technology continues to evolve, there is a growing need to reassess democratic principles and structures to ensure that they remain relevant and effective. Digital tools and platforms are reshaping the political landscape, but they are also challenging the traditional mechanisms of democratic accountability, representation, and participation. For example, the rise of digital lobbying, the manipulation of public opinion through social media, and the growing influence of tech companies on policy decisions all suggest that new approaches to governance may be needed to preserve the integrity of democratic institutions (European Parliament & Council, 2024; Council of Europe, 2024).

Governments must also be mindful of the ethical considerations that come with the use of technology in governance. The rapid pace of technological change often outpaces the development of laws and regulations designed to safeguard the public interest. As new technologies emerge, such as AI, machine learning, and blockchain, governments must consider how these innovations can be harnessed for the public good while mitigating the potential risks they pose. This requires a proactive approach to policy-making that balances innovation with accountability and ethical responsibility (Council of Europe, 2024; OMB, 2024).

In conclusion, technology has fundamentally transformed the way governments operate, make decisions, and engage with citizens. It has opened new avenues for transparency, accountability, and participation, but it has also introduced significant challenges related to privacy, power dynamics, and security. As we move forward into an increasingly digital age, governments must navigate the complexities of integrating technology into political systems while ensuring that democratic values and citizens' rights are upheld. By doing so, they can leverage the power of technology to strengthen governance and foster more inclusive, transparent, and participatory political structures (UN DESA, 2024; World Bank, 2024).

TECHNOLOGY AND POLICY-MAKING: DATA-DRIVEN DECISION MAKING

The role of technology in policymaking has undergone a significant transformation with the advent of big data, machine learning, and artificial intelligence. These tools are revolutionizing how governments and policymakers approach the formulation and evaluation of policies. In the past, policy decisions were often based on limited data, expert opinions, or political ideologies, which could be subject to bias or incomplete information. Today, these advanced technologies enable policymakers to analyze complex social, economic, and environmental issues with greater precision, speed, and efficiency. The availability of vast datasets allows governments to identify emerging trends, monitor real-time events, and make informed decisions that are grounded in empirical evidence (OMB, 2024; World Bank, 2024).

For example, data analytics allows governments to identify trends in public health, predict economic outcomes, and allocate resources more effectively. In public health, data analytics can track the spread of diseases, helping authorities respond swiftly to outbreaks and implement targeted health interventions. Similarly, economic data can be analyzed to predict economic recessions or booms, enabling proactive policy measures to mitigate negative impacts. Governments can also use data to optimize resource allocation, ensuring that public funds are directed toward the most pressing needs, whether that be education, infrastructure, or welfare services. With data-driven insights, policymakers are better equipped to address complex societal challenges (UN DESA, 2024; OECD, 2024a).

In addition to data analytics, AI-driven simulations are increasingly being used to model the potential outcomes of various policy options. These simulations can replicate real-world scenarios and predict the long-term consequences of different courses of action. For example, AI can simulate the effects of climate change mitigation strategies on local economies, taking into account variables such as industry shifts, employment patterns, and environmental impacts. Similarly, AI can predict the impact of taxation changes on business growth or employment rates, providing policymakers with valuable insights into the potential economic consequences of their decisions (OECD, 2024b; OMB, 2024).

AI-powered simulations are also proving invaluable in areas such as transportation planning, urban development, and environmental regulation. These models allow policymakers to explore different scenarios and assess the risks and benefits of various policy interventions. By using machine learning algorithms, governments can continually refine their models to improve their accuracy and predictiveness, creating more robust policy frameworks. The ability to model and forecast the impact of policies before they are implemented represents a major leap forward in evidence-based policymaking, offering a more proactive and informed approach to governance (World Bank, 2024).

As these technologies advance, they promise to create more sophisticated, evidence-based policy frameworks. The integration of AI, big data, and machine learning into the policy-making process allows for more nuanced and tailored solutions that are better aligned with the needs of specific communities or regions. For instance, policies related to climate change, social justice, or public health can be customized based on real-time data, ensuring that interventions are as effective as possible. The promise of personalized policy interventions also holds potential for reducing disparities between different demographic groups, as policies can be designed to specifically address the unique challenges faced by marginalized populations (Council of Europe, 2024; OMB, 2024).

However, the use of technology in policymaking also introduces several challenges. One of the primary concerns is the potential for bias in AI algorithms. Machine learning algorithms rely on data to make predictions, and if the data used to train these algorithms is flawed or biased, the resulting policies may disproportionately benefit certain groups while disadvantaging others. For example, if the data used to develop AI models reflects existing social inequalities—such as racial or economic disparities—the AI may perpetuate these biases in its predictions and recommendations. This could lead to policies that unfairly impact marginalized communities or reinforce systemic inequalities (European Parliament & Council, 2024; USCCR, 2024).

The risk of algorithmic bias is particularly concerning in areas such as criminal justice, healthcare, and housing, where decisions made by AI systems can have life-altering consequences. In the criminal justice system, for instance, predictive algorithms are used to assess the risk of recidivism, which can influence sentencing, parole decisions, and probation. If these algorithms are trained on biased data, they may unfairly target certain racial or socioeconomic groups, leading to disproportionate incarceration rates. Similarly, AI systems used in healthcare may not account for the specific needs of minority populations, potentially exacerbating health disparities (Council of Europe, 2024).

In response to these concerns, policymakers must ensure that AI algorithms are transparent, equitable, and accountable. Transparency is key to understanding how algorithms make decisions and identifying potential sources of bias. Governments must establish clear guidelines for how data is collected, processed, and used to train AI models, ensuring that these practices are consistent with principles of fairness and justice. In addition, independent audits of AI systems should be conducted regularly to identify and mitigate bias, ensuring that the technology is functioning as intended and not reinforcing existing inequalities (OMB, 2024; European Parliament & Council, 2024).

Another challenge with the integration of technology into policymaking is the accessibility of these systems. As AI and data-driven decision-making become increasingly central to the policymaking process, there is a risk that these technologies could deepen the digital divide, particularly in low-income or rural areas where access to high-speed internet, digital tools, and technical skills is limited. If policymakers rely solely on these advanced technologies to inform their decisions, they may inadvertently exclude certain groups from the political process, leading to policies that do not reflect the needs or perspectives of all citizens (ITU, 2024; UN DESA, 2024).

To address this challenge, governments must prioritize digital inclusion and ensure that all citizens have access to the tools and knowledge needed to participate in the digital economy. This includes investing in broadband infrastructure, providing digital literacy programs, and promoting policies that ensure equitable access to technology. By ensuring that technology is accessible to all, governments can help ensure that the benefits of data-driven policymaking are distributed fairly across society (ITU, 2024).

The increased reliance on technology in policymaking also raises questions about accountability. While AI and big data can enhance decision-making, they also introduce a level of complexity that can make it more difficult for the public to understand and scrutinize policy decisions. As policies are increasingly informed by AI models and data-driven simulations, it may become more challenging for citizens to assess the reasoning behind specific policy choices. This lack of transparency can undermine public trust in government institutions, especially if people feel that decisions are being made by faceless algorithms rather than elected representatives (OMB, 2024; OECD, 2024a).

To address these concerns, governments must ensure that their use of technology in policymaking is transparent and accountable. This includes providing clear explanations of how data and algorithms are being used to inform policy decisions, as well as engaging citizens in the policymaking process through consultations and public feedback mechanisms. By promoting transparency and accountability, governments can build trust with their constituents and ensure that technology is being used in a way that reflects the public's best interests (UN DESA, 2024).

Additionally, policymakers must be mindful of the ethical implications of using AI and big data in decision-making. As these technologies become more integrated into the policy process, there is an increasing need to establish ethical guidelines that govern their use. These guidelines should address issues such as privacy, consent, and fairness, and ensure that AI systems are designed to respect human rights and uphold democratic values. For example, when using AI to make decisions about welfare distribution or healthcare access, policymakers must ensure that these systems do not discriminate against vulnerable populations or violate individuals' privacy rights (Council of Europe, 2024).

Ethical concerns also arise when considering the potential for AI to replace human decision-makers in certain policy areas. While AI systems can process vast amounts of data and generate insights quickly, they lack the empathy, judgment, and moral reasoning that human decision-makers bring to the table. In areas such as healthcare or criminal justice, where decisions often involve deeply personal or ethical considerations, it is important to ensure that technology supplements human judgment rather than replacing it entirely. Policymakers must strike a balance between leveraging the power of AI and ensuring that human values and ethical principles are upheld in the decision-making process (European Parliament & Council, 2024; Council of Europe, 2024).

The rapid pace of technological innovation also presents challenges for lawmakers and regulators. As new technologies such as AI, blockchain, and quantum computing continue to emerge, governments must adapt their regulatory frameworks to keep up with the evolving landscape. This requires policymakers to have a deep understanding of the technologies they are regulating and the potential risks they pose. In many cases, regulators are playing catch-up, working to establish policies and regulations that can address the complex issues raised by new technologies. This can create a sense of uncertainty for businesses and citizens, who may be unsure about how new technologies will be regulated or whether they will face new restrictions or requirements (European Commission, 2024a; European Parliament & Council, 2024).

Despite these challenges, the integration of technology into policymaking offers immense potential to improve governance and public service delivery. By leveraging big data, AI, and machine learning, governments can create more efficient, effective, and responsive policies that are based on evidence rather than ideology. The potential for AI to model the long-term effects of policies, predict future trends, and identify emerging issues represents a significant step forward in the policymaking process. However, to fully realize the benefits of these technologies, policymakers must address the ethical, social, and economic challenges they present. By ensuring that technology is used in a transparent, equitable, and accountable manner, governments can harness the power of innovation to create a more just and prosperous society for all citizens (OMB, 2024; Council of Europe, 2024).

As we move into the future, the role of technology in policymaking will only continue to grow. New advancements in AI, big data, and machine learning will enable governments to make even more accurate predictions and optimize policy decisions. However, these technologies must be used responsibly and ethically to ensure that they serve the public good. The key to successful integration lies in balancing the power of technology with the values of transparency, fairness, and accountability, ensuring that technology is a force for positive change rather than a tool for exploitation or inequality (European Parliament & Council, 2024; OECD, 2024a).

TECHNOLOGY AND CIVIC ENGAGEMENT: EMPOWERING OR ERODING DEMOCRACY?

The intersection of technology and civic engagement has been one of the most profound developments in contemporary politics. Technology has fundamentally transformed how citizens participate in the political process, enabling new forms of political engagement that were once unimaginable. The digital age has introduced online petitions, social media activism, crowdfunding for political causes, and virtual town halls, among other innovative platforms for political participation. These digital tools provide individuals with unprecedented opportunities to voice their opinions, organize communities, and influence public policy in ways that were previously reserved for a select few (UN DESA, 2024; World Bank, 2024).

Social media platforms, in particular, have become vital tools for organizing political campaigns, spreading political ideologies, and creating global movements. The ability to instantly connect with millions of people, share ideas, and mobilize resources has allowed ordinary citizens to participate in political discourse on a global scale. Platforms such as Twitter, Facebook, Instagram, and

TikTok have become central to political communication, where ideas are shared, debates unfold, and calls to action are amplified. The ease of communication has made it possible for social movements to gain momentum rapidly, while also offering a space for political leaders and public figures to connect directly with their constituents (Pew Research Center, 2024; Freedom House, 2024).

The rise of online activism, from the Arab Spring to Black Lives Matter, demonstrates the power of technology to amplify marginalized voices and mobilize large-scale movements. These movements have showcased how technology can be harnessed to challenge oppressive regimes, advocate for social justice, and hold governments accountable. The use of social media during the Arab Spring, for example, enabled protestors to circumvent censorship and organize mass demonstrations, contributing to political revolutions across the Middle East and North Africa. Similarly, the Black Lives Matter movement, fueled by social media, has raised awareness of police brutality and systemic racism, leading to protests and policy changes around the world. The ability to reach a global audience, organize protests in real-time, and share powerful messages of social change has made technology an indispensable tool for modern activism (Freedom House, 2024).

Social media platforms have also played a pivotal role in the creation of virtual communities and the formation of collective identities. Activists can now come together online to share their experiences, discuss issues, and coordinate actions in ways that transcend geographical boundaries. This digital solidarity is especially significant for marginalized groups who may face barriers to participation in traditional political spaces. For instance, individuals who live in regions with limited access to physical protests or demonstrations can still participate in online activism and engage with global conversations. Technology has democratized political participation, allowing people from diverse backgrounds to come together, amplify their voices, and advocate for change (UN DESA, 2024; Freedom House, 2024).

By bypassing traditional media channels, activists can directly communicate with their audiences, build solidarity, and organize protests or campaigns in real-time. The ability to reach large audiences instantly without relying on journalists or broadcasters has given activists unprecedented control over the narrative. This direct communication allows movements to shape their own stories, frame issues in their own terms, and resist the framing of traditional media outlets. Social media, as a result, has become a powerful counterbalance to institutionalized power structures and a tool for challenging the status quo (Pew Research Center, 2024).

However, technology's role in civic engagement is not without its drawbacks. The increasing prevalence of online echo chambers and filter bubbles, where users are exposed only to information that aligns with their existing beliefs, has led to political polarization. Echo chambers limit the diversity of information individuals receive, reinforcing their preexisting biases and discouraging open dialogue. The ability to filter content based on personal preferences or algorithmic recommendations exacerbates this issue, as users are often presented with information that confirms their worldview while avoiding content that challenges their perspectives (Milli et al., 2025; Pew Research Center, 2024).

The polarization driven by echo chambers is further amplified by the algorithms used by social media platforms, which prioritize sensational content that generates high levels of engagement. This algorithmic bias leads to the spread of divisive rhetoric, misinformation, and fake news, as content that is emotionally charged or controversial is more likely to be shared and go viral. The proliferation of false information can distort public understanding of key political issues, making it harder for citizens to make informed decisions and participate in meaningful political discourse. As a result, technology that was initially intended to enhance civic engagement may inadvertently contribute to a more fragmented and misinformed public sphere (Milli et al., 2025; Freedom House, 2024).

Moreover, the rapid spread of misinformation can have serious consequences for democracy. False claims and misleading narratives can undermine trust in institutions, skew election outcomes, and polarize political debates. In some cases, misinformation can even incite violence or lead to civil unrest, as seen in the aftermath of elections or political events where false narratives have fueled protests, riots, or other forms of civil disobedience. As the digital space becomes increasingly central to political engagement, addressing the challenges of misinformation and disinformation is crucial to ensuring the integrity of democratic processes (Freedom House, 2024; European Commission, 2024).

Online harassment is another significant issue in digital civic engagement. The anonymity and distance provided by the internet have made it easier for individuals to engage in harmful behavior, including cyberbullying, trolling, and hate speech. Public figures, activists, and ordinary citizens alike can become targets of online abuse, which can deter them from participating in political discourse or activism. In some cases, online harassment can escalate into physical threats or intimidation, creating a hostile environment for those who seek to engage in political discussions. The rise of cyberattacks, such as hacking and data breaches, also raises concerns about the security of online political engagement, as malicious actors can undermine the integrity of political campaigns or steal sensitive information (Inter-Parliamentary Union, 2025; Freedom House, 2024).

The digital space has also introduced new challenges related to privacy and surveillance. While technology enables greater participation, it also raises concerns about the collection and use of personal data. Governments, corporations, and other organizations now have access to vast amounts of data about citizens' online activities, including their political views, affiliations, and behaviors. This data can be used to target individuals with personalized political ads, manipulate opinions, or suppress dissent. The use of personal data for political purposes has sparked debates about consent, transparency, and the ethical implications of digital surveillance. The balance between enabling civic engagement and protecting privacy is an ongoing challenge in the digital age (European Parliament & Council, 2024; OECD, 2024).

Furthermore, while technology has empowered individuals to participate in political processes, it has also created barriers for some groups. Digital literacy remains a significant challenge for older adults, marginalized communities, and those without access to the internet or modern technology. The digital divide, therefore, poses a significant threat to equitable political participation. Those who lack access to technology are excluded from digital political spaces, which can result in unequal representation and political power. This divide is particularly pronounced in rural or low-income areas, where internet connectivity and technological infrastructure are limited. The rise of digital politics, if left unaddressed, may deepen existing inequalities and marginalize those who are already disadvantaged (ITU, 2024; UN DESA, 2024).

Ensuring that all citizens have access to the tools and knowledge necessary to engage in digital democracy is crucial to maintaining an inclusive political system. This requires investment in digital literacy programs, public internet access, and policies that reduce the cost of technology for low-income households. Governments, civil society organizations, and the private sector must work together to bridge the digital divide and ensure that no one is left behind in the digital political landscape (OECD, 2024; ITU, 2024).

The digital divide also extends to the accessibility of online platforms themselves. While many people have access to the internet, they may not have the necessary devices or skills to navigate digital platforms effectively. For instance, individuals with disabilities may face challenges when trying to engage in online political activities if platforms are not designed to be accessible. Ensuring that online spaces are inclusive and accessible to all citizens is essential for fostering a truly participatory democracy (UN DESA, 2024).

Moreover, the rapid pace of technological innovation means that the landscape of digital civic engagement is constantly evolving. New technologies, such as artificial intelligence, virtual reality, and blockchain, have the potential to further transform political participation and engagement. For example, AI could be used to analyze political trends or predict voter behavior, while virtual reality could allow citizens to participate in immersive political debates or town halls from anywhere in the world. These technologies could create new opportunities for political engagement, but they also raise questions about privacy, ethics, and the potential for manipulation (OECD, 2024; Freedom House, 2024).

In response to these challenges, policymakers must ensure that regulations and policies are put in place to protect the integrity of digital democracy. This includes addressing issues such as misinformation, online harassment, data privacy, and digital accessibility. Governments must strike a balance between fostering innovation and protecting citizens' rights, ensuring that technology serves the public good rather than undermining democratic principles (European Commission, 2024; European Parliament & Council, 2024).

As technology continues to shape civic engagement, it is important to remember that it is only one tool among many in the political process. While digital platforms offer new opportunities for participation, they are not a panacea for the challenges facing democracy. The quality of civic engagement depends not only on access to technology but also on the willingness of citizens to engage in meaningful dialogue, the commitment of political leaders to represent diverse voices, and the ability of institutions to ensure that all citizens are treated fairly and with respect. The future of civic engagement lies in harnessing the power of technology while upholding the values of inclusivity, transparency, and accountability (Pew Research Center, 2024; Freedom House, 2024).

THE FUTURE OF TECHNOLOGY AND POLITICAL STRUCTURES: CHALLENGES AND OPPORTUNITIES

As technology continues to evolve, its impact on political structures will likely grow even more profound. Emerging technologies such as blockchain, virtual reality, and augmented reality hold immense potential to reshape the political landscape. These innovations are expected to transform how political campaigns are conducted, how policies are implemented, and how citizens interact with their governments. As we enter an era where technology becomes even more integrated into political processes, the question arises: how can we harness these advancements while safeguarding democratic values? (OECD, 2024).

Blockchain, for example, could revolutionize voting systems by ensuring secure, transparent, and tamper-proof elections. This technology enables the creation of decentralized ledgers that are immutable, meaning once information is recorded, it cannot be altered. By using blockchain, elections could be more secure, reducing the risk of fraud or tampering with votes. This increased transparency could help rebuild public trust in democratic processes, especially in countries where election integrity is often questioned. Blockchain's potential to ensure the authenticity of votes would be a significant step forward in the fight against corruption and electoral fraud, offering a new level of accountability in democratic systems. At the same time, leading election-security authorities continue to caution that remote electronic ballot return remains high-risk even with controls, underscoring the need for careful design and strong paper trails (CISA, EAC, FBI, & NIST, 2020; NIST, 2024).

Furthermore, blockchain could also create a more accessible and convenient voting experience, particularly for citizens who have difficulty accessing traditional polling stations. Blockchain-based systems could enable remote voting, allowing citizens to cast their ballots securely from anywhere in the world. This could increase voter turnout, especially among underserved populations such as expatriates, people with disabilities, or those living in conflict zones. By making voting more accessible, blockchain technology could strengthen democratic engagement and ensure that all voices are heard, regardless of geographic or social barriers (CISA, EAC, FBI, & NIST, 2020).

Virtual reality (VR) and augmented reality (AR) also have the potential to transform political engagement by creating immersive and interactive environments. VR could allow citizens to experience political events, debates, and town halls in a fully immersive virtual space, making it easier for individuals to participate in political discussions regardless of their physical location. Citizens could engage with political campaigns in ways that were previously unimaginable, such as attending virtual rallies, touring virtual representations of government projects, or experiencing the impact of policies in real-time simulations. VR could also be used as a tool for political education, offering interactive learning experiences that help citizens understand complex political issues, historical events, or government procedures (OECD, 2024).

Augmented reality, on the other hand, could enhance the way citizens interact with their physical environment and access political information. By overlaying digital information onto the real world, AR could provide citizens with real-time access to data about political events, government actions, or public policies as they move through public spaces. For instance, a citizen could point their smartphone at a government building and receive detailed information about the representatives who work there, the policies they support, or upcoming legislative votes. This integration of AR into everyday life would make it easier for citizens to stay informed and engaged with the political process, fostering a more informed and active electorate (UN DESA, 2024; OECD, 2024).

While these technologies offer immense potential, they also bring new challenges that need to be addressed. One significant concern is the centralization of technological power in the hands of a few large corporations. Companies such as Facebook, Google, and Amazon dominate the digital space, controlling vast amounts of personal data and influencing political discourse on an unprecedented scale. These companies have the resources and influence to shape public opinion, interfere with political processes, and even sway election outcomes. The concentration of digital power in the hands of a few entities raises important questions about the role of private companies in governing public discourse (Freedom House, 2024; European Parliament & Council, 2024).

The control that these corporations wield over digital platforms has serious implications for democracy. Large tech companies are able to shape the narratives that circulate online, controlling the flow of information and determining what content is prioritized in users' feeds. This level of control can be used to promote certain political ideologies, suppress dissenting voices, or manipulate public opinion through targeted advertising. With their vast databases of user information, these corporations have the ability to tailor content to specific individuals, influencing their political views and behaviors without their full awareness or consent. This creates a situation where public discourse is heavily mediated by corporate interests rather than a free exchange of ideas (Pew Research Center, 2024; European Commission, 2024).

Moreover, the concentration of technological power also raises concerns about privacy and data security. With personal data being collected and analyzed on such a massive scale, there is an increased risk of data breaches, identity theft, and surveillance. Citizens may not always be aware of the extent to which their information is being tracked and used, making it difficult for individuals to fully understand the implications of their online actions. Governments must consider how to regulate the use of personal data while ensuring that the benefits of technology, such as personalized political engagement or targeted advocacy, are not overshadowed by privacy concerns (OECD, 2024; European Parliament & Council, 2024).

The centralization of power in the tech industry also creates challenges in terms of accountability. Private companies often have more influence over public discourse than elected officials, leading to a situation where corporate interests may take precedence over democratic values. This concentration of power makes it difficult for citizens to hold these companies accountable for their actions, as they are not directly answerable to the public in the same way that politicians are. In some cases, these companies may operate outside the jurisdiction of national governments, making it difficult to impose regulations or enforce laws that protect public interests. To address this issue, governments may need to introduce new regulatory frameworks that ensure tech companies are held accountable for their actions and that their practices align with democratic principles (European Commission, 2024; European Parliament & Council, 2024).

As emerging technologies continue to reshape the political landscape, governments must consider how to regulate their use without stifling innovation. Striking the right balance between encouraging technological progress and protecting democratic values will be one of the key challenges of the coming decades. Policymakers will need to establish frameworks for ethical technology use that promote transparency, fairness, and accountability while fostering innovation. This requires careful consideration of the potential risks and benefits of emerging technologies, as well as a commitment to ensuring that these technologies serve the public interest rather than corporate profits (European Parliament & Council, 2024; OECD, 2024).

One important area for regulation is the ethical use of artificial intelligence (AI) and machine learning in political campaigns. AI has the potential to revolutionize how political campaigns are run, enabling more targeted outreach and more efficient voter engagement. However, the use of AI in politics also raises concerns about privacy, manipulation, and the fairness of elections. AI-powered algorithms can be used to analyze vast amounts of voter data, predict behavior, and craft personalized messages that may sway individuals' opinions or voting decisions. Without proper oversight, AI could be used to manipulate voters or spread misinformation, undermining the integrity of democratic processes. Governments must establish ethical guidelines for the use of AI in politics to ensure that it is used responsibly and transparently (European Parliament & Council, 2024; European Commission, 2024).

Another challenge related to emerging technologies is the potential for technological monopolies to form. As tech companies continue to consolidate and acquire smaller startups, the market becomes increasingly dominated by a few powerful players. This raises concerns about competition and market fairness, as monopolies can stifle innovation, limit consumer choice, and lead to the concentration of political power. To address this, governments may need to strengthen antitrust laws and promote policies that encourage competition and innovation in the tech sector. By fostering a more competitive market, governments can help ensure that technological advancements benefit society as a whole, rather than just a select few corporations (OECD, 2024).

In addition to regulating the use of technology, governments must also invest in digital infrastructure and education to ensure that all citizens have access to the tools and knowledge necessary to participate in the digital political landscape. As digital engagement becomes increasingly central to political participation, it is crucial to address the digital divide and ensure that marginalized communities have access to the resources they need. This includes improving access to high-speed internet, providing digital literacy programs, and ensuring that all citizens can navigate digital platforms effectively (ITU, 2024; UN DESA, 2024).

Digital literacy will become an essential skill for future generations, as the ability to understand, analyze, and critically engage with digital content will be necessary for informed political participation. Governments and educational institutions must work together to integrate digital literacy into curricula and provide citizens with the tools they need to engage with political content responsibly. This will help citizens make informed decisions, recognize misinformation, and participate in meaningful political discourse (OECD, 2024; Pew Research Center, 2024).

In conclusion, emerging technologies such as blockchain, virtual reality, and augmented reality offer tremendous opportunities for political engagement, but they also pose significant challenges. To ensure that these technologies serve the public good, governments must regulate their use responsibly, address issues related to privacy and data security, and promote equitable access to digital tools. By striking the right balance between innovation and regulation, policymakers can help create a more inclusive, transparent, and accountable political system that leverages the power of technology to strengthen democracy and foster greater civic participation (European Parliament & Council, 2024; OECD, 2024).

CONCLUSION

The relationship between technology and political structures is complex and multifaceted, with both significant opportunities and profound challenges. Technology, when effectively harnessed, has the potential to enhance governance, streamline policymaking, and increase civic participation, fundamentally transforming how societies function. From making decision-making processes more data-driven and transparent to enabling more direct engagement between governments and citizens, the digital age has opened new possibilities for political systems worldwide. At the same time, the proliferation of new technologies has raised concerns related to power concentration, equity, and accountability—issues that must be carefully navigated to preserve democratic values (UN DESA, 2024; OECD, 2024).

One of the greatest promises of technology lies in its ability to enhance the efficiency and effectiveness of governance. The advent of tools such as big data analytics, artificial intelligence, and blockchain has enabled governments to make more informed and evidence-based decisions. Data analytics allows policymakers to examine vast quantities of information, recognize trends, and make predictions, leading to more precise policy recommendations and better-targeted resource allocation. Furthermore, the transparency and security offered by technologies like blockchain could revolutionize systems such as voting, providing a more reliable and tamper-proof means of ensuring the integrity of democratic processes. However, while these technologies offer new ways to manage societal challenges, their full potential can only be realized if they are implemented in ways that uphold the principles of justice and fairness (European Parliament & Council, 2024; CISA, EAC, FBI, & NIST, 2020).

Another key benefit of technological advancements is the expansion of civic engagement. The rise of social media and digital platforms has empowered citizens to participate in political life like never before. Individuals now have more opportunities to voice their opinions, mobilize communities, and influence public policies, often bypassing traditional media or institutional gatekeepers. Digital platforms have played pivotal roles in large-scale political movements such as the Arab Spring and Black Lives Matter, where activists have used these platforms to raise awareness, organize protests, and challenge political systems. These technologies allow for real-time communication, democratizing access to information and giving voice to marginalized communities that might otherwise remain unheard. This represents a significant shift toward more participatory models of democracy, where citizens are not just passive recipients of government decisions but active participants in shaping policy and societal norms (Freedom House, 2024; Pew Research Center, 2024).

Despite these benefits, the rise of technology also presents substantial risks. One of the most pressing concerns is the concentration of power in the hands of a few large corporations, which control much of the digital landscape. Companies like Facebook, Google, and Amazon have accumulated vast amounts of data, giving them considerable influence over public discourse and political outcomes. This concentration of power raises important questions about how to balance corporate interests with the public good. These companies often act as arbiters of information, controlling the flow of news, political discourse, and even determining which ideas are amplified or suppressed. The potential for these companies to influence or manipulate public opinion, through algorithms that prioritize sensational or divisive content, is a growing concern. Moreover, the commercialization of data leads to situations where citizens' personal information is exploited for profit, often without their informed consent. The unchecked power of these digital giants could erode trust in democratic institutions and undermine the very principles upon which democratic societies are built (OECD, 2024; European Commission, 2024).

Furthermore, the increasing reliance on technology in political processes exacerbates issues of digital inequity. The digital divide remains a significant barrier to equal participation in the political system. For many individuals, especially those in rural or marginalized communities, access to the internet, smartphones, or digital tools is limited. Without access to these technologies, many are excluded from the political conversation, missing out on opportunities for engagement and representation. Digital literacy, too, becomes a crucial factor in determining who can fully participate in digital democracy. Older adults, individuals in lower-income households, and those without formal education may lack the skills needed to navigate online platforms effectively, thus creating a new form of political disenfranchisement. To ensure that technology works for all, it is essential to bridge these gaps in access and education, providing all citizens with the tools they need to engage in the digital political landscape (ITU, 2024; UN DESA, 2024).

As the role of technology in politics continues to expand, the need for thoughtful regulation and governance has never been more urgent. Governments must play an active role in ensuring that technological advancements align with democratic values, protect citizens' privacy, and safeguard against abuses of power. This requires not only creating laws that regulate the use of technology but also fostering a culture of ethical innovation. Policymakers must develop frameworks that prioritize the public interest over corporate profits, ensuring that technologies are designed and implemented with fairness, transparency, and accountability in mind. These regulations should address issues like data privacy, algorithmic fairness, and the ethical implications of emerging technologies such as AI and machine learning. Ensuring that these technologies are used responsibly will help prevent them from exacerbating existing social inequalities or undermining democratic processes (European Parliament & Council, 2024; European Commission, 2024).

Ethical innovation is critical to ensuring that technology remains a tool for empowerment rather than a mechanism for oppression. Companies developing new technologies should be held to high ethical standards, ensuring that their products serve the public good and respect human rights. This includes creating technologies that promote inclusivity, accessibility, and equity, and actively working to eliminate bias from algorithms and data models. Companies and governments must also collaborate to address the environmental and societal impacts of technological development, ensuring that innovation does not come at the expense of public well-being or environmental sustainability. By embedding ethical considerations into the design and implementation of technology, we can ensure that technological progress supports rather than undermines democratic principles (OECD, 2024; Freedom House, 2024).

Public engagement also plays a vital role in shaping how technology is used in politics. Citizens must be actively involved in the conversation around technological development and its implications for democracy. This includes fostering a public understanding of how digital platforms work, how algorithms shape information, and how personal data is collected and used. Education and outreach programs can help empower citizens to navigate the digital landscape responsibly, enabling them to participate fully in the political process. Moreover, public discourse around the ethical implications of technology can help ensure that these technologies are developed with the broader public interest in mind, rather than just the interests of powerful corporations or governments. It is essential for citizens to hold both corporations and governments accountable for how technology is used, ensuring that these entities act in the best interest of society as a whole (Pew Research Center, 2024; OECD, 2024).

In conclusion, while technology offers enormous potential to enhance governance, increase civic participation, and streamline policymaking, it also presents new challenges related to power, equity, and accountability. As we continue to navigate the digital age, it is essential for governments, businesses, and citizens to work together to ensure that technology is used in ways that strengthen democratic values, protect individual rights, and foster inclusive political systems. Thoughtful regulation, ethical innovation, and active public engagement will be key to harnessing the full potential of technology while safeguarding the principles of democracy and justice. Through collective effort and a commitment to fairness and equity, we can shape a future where technology serves as a force for good, supporting a more just, transparent, and participatory political system (European Parliament & Council, 2024; UN DESA, 2024).

CHAPTER 4 ESSAY PROMPTS

Technology and Political Structures—Power, Policy, and Participation

1. How has the advent of digital technologies, such as big data and artificial intelligence, transformed decision-making in governmental bodies? Discuss both the benefits and the challenges these technologies present to political structures.
2. In what ways can technology improve the transparency and accountability of governments? Provide specific examples of digital tools or platforms that have enhanced government operations.
3. Discuss the impact of social media on political participation. How have platforms like Twitter, Facebook, and Instagram empowered citizens, and what potential risks do these platforms pose to the democratic process?
4. Analyze the role of technology in fostering civic engagement. How have tools like online petitions, virtual town halls, and crowdfunding campaigns altered the landscape of political activism?
5. What are the ethical implications of using artificial intelligence in policymaking? Discuss the potential for bias in AI algorithms and its impact on marginalized communities and the fairness of policy outcomes.
6. Examine the role of blockchain technology in enhancing electoral processes. How could blockchain ensure secure, transparent, and tamper-proof elections, and what are the possible limitations of such a system?
7. How has the rise of digital platforms contributed to political polarization? Discuss how filter bubbles, echo chambers, and algorithm-driven content can influence public opinion and political discourse.
8. What are the challenges of regulating emerging technologies like AI, blockchain, and virtual reality in the context of democratic governance? How can governments balance innovation with the need to protect citizens' rights?
9. How do digital divides impact political participation and equity? Discuss the challenges faced by marginalized communities in accessing the tools and knowledge necessary for engaging in digital democracy.
10. In the age of technology-driven political structures, how can policymakers ensure that emerging technologies are used ethically and responsibly to benefit society at large? What frameworks or guidelines could be implemented to safeguard democratic principles?

REFERENCES

Council of Europe. (2024). *Framework Convention on Artificial Intelligence and Human Rights, Democracy and the Rule of Law*. Author.

Estonian National Electoral Committee. (n.d.). *Introduction to i-voting*. (Accessed November 10, 2025). Author.

European Commission. (2024a, March 26). *Commission publishes guidelines under the Digital Services Act for the mitigation of systemic risks online for elections* [Press release]. Author.

European Commission. (2024b, April 30). *Commission opens formal proceedings to assess possible breaches of the Digital Services Act by Meta* [Press release]. Author.

European Parliament & Council of the European Union. (2024). *Regulation (EU) 2024/1689 of 13 June 2024 laying down harmonised rules on artificial intelligence (AI Act). Official Journal of the European Union*.

Freedom House. (2024). *Freedom on the Net 2024: The struggle for trust online*. Author.

Inter-Parliamentary Union. (2025, March). *Sexism, harassment and violence against women in parliaments in the Asia-Pacific region* [Issue brief]. IPU.

International Telecommunication Union. (2024). *Measuring digital development: Facts and figures 2024*. ITU.

Milli, S., Carroll, M., Wang, Y., Pandey, S., Zhao, S., & Dragan, A. D. (2025). Engagement, user satisfaction, and the amplification of divisive content on social media. *PNAS Nexus, 4*(3), pgaf062.

NATO. (2024, July 10). *Summary of NATO's revised Artificial Intelligence strategy*. Author.

National Institute of Standards and Technology. (2024). *Election security* [Program overview and resources]. U.S. Department of Commerce.

Organisation for Economic Co-operation and Development. (2024a). *OECD digital economy outlook 2024* (Vol. 2). OECD Publishing.

Organisation for Economic Co-operation and Development. (2024b). *Artificial intelligence, data and competition* (DAF/COMP/WD(2024)49). OECD Publishing.

Office of Management and Budget. (2024, March 28). *M-24-10: Advancing governance, innovation, and risk management for agency use of AI*. Executive Office of the President.

Pew Research Center. (2024). *Social media fact sheet*. Author.

Saeva, E., & Tasheva, I. (2024). The 2024 EU elections and cybersecurity: A retrospective and lessons learned. *European View, 23*(2), 226–234.

U.S. Digital Service. (2024). *Impact report 2024*. Author.

World Bank. (2024). *GovTech Maturity Index 2024*. Author.

Ethical Frameworks for Emerging Technologies

5

ABSTRACT

Chapter 5 explores the vital role of ethical frameworks in navigating the complexities of emerging technologies. It introduces foundational ethical theories, such as deontology, consequentialism, virtue ethics, and care ethics, and demonstrates their practical application to contemporary technological dilemmas like AI bias, autonomous weapons, and data privacy. The chapter emphasizes the importance of ethical guidelines, global cooperation, and leadership in fostering responsible innovation. Through case studies and theoretical insights, it highlights the need for adaptable, inclusive, and forward-thinking approaches to ethical decision-making, ensuring technological advancements align with societal values and human dignity.

INTRODUCTION

The rapid advancement of technology presents unprecedented ethical challenges that demand robust frameworks for decision-making. As technologies evolve and integrate deeper into our lives, ethical considerations are essential to guide their development and use. This chapter introduces foundational ethical theories and examines their practical applications in addressing contemporary technological dilemmas (UNESCO, 2021; NIST, 2023; European Parliament & Council, 2024).

Ethical dilemmas in technology are not new but have gained prominence with the acceleration of innovations such as artificial intelligence, biotechnology, and blockchain. These technologies influence every aspect of human life—from healthcare and education to governance and social interactions. The increasing complexity and reach of these technologies underscore the urgency for ethical guidelines that can anticipate potential consequences while fostering innovation responsibly (WHO, 2023; ISO/IEC, 2023).

One of the significant ethical concerns is the dual-use nature of many emerging technologies. For example, artificial intelligence can enhance medical diagnostics and streamline administrative tasks, but it can also perpetuate bias, violate privacy, or enable mass surveillance. Similarly, biotechnology holds the promise of curing genetic disorders but raises questions about designer babies, eugenics, and unequal access. These dual-use scenarios highlight the need for proactive ethical frameworks that weigh the benefits against potential risks (UNESCO, 2021; WHO, 2023).

The foundation of ethical decision-making lies in well-established ethical theories such as deontology, consequentialism, virtue ethics, and care ethics. Each offers a unique lens to evaluate technological dilemmas. Deontology, for instance, emphasizes adherence to moral principles and rules, making it particularly relevant in ensuring technologies respect fundamental rights. Consequentialism focuses on outcomes, pushing developers and policymakers to consider the broader implications of technological deployment. Virtue ethics highlights the character and intentions of those creating and using technology, while care ethics prioritizes relational responsibilities and the impact on vulnerable populations (Alexander & Moore, 2021; Sinnott-Armstrong, 2023; Hursthouse & Pettigrove, 2020; Held, 2020).

The practical application of these ethical theories becomes evident when addressing contemporary issues like AI bias. Machine learning algorithms, for example, are often trained on historical data that may contain biases, perpetuating discrimination in areas such as hiring, law enforcement, and credit scoring. Ethical frameworks can guide the development of unbiased algorithms and establish accountability mechanisms to prevent harm (NIST, 2023; OECD, 2024).

Another pressing ethical dilemma is the rise of autonomous systems, such as self-driving cars and drones. These technologies challenge traditional moral paradigms, requiring decisions about life-and-death scenarios, liability in accidents, and regulatory oversight. Ethical theories provide tools for examining these situations, helping stakeholders balance innovation with public safety and equity (IEEE, 2021; ISO/IEC, 2023).

Data privacy and surveillance are also central to ethical discussions in the digital age. The widespread collection and analysis of personal data have sparked debates over consent, ownership, and misuse. Ethical frameworks can inform policies that protect individuals' rights while allowing businesses and governments to harness data responsibly for societal benefit (European Parliament & Council, 2016; OECD, 2022; Freedom House, 2024).

Beyond individual technologies, the ethical implications of systemic shifts such as digital transformation and automation require attention. Automation, for instance, has the potential to increase efficiency and reduce costs but threatens job displacement and economic inequality. Ethical considerations must address how to mitigate these impacts, ensuring that technological progress benefits all sections of society equitably (OECD, 2024; World Economic Forum, 2025).

In addition to established ethical theories, emerging fields like technoethics and AI ethics are contributing to the discourse. These interdisciplinary approaches integrate insights from philosophy, sociology, and computer science to address the unique challenges posed by rapidly evolving technologies. They emphasize the importance of inclusivity, transparency, and adaptability in developing ethical guidelines (UNESCO, 2021; NIST, 2023; IEEE, 2021).

Global cooperation is another crucial aspect of navigating ethical dilemmas in technology. As technological advancements often transcend national borders, ethical frameworks must reflect diverse cultural values and perspectives. International agreements and collaborative research efforts can ensure that ethical standards are consistent and inclusive, fostering trust and reducing conflicts (Department for Science, Innovation & Technology, 2023; AI Seoul Summit Leaders, 2024).

The role of leadership in promoting ethical practices cannot be overstated. Industry leaders, policymakers, and educators have a responsibility to champion ethical innovation by integrating ethical principles into research, development, and deployment processes. This includes creating mechanisms for accountability, encouraging public discourse, and prioritizing long-term societal well-being over short-term gains (ACM, 2018; IEEE, 2021; NIST, 2023).

Education and awareness are foundational to cultivating a culture of ethical technology use. By embedding ethics in STEM curricula and providing training for professionals, societies can equip individuals with the tools to make informed decisions. Public campaigns and accessible resources can also raise awareness among non-experts, empowering them to advocate for ethical practices in their communities (UNESCO, 2021).

The inclusion of marginalized voices is essential in shaping ethical frameworks. Vulnerable populations often bear the brunt of technological misuses, whether through algorithmic bias, digital divides, or exploitative labor practices. Ensuring that their perspectives are heard can lead to more equitable and humane outcomes, addressing disparities and promoting social justice (ITU, 2024; Freedom House, 2024).

Case studies are instrumental in illustrating the practical application of ethical frameworks. Examples like the Facebook-Cambridge Analytica scandal, controversies surrounding facial recognition technology, and debates over gene-editing tools like CRISPR demonstrate the complexities of ethical decision-making. These cases highlight the need for ethical foresight, stakeholder engagement, and iterative policy development (Information Commissioner's Office, 2018; Amnesty International, 2024; WHO, 2023).

This chapter also explores the limitations of existing ethical frameworks and the need for continuous evolution. As technologies evolve, so too must the ethical considerations that govern them. Adaptive and dynamic ethical models can address unforeseen challenges, fostering resilience and innovation without compromising fundamental values (NIST, 2023; OECD, 2022; Alexander & Moore, 2021).

In conclusion, the interplay between ethics and emerging technologies is a dynamic and multifaceted domain that requires collective effort and foresight. By integrating ethical theories into practical decision-making, fostering collaboration, and prioritizing inclusivity, societies can navigate the challenges

Ethics for emerging tech: deontology, consequentialism, virtue, and care to steer AI, data, and autonomous systems toward human dignity

Read through an ethics lens, the image invites viewers to weigh four complementary tests for emerging tech: a deontological check ("are we honoring rights, duties, and consent?"), a consequentialist check ("do expected benefits outweigh harms, including to the most vulnerable?"), a virtue check ("does this cultivate trustworthy character and professional integrity in its makers and users?"), and a care check ("are relationships, context, and long-term human dignity being protected?"). It suggests that responsible AI and data systems braid these lenses into practice—risk assessment before deployment, transparency and redress after, and continuous monitoring to catch bias, drift, and unintended effects.

of technological innovation responsibly. This chapter aims to provide a comprehensive understanding of ethical frameworks and their application, equipping readers with the tools to address contemporary dilemmas with clarity and integrity (UNESCO, 2021; NIST, 2023; ISO/IEC, 2023).

THE IMPORTANCE OF ETHICS IN TECHNOLOGY

Technological innovations influence every aspect of society, from individual privacy to global sustainability. The rapid pace of these advancements amplifies both the benefits and risks, making ethical considerations a cornerstone in the development and application of new technologies. As tools like artificial intelligence (AI), blockchain, and genetic engineering redefine human capabilities, it is crucial to ask how they impact fundamental rights such as privacy, equity, and security. Equally important is the question of who bears responsibility for mitigating harm in the face of these transformative changes (European Parliament & Council, 2016; European Parliament & Council, 2024; UNESCO, 2021).

Ethical frameworks serve as critical tools for addressing such questions, providing structured approaches to balance innovation with societal well-being. By prioritizing ethical reflection, developers, policymakers, and end-users can work collaboratively to protect vulnerable populations from unintended harm while fostering technologies that contribute to the public good. The absence of such considerations risks creating technologies that exacerbate social inequalities, infringe on individual rights, or introduce unforeseen dangers (NIST, 2023; IEEE, 2021).

One of the most significant challenges in emerging technologies is their unpredictability. For instance, AI systems often operate as "black boxes," producing outcomes that even their creators struggle to fully understand or control. This opacity raises ethical concerns about accountability, fairness, and trust. Ethical frameworks can guide developers in making transparent and explainable AI systems that align with societal values and avoid perpetuating harm (NIST, 2023; OECD, 2024).

Biotechnology, another field undergoing rapid growth, presents equally complex ethical dilemmas. Innovations like CRISPR gene editing hold the promise of curing hereditary diseases but also introduce concerns about genetic discrimination and the potential misuse of technology for eugenics. Ethical frameworks in this domain emphasize informed consent, equitable access, and the need to consider long-term societal implications (WHO, 2023).

The dual-use nature of many technologies complicates ethical decision-making further. A single innovation can serve both beneficial and harmful purposes, depending on its application. For example, while drones enhance disaster response efforts and agricultural monitoring, they also enable invasive surveillance and military strikes. Robust ethical frameworks help policymakers weigh these dual-use implications, promoting uses that align with human rights and social equity (Freedom House, 2024; Amnesty International, 2024).

As society becomes increasingly interconnected through digital platforms, data privacy has emerged as a critical ethical concern. The widespread collection, storage, and analysis of personal data raise questions about consent, ownership, and misuse. Data breaches, algorithmic profiling, and targeted misinformation illustrate the risks posed by inadequate data governance. Ethical principles, such as respect for autonomy and justice, provide the foundation for policies that safeguard individual rights while supporting innovation (European Parliament & Council, 2016; OECD, 2022; Freedom House, 2024).

Automation, driven by advancements in AI and robotics, has introduced another layer of ethical complexity. While automation increases productivity and efficiency, it also disrupts labor markets, threatening job security for millions. Ethical frameworks in this context guide stakeholders in addressing the socioeconomic impacts of automation, advocating for measures such as reskilling programs and equitable distribution of benefits (OECD, 2024; World Economic Forum, 2025).

Autonomous systems, including self-driving vehicles and drones, present unique ethical challenges that defy traditional decision-making models. These systems often operate in life-and-death scenarios, requiring real-time moral judgments. Ethical theories like utilitarianism, which seeks to maximize overall well-being, and deontology, which emphasizes adherence to moral rules, offer contrasting approaches to addressing these dilemmas. Integrating such frameworks ensures that autonomous technologies operate within clear ethical boundaries (IEEE, 2021; ISO/IEC, 2023).

Globalization adds another dimension to the ethical considerations surrounding technology. As innovations cross borders, they interact with diverse cultural, legal, and ethical norms. A technology deemed acceptable in one context may provoke controversy in another. Ethical frameworks must therefore be adaptable and inclusive, incorporating a wide range of perspectives to ensure global fairness and mutual respect (UNESCO, 2021; Department for Science, Innovation & Technology, 2023; AI Seoul Summit Leaders, 2024).

The role of leadership in fostering ethical technology use cannot be overstated. Industry leaders, policymakers, and educators are uniquely positioned to promote ethical practices by embedding principles like transparency, accountability, and inclusivity into their strategies. Leaders who prioritize long-term societal well-being over short-term profit can drive meaningful change, setting ethical standards that resonate across sectors (ACM, 2018; NIST, 2023; IEEE, 2021).

Education plays a pivotal role in cultivating a culture of ethical technology development and use. By integrating ethics into STEM curricula and providing specialized training for technology professionals, societies can equip individuals with the skills to navigate complex moral landscapes. Public education campaigns can further empower non-experts, enabling them to advocate for responsible practices and hold decision-makers accountable (UNESCO, 2021).

One of the most compelling reasons for integrating ethics into technological decision-making is the potential for harm reduction. From mitigating biases in AI systems to preventing the misuse of genetic data, ethical frameworks provide actionable pathways to address vulnerabilities and protect marginalized communities. These frameworks challenge developers to anticipate unintended consequences and design technologies that prioritize equity and inclusivity (NIST, 2023; WHO, 2023).

The inclusion of diverse voices is critical in shaping ethical considerations for emerging technologies. Historically marginalized groups, including women, minorities, and individuals from low-income countries, often face the greatest risks from technological misuse. Ensuring their perspectives are represented in decision-making processes leads to more equitable outcomes and minimizes the risk of reinforcing systemic inequalities (ITU, 2024; Freedom House, 2024).

Case studies offer valuable insights into the practical application of ethical frameworks. Events like the Facebook-Cambridge Analytica scandal highlight the consequences of prioritizing profit over privacy. Similarly, debates over facial recognition technology illustrate the tension between public safety and individual rights. These examples underscore the importance of ethical foresight and collaborative governance (Information Commissioner's Office, 2018; Amnesty International, 2024).

The emergence of technoethics and AI ethics as distinct fields reflects the growing recognition of the need for specialized approaches to modern challenges. These interdisciplinary areas draw from philosophy, sociology, and computer science to develop context-specific ethical guidelines. By emphasizing adaptability, inclusivity, and transparency, they help bridge the gap between theoretical principles and real-world applications (UNESCO, 2021; NIST, 2023).

Despite their strengths, existing ethical frameworks are not without limitations. Many are rooted in historical contexts that may not fully address the complexities of today's technologies. Evolving these frameworks requires ongoing dialogue and research, ensuring they remain relevant and effective in the face of rapid innovation (Sinnott-Armstrong, 2023; Hursthouse & Pettigrove, 2020).

Adaptive ethical models are essential for addressing unforeseen challenges. Unlike rigid systems, adaptive frameworks evolve alongside technological advancements, incorporating new insights and perspectives. This flexibility allows stakeholders to respond proactively to emerging dilemmas, fostering resilience and innovation without compromising core values (NIST, 2023).

Ethical considerations also extend to the environmental impacts of technology. From the energy consumption of data centers to the electronic waste generated by consumer devices, technology's ecological footprint raises urgent questions about sustainability. Ethical frameworks encourage developers and policymakers to adopt green practices, ensuring that innovation does not come at the expense of the planet (IEA, 2024; Baldé et al., 2024).

International cooperation is vital for navigating the ethical dimensions of emerging technologies. Global challenges like cybersecurity threats and climate change require coordinated responses that transcend national boundaries. By fostering collaboration and aligning ethical standards, countries can address shared risks while respecting cultural diversity (Department for Science, Innovation & Technology, 2023; AI Seoul Summit Leaders, 2024).

Transparency is a cornerstone of ethical technology use. Open communication about the capabilities, limitations, and potential risks of new technologies builds trust among stakeholders. Transparency also enables informed decision-making, allowing users to understand and mitigate risks associated with technology adoption (GDPR: European Parliament & Council, 2016; OECD, 2022).

Accountability mechanisms are equally important in ensuring ethical compliance. From regulatory frameworks to self-governance initiatives, these mechanisms hold developers and organizations responsible for their actions. Accountability fosters a culture of integrity, discouraging unethical behavior and promoting public trust in technological advancements (ACM, 2018; IEEE, 2021).

Public engagement is a critical component of ethical governance. By involving citizens in discussions about the development and use of new technologies, policymakers can align decisions with societal values and expectations. Participatory approaches ensure that ethical considerations reflect the priorities of diverse communities, enhancing legitimacy and effectiveness (UNESCO, 2021).

Ethical foresight is essential for anticipating and addressing future challenges. By exploring potential scenarios and their implications, stakeholders can identify risks and opportunities before they materialize. This proactive approach minimizes harm and maximizes the benefits of technological innovation (NIST, 2023; OECD, 2024).

In conclusion, ethical frameworks are indispensable for navigating the complexities of emerging technologies. By integrating theoretical insights with practical applications, they provide a roadmap for responsible innovation. As technologies continue to shape society, fostering a culture of ethical awareness and accountability will be critical for ensuring that progress benefits all (UNESCO, 2021; NIST, 2023).

OVERVIEW OF ETHICAL THEORIES

To apply ethics to technology effectively, it is essential to understand the foundational theories that underpin ethical reasoning. These theories guide individuals and organizations in making decisions that align with moral principles and contribute to the greater good. By examining the key ethical frameworks, we can better navigate the complexities of emerging technologies and their potential societal impact. This section provides a concise overview of four major ethical theories: Deontological Ethics, Consequentialism, Virtue Ethics, and Care Ethics. Each of these frameworks offers a unique lens through which to evaluate the moral implications of technological innovations (Alexander & Moore, 2021; Sinnott-Armstrong, 2023; Hursthouse & Pettigrove, 2020; Held, 2020).

Deontological Ethics, or duty-based ethics, is grounded in the belief that actions should be judged by their adherence to moral rules or duties, regardless of the consequences. The term "deontological" comes from the Greek words "deon," meaning duty, and "logos," meaning study or discourse. This ethical framework emphasizes the intrinsic value of moral actions, asserting that some actions are morally obligatory, even if they result in unfavorable outcomes. For example, one might impose strict regulations on the use of artificial intelligence to prevent misuse, even if these regulations slow down technological innovation. The key tenet of deontological ethics is that it prioritizes the rule or duty over the consequences, making it a useful approach when establishing non-negotiable standards for technology development (Alexander & Moore, 2021).

In the context of emerging technologies, deontological ethics helps ensure that developers, users, and policymakers uphold ethical principles that protect society. It serves as a safeguard against the potential misuse of powerful technologies such as AI, genetic engineering, and surveillance systems. By setting clear rules or guidelines for their use, deontological ethics can prevent harmful outcomes, such as privacy violations or the reinforcement of biases in algorithmic decision-making. However, one challenge of deontological ethics is that it may be too rigid in some contexts, as it does not consider the complexities of balancing moral rules with real-world consequences (Alexander & Moore, 2021).

Consequentialism, on the other hand, evaluates actions based on their outcomes or utility. This ethical theory holds that the morality of an action is determined by its consequences, with the aim of maximizing overall good or minimizing harm. The most well-known form of consequentialism is utilitarianism, which advocates for actions that provide the greatest benefit to the greatest number of people. For instance, when developing autonomous vehicles, a consequentialist approach might prioritize minimizing harm in accidents, even if it means sacrificing one individual's life to save multiple others. The focus here is on outcomes rather than the morality of the action itself (Sinnott-Armstrong, 2023).

In technological development, consequentialism provides a framework for making decisions that maximize benefits and minimize risks. It offers a practical approach to evaluating the societal impact of emerging technologies, particularly when there are trade-offs between different stakeholders. For example, in the case of AI decision-making systems, a consequentialist might argue that if the algorithm results in fewer discriminatory practices or better allocation of resources, then its development is morally justified, despite potential challenges or risks. However, a limitation of consequentialism is its potential to justify harmful actions if they result in perceived greater good.

For example, prioritizing efficiency in an autonomous vehicle system may lead to ethical dilemmas in deciding whose life to prioritize in a life-and-death situation (Sinnott-Armstrong, 2023).

Virtue Ethics, by contrast, emphasizes the importance of an individual's character and moral virtues in guiding ethical behavior. Unlike deontology and consequentialism, which focus on rules and outcomes, virtue ethics is concerned with the qualities that make a person morally good. Developed by Aristotle, virtue ethics advocates for the cultivation of virtues such as honesty, courage, empathy, and integrity, believing that a virtuous individual will naturally make ethical decisions. For example, in the context of technology leadership, virtue ethics encourages tech leaders to adopt humility, transparency, and integrity in their decision-making, ensuring that they prioritize the well-being of others in their pursuit of innovation (Hursthouse & Pettigrove, 2020).

Virtue ethics encourages a more holistic approach to technological development, one that is not solely concerned with outcomes or following rules but also with the intentions and character of the individuals involved in the decision-making process. Tech leaders, engineers, and developers who practice virtues like empathy and responsibility will be more likely to create technologies that consider societal needs and promote well-being. The focus on moral character makes virtue ethics particularly valuable when addressing issues like accountability and trust in technology. However, one challenge of virtue ethics is its subjectivity, as different cultures and individuals may have varying views on what constitutes a virtuous character, potentially leading to disagreements on what constitutes ethical behavior (Hursthouse & Pettigrove, 2020).

Care Ethics, another key ethical framework, focuses on the importance of relationships and responsibilities to others. It emphasizes empathy, compassion, and the moral obligation to care for those who are vulnerable or in need. Unlike deontology and consequentialism, which often focus on abstract principles or outcomes, care ethics is deeply relational, stressing the importance of maintaining and nurturing human connections. For example, designing assistive technologies with input from end-users is a core principle of care ethics, ensuring that the needs and preferences of those who will use the technology are genuinely met (Held, 2020).

In the realm of technology, care ethics encourages the development of technologies that promote human flourishing, taking into account the social and emotional needs of individuals. It highlights the importance of user-centered design and the necessity of engaging with marginalized or vulnerable groups to ensure that technological advancements are inclusive and responsive to their unique needs. Care ethics also advocates for the creation of technologies that enhance, rather than exploit, human relationships. For instance, technologies designed to assist the elderly or people with disabilities should be built with a deep understanding of the users' specific needs and preferences. However, care ethics can be difficult to implement in large-scale technological projects, where the relational aspects of design may be overlooked in favor of efficiency or profit (Held, 2020).

While each of these ethical frameworks offers distinct perspectives, they are not mutually exclusive. In fact, a well-rounded approach to ethical decision-making in technology often involves integrating multiple frameworks to address complex dilemmas. For example, when developing AI systems for healthcare, a developer might consider deontological principles to ensure that privacy is protected, consequentialist principles to evaluate the system's overall benefits and risks, virtue ethics to guide the developer's conduct, and care ethics to prioritize patient well-being. By drawing on these frameworks, developers and policymakers can create a more comprehensive ethical strategy that takes into account the multifaceted nature of technological innovation (UNESCO, 2021; NIST, 2023).

Understanding these ethical frameworks is essential for anyone involved in the development, regulation, or use of emerging technologies. Whether it is addressing the ethical implications of AI, data privacy, or biotechnology, these frameworks provide the tools necessary to navigate the complex moral landscape of technological progress. By applying these frameworks thoughtfully, we can ensure that emerging technologies serve the greater good and contribute to a more equitable, just, and sustainable society (UNESCO, 2021; OECD, 2022).

However, the application of these ethical frameworks is not without challenges. One major issue is the fast-paced nature of technological innovation, which often outpaces the development of ethical guidelines and regulations. In many cases, technological advancements are already in place before ethical considerations are adequately addressed, creating a reactive rather than proactive approach to ethical dilemmas. As a result, it is important for stakeholders to work together—developers, policymakers, ethicists, and the public—to ensure that ethical standards evolve alongside technological advancements (NIST, 2023; Department for Science, Innovation & Technology, 2023).

Another challenge is the global nature of technology, which often raises questions about the compatibility of different cultural, legal, and ethical norms. A technology that is deemed ethical in one country or culture may not align with the values of another. For example, data privacy regulations differ significantly across regions, such as the European Union's General Data Protection Regulation (GDPR) and the more relaxed data protection laws in some parts of the United States. Navigating these differences requires international cooperation and the development of shared ethical standards that respect cultural diversity while promoting universal principles of justice and human rights (European Parliament & Council, 2016; OECD, 2022).

In addition to these external challenges, there are internal challenges within the tech industry itself. Many technology companies prioritize profitability and market dominance over ethical considerations, which can lead to the development of technologies that exploit users or violate their rights. For example, social media platforms that prioritize engagement often use algorithms that promote divisive content, leading to harmful societal consequences. In such cases, ethical frameworks like deontological ethics and care ethics can provide a counterbalance, urging companies to prioritize user well-being and long-term societal benefits over short-term profits (Freedom House, 2024; Amnesty International, 2024).

Despite these challenges, the growing awareness of the ethical implications of technology offers hope for a more responsible and equitable future. As the tech industry becomes increasingly aware of the importance of ethical decision-making, there is potential for meaningful change. Initiatives like AI ethics boards, corporate social responsibility programs, and the inclusion of ethicists in technology development teams are all steps in the right direction. By integrating ethical frameworks into the design, development, and deployment of emerging technologies, we can ensure that technological progress aligns with moral principles and serves the common good (ACM, 2018; IEEE, 2021; NIST, 2023).

As we move forward, it is essential to continue refining these ethical frameworks to account for new technological developments and emerging dilemmas. The integration of ethics into technology is not a one-time effort but an ongoing process that requires constant reflection, adaptation, and collaboration. Only by doing so can we ensure that emerging technologies contribute to a more just, equitable, and sustainable world (UNESCO, 2021; AI Seoul Summit Leaders, 2024).

APPLYING ETHICAL THEORIES TO TECHNOLOGICAL DILEMMAS

This section explores how ethical frameworks can be applied to emerging technologies, providing case studies that illustrate the complexity of ethical decision-making. As technology evolves at a rapid pace, the ethical implications of its use become increasingly important. In particular, technologies like artificial intelligence (AI) and autonomous weapons systems present novel ethical dilemmas that require thoughtful analysis and a careful balance of competing interests. The application of various ethical theories—deontological ethics, consequentialism, virtue ethics, and care ethics—can guide decision-making in these areas and help navigate the challenges that arise (Alexander & Moore, 2021; Sinnott-Armstrong, 2023; Hursthouse & Pettigrove, 2020; Held, 2020).

Artificial Intelligence (AI) has become an integral part of many sectors, from healthcare to finance, education, and even law enforcement. However, one of the most pressing ethical concerns surrounding AI is the issue of bias. AI systems are often trained on vast datasets, which may contain biases that reflect historical prejudices, societal inequalities, or skewed representations. When these systems are deployed, they can perpetuate or even amplify these biases, leading to discriminatory outcomes. For example, facial recognition software has been shown to have higher error rates for people of color and women, leading to concerns about racial profiling and unequal treatment. Similarly, predictive policing algorithms have been criticized for disproportionately targeting marginalized communities based on biased crime data (Buolamwini & Gebru, 2018; Grother et al., 2019; Lum & Isaac, 2016; Ensign et al., 2018).

The application of deontological ethics to AI bias mandates that developers adhere to strict rules and guidelines that prevent the use of biased training data. From this perspective, it is the duty of AI creators to ensure that their systems are fair and just. For instance, deontology would require the implementation of rigorous auditing mechanisms to detect and correct biases in training datasets before AI systems are deployed. Transparency in the design and functioning of AI algorithms is also essential to prevent harm and uphold ethical duties. Developers should be held accountable for ensuring that their systems comply with these ethical duties, regardless of the potential delays or costs that may arise from extensive oversight (NIST, 2023; IEEE, 2021; European Parliament & Council, 2024; Gebru et al., 2021; Mitchell et al., 2019).

From a consequentialist perspective, the ethical evaluation of AI systems involves assessing the outcomes of their deployment. A consequentialist would weigh the benefits of AI technologies—such as efficiency, cost savings, and improved decision-making—against the potential harms, such as exacerbating social inequalities or perpetuating discrimination. The goal is to maximize overall societal benefit while minimizing harm. In the case of biased AI, a consequentialist approach would involve addressing bias in training data to ensure that AI systems provide fair and equitable outcomes for all users. For example, revising biased algorithms to ensure they do not disproportionately affect specific groups would be considered an ethical response that aligns with the maximization of social utility (Sinnott-Armstrong, 2023; Obermeyer et al., 2019; OECD, 2019).

Virtue ethics focuses on the moral character of individuals involved in AI development. From this standpoint, fostering a culture of fairness, accountability, and empathy among AI developers is essential to addressing bias. Virtue ethics suggests that the creators of AI systems should cultivate virtues like humility, integrity, and openness to feedback. For instance, AI developers should be encouraged to consider the social impact of their work and to engage in ongoing education about the potential biases inherent in their algorithms. By prioritizing virtues like fairness and responsibility, the technology industry can help prevent the development and deployment of harmful biased systems (Hursthouse & Pettigrove, 2020; Raji et al., 2020; Benjamin, 2019).

Care ethics, which emphasizes the importance of relationships and empathy, advocates for prioritizing the needs of marginalized communities that are disproportionately affected by biased AI systems. From a care ethics perspective, developers and policymakers should recognize the vulnerabilities of these communities and ensure that their voices are included in the development of AI technologies. This may involve conducting outreach to underserved groups, soliciting their input on how AI systems affect their lives, and designing systems that are more inclusive and responsive to their needs. By considering the humanitarian aspects of AI, care ethics ensures that technology serves to enhance, rather than harm, the well-being of vulnerable populations (Held, 2020; Noble, 2018; Eubanks, 2018).

Autonomous weapons systems present another significant ethical dilemma. These technologies, which include unmanned drones and robotic soldiers, have the ability to make life-and-death decisions autonomously, without human intervention. The development of such systems raises profound moral and legal questions about accountability, human oversight, and the nature of warfare. Autonomous weapons systems could theoretically reduce human casualties on the battlefield by allowing machines to carry out military operations. However, the idea of machines making decisions about life and death without human oversight is deeply troubling for many ethicists and policymakers, as it removes accountability from the decision-making process (ICRC, 2023; UN Secretary-General & ICRC, 2023; U.S. DoD, 2023).

Deontological ethics would argue against the development and deployment of autonomous weapons systems, as they violate the moral duty of human oversight in matters of life and death. According to deontological principles, the duty to protect human lives and exercise moral judgment in the context of war cannot be delegated to machines. Prohibiting autonomous weapons is seen as necessary to ensure that humans remain in control of military decisions and that ethical principles guide actions in times of conflict. From this perspective, allowing machines to make decisions about the use of force undermines the moral responsibility of military personnel and the legal frameworks that govern warfare, such as the Geneva Conventions (Alexander & Moore, 2021; ICRC, 2006; European Parliament, 2021).

Consequentialism, however, evaluates autonomous weapons systems based on their potential outcomes. A consequentialist would examine the risks and benefits of deploying autonomous weapons, considering factors such as the reduction of civilian casualties, the speed and efficiency of military operations, and the strategic advantages of using autonomous systems. From this perspective, the use of autonomous weapons might be justified if the overall benefits outweigh the risks, such as minimizing harm to soldiers and non-combatants. However, a consequentialist would also need to consider the potential negative consequences, such as the risk of malfunction, the lack of accountability in decision-making, and the possibility that autonomous weapons could be used in ethically questionable ways (Sparrow, 2007; Sharkey, 2016; Roff, 2019).

Virtue ethics in the context of autonomous weapons calls for restraint and wisdom among nations developing these technologies. Virtue ethics stresses that the moral character of those involved in decision-making is crucial, and in the case of autonomous weapons, it calls for nations to exercise caution and responsibility. Developers and military leaders should act with prudence, empathy, and respect for human dignity when creating and deploying autonomous weapons. This approach encourages ongoing dialogue about the ethical implications of these systems and advocates for careful consideration of their potential impact on international security and global peace (Hursthouse & Pettigrove, 2020; UN Secretary-General & ICRC, 2023).

Care ethics, with its focus on relationships and human well-being, underscores the humanitarian consequences of using autonomous weapons in conflict zones. In particular, care ethics emphasizes the need to consider the effects of such weapons on civilian populations, particularly those who are already vulnerable due to poverty, displacement, or ongoing conflict. From a care ethics perspective, the development and use of autonomous weapons should be guided by a deep concern for the safety and dignity of civilians, especially those in war-torn regions. Additionally, care ethics calls for attention to the psychological and emotional impacts of warfare, both on soldiers and on civilian populations. Ensuring that technology is used in a way that minimizes harm and fosters compassion is key in addressing the ethical challenges posed by autonomous weapons (Held, 2020; ICRC, 2023).

The application of these ethical frameworks to the dilemmas posed by AI and autonomous weapons highlights the complexity of decision-making in the age of advanced technology. In both cases, no single ethical theory offers a complete solution to the problems at hand. Instead, a combination of deontological, consequentialist, virtue ethics, and care ethics provides a more nuanced approach that takes into account the moral responsibilities of technology developers, the potential outcomes of their actions, and the broader societal implications of emerging technologies. By applying these frameworks in a thoughtful and balanced way, we can better navigate the ethical challenges that arise as technology continues to advance (UNESCO, 2021; OECD, 2019).

As we move forward, it is essential to continue refining and expanding our understanding of ethical decision-making in relation to emerging technologies. The challenges posed by AI and autonomous weapons are just the beginning, as new technologies—such as biotechnology, nanotechnology, and quantum computing—raise additional ethical questions that will require careful consideration and responsible action. By applying ethical frameworks in a systematic and consistent way, we can ensure that technological advancements contribute to the common good and promote justice, equity, and human dignity. Ultimately, the responsible development and use of technology will depend on our collective ability to engage with these ethical dilemmas thoughtfully and with a commitment to ensuring that technology serves humanity rather than undermines it (Floridi et al., 2018; Friedman & Hendry, 2019).

DEVELOPING ETHICAL GUIDELINES FOR EMERGING TECHNOLOGIES

Building on the theoretical foundations, this section outlines how organizations and governments can develop ethical guidelines tailored to technological advancements. As technology continues to evolve at an accelerating pace, the ethical dilemmas it generates become increasingly complex. To address these challenges, it is essential that ethical frameworks and policies are crafted with careful consideration and foresight. The development of these ethical guidelines is not solely the responsibility of technologists and engineers but also involves input from various disciplines, stakeholders, and the public. This collaborative, multidisciplinary approach is crucial for creating ethical policies that are both comprehensive and adaptable to future technological innovations (Stilgoe et al., 2013; UNESCO, 2021).

A key aspect of developing robust ethical guidelines for technology is the role of multidisciplinary committees. These committees should consist of experts from a diverse range of fields, including technology, ethics, law, public policy, and social sciences. By bringing together people with different perspectives and areas of expertise, organizations and governments can ensure that ethical considerations are fully explored and that all potential impacts of new technologies are considered. For example, when designing AI systems or developing new medical technologies, it is essential to involve ethicists who can guide decision-making on moral issues, legal experts who can assess regulatory compliance, and technologists who can offer insight into the technical feasibility and limitations of these technologies (Friedman & Hendry, 2019; National Academies, 2022).

Multidisciplinary committees can help to identify and address potential risks and unintended consequences that may arise from the implementation of new technologies. These committees are particularly important in the context of complex technologies like artificial intelligence, gene editing, and autonomous systems, where the ethical implications can be difficult to foresee. The involvement of professionals from various fields ensures that no critical aspect of technology's impact is overlooked, whether it involves privacy, security, equity, or environmental concerns. The committee can also ensure that ethical guidelines are continually updated to reflect new technological developments and the evolving needs of society (ISO/IEC, 2023; NIST, 2023).

The importance of public engagement and stakeholder involvement in ethical policymaking cannot be overstated. While experts and policymakers play a vital role in shaping ethical guidelines, the views and concerns of the public should also be taken into account. This ensures that the policies developed are not only technically sound but also socially acceptable and aligned with the values and expectations of the broader community. Public engagement can take various forms, including consultations, surveys, public forums, and town hall meetings, where citizens can voice their opinions on emerging technologies and their potential ethical implications (Rowe & Frewer, 2000; OECD, 2019).

Stakeholder involvement is also critical, particularly in industries where technology has far-reaching consequences. For example, in the healthcare sector, the development of new medical technologies should involve not only healthcare professionals but also patients, advocacy groups, and policymakers. Similarly, when developing AI systems, it is important to include input from diverse communities who may be affected by these technologies, including marginalized groups. This inclusive approach ensures that ethical guidelines are not only comprehensive but also equitable, ensuring that the rights and concerns of all stakeholders are considered (Benjamin, 2019; UNESCO, 2021).

Moreover, stakeholder engagement is essential in ensuring that technological advancements are aligned with broader societal goals, such as promoting social justice, equity, and environmental sustainability. Engaging with the public and stakeholders allows policymakers to better understand the social implications of new technologies and to develop ethical guidelines that promote the common good. By involving a diverse array of voices in the policymaking process, governments and organizations can prevent the exclusion of certain groups and mitigate the risk of technological development exacerbating social inequalities (Floridi et al., 2018; O'Neil, 2016).

Another critical factor in developing ethical guidelines for emerging technologies is the need for adaptability. As technology evolves rapidly, ethical frameworks and guidelines must also be flexible and responsive to new challenges and risks. What may be considered ethical in one era may no longer be appropriate in the context of new technological advancements. For example, the ethical guidelines for data privacy that were developed a decade ago may no longer be sufficient given the proliferation of big data and advances in AI, which enable the collection and analysis of vast amounts of personal information. As such, it is crucial for ethical guidelines to be regularly updated and adapted to reflect new technological realities (European Parliament & Council, 2016; OECD, 2022).

To ensure that ethical guidelines remain relevant, organizations and governments must create systems for continuous monitoring and review. This may involve setting up specialized ethics review boards or committees that are tasked with evaluating emerging technologies and their associated ethical risks on an ongoing basis. Additionally, technological advancements should be subject to regular ethical audits, where the impacts of a technology are assessed throughout its lifecycle—from development and deployment to long-term societal consequences. This ensures that any unforeseen ethical challenges are identified and addressed before they become significant issues (Reisman et al., 2018; NIST, 2023).

Moreover, the adaptability of ethical guidelines extends to their ability to respond to rapid changes in public attitudes and societal values. Ethical decision-making is not static; it must evolve in response to shifting norms and concerns. For instance, as societal attitudes toward privacy, surveillance, and data security change, ethical guidelines for AI and data use must evolve accordingly. The challenge for policymakers and organizations is to stay ahead of these changes while ensuring that their guidelines remain grounded in core ethical principles (UNESCO, 2021; OECD, 2019).

In developing ethical guidelines, it is also important to recognize the diverse cultural and social contexts in which technologies are deployed. Ethical guidelines that work well in one country or region may not be applicable in another. For example, the ethical considerations surrounding genetic engineering may vary significantly between cultures with different religious beliefs and values. In some countries, biotechnology may be viewed as a tool for progress and scientific advancement, while in others, it may raise concerns about moral and ethical boundaries. As such, global cooperation and dialogue are essential in developing international ethical standards for emerging technologies (Council of Europe, 2024; UN General Assembly, 2024).

Global collaboration allows countries to share best practices and learn from each other's experiences in developing ethical frameworks for technology. International organizations, such as the United Nations or the World Health Organization, can play an important role in facilitating these discussions and helping to establish global ethical guidelines. At the same time, it is essential for local and national contexts to be taken into account, ensuring that ethical policies reflect the unique needs and values of different communities (UNESCO, 2021; WHO, 2023).

Ethical guidelines for emerging technologies must also address the potential for unequal access to technological advancements. In many cases, new technologies exacerbate existing inequalities, creating divisions between those who have access to the benefits of these technologies and those who do not. For example, access to AI-powered healthcare technologies may be limited to wealthier countries or populations, leaving disadvantaged communities without critical resources. Ethical guidelines should ensure that emerging technologies are developed and deployed in ways that promote equity and inclusivity, ensuring that the benefits of innovation are shared broadly (Noble, 2018; Eubanks, 2018; UNESCO, 2021).

Governments and organizations must also consider the environmental impact of new technologies when developing ethical guidelines. Technological advancements often have unintended consequences on the environment, such as increased energy consumption, pollution, or resource depletion. Ethical guidelines should require technology developers to assess the environmental impacts of their innovations and prioritize sustainability in their designs. This can include the development of energy-efficient technologies, the reduction of electronic waste, and the promotion of environmentally friendly materials and practices (IEA, 2024; Baldé et al., 2024).

In addition to promoting ethical behavior among developers, organizations must also take steps to ensure that their employees are well-versed in ethical decision-making. Training programs on ethics in technology can help individuals in various sectors—whether they are engineers, designers, or policymakers—understand the ethical implications of their work and how to navigate complex dilemmas. Ethical training should be an ongoing process, embedded into the culture of technology organizations and reinforced throughout employees' careers (National Academies, 2022; ACM, 2018).

The integration of ethics into technological decision-making processes requires strong leadership and a commitment to responsible innovation. Leaders within organizations must set an example by prioritizing ethical considerations in their decision-making and ensuring that ethical frameworks are not only developed but actively applied. In practice, this means fostering a culture of transparency, accountability, and responsibility, where the ethical implications of decisions are carefully weighed and considered. By establishing ethical guidelines that are embedded into organizational processes, leaders can ensure that ethical behavior is not merely an afterthought but a fundamental part of the development and deployment of emerging technologies (IEEE, 2021; NIST, 2023).

As technology continues to shape every facet of modern life, the role of ethics in guiding technological development will only grow more important. In this context, the development of ethical guidelines that are grounded in multidisciplinary collaboration, public engagement, adaptability, and global cooperation will be crucial. By ensuring that technology is developed and deployed responsibly, governments and organizations can help ensure that emerging technologies contribute to the common good, respect fundamental rights, and foster a more equitable, just, and sustainable future for all (UNESCO, 2021; OECD, 2019).

ETHICAL CHALLENGES IN A GLOBALIZED TECHNOLOGICAL LANDSCAPE

Emerging technologies are not confined by borders, and their ethical implications often span cultures, laws, and value systems. The rapid pace at which technological innovations are being developed and deployed has made it clear that ethical considerations surrounding these advancements cannot be isolated to a single country, region, or community. The effects of technologies like artificial intelligence, biotechnology, and autonomous systems ripple across the globe, raising concerns that require collaborative, international responses. This section discusses the need for global cooperation to address cross-border ethical dilemmas, the role of international organizations in establishing universal ethical standards, and the challenges of reconciling diverse cultural perspectives on ethics and technology (UNESCO, 2021; Council of Europe, 2024).

The need for global cooperation to address cross-border ethical dilemmas is becoming increasingly urgent as the world becomes more interconnected through digital technologies. For example, data privacy is a global issue, with personal information easily crossing national borders through cloud services and social media platforms. The rise of AI and automation has led to questions about the displacement of workers worldwide, with some countries benefiting from technological advancements while others may face significant economic challenges. Furthermore, the deployment of autonomous weapons systems and the use of genetic engineering raise moral questions that extend beyond any one nation's jurisdiction, affecting global security, public health, and human rights. These are all examples of ethical challenges that transcend borders and require coordinated efforts across nations (European Parliament & Council, 2016, 2018; UN General Assembly, 2024).

To effectively address these complex ethical dilemmas, there is a need for global cooperation. National governments, international organizations, corporations, and civil society must work together to create ethical guidelines that consider the broader, cross-border implications of new technologies. For instance, when addressing the ethical use of AI, countries need to collaborate to ensure that AI technologies are developed and deployed in ways that promote fairness, transparency, and accountability, while also minimizing the potential harms of bias and discrimination. By fostering dialogue and cooperation between governments and international institutions, countries can align their policies and regulations on emerging technologies, ensuring that they address ethical challenges in a cohesive and comprehensive manner (OECD, 2019; UNESCO, 2021).

One of the key challenges in establishing global cooperation is the need to harmonize ethical guidelines across countries with different legal systems, cultural values, and societal norms. The ethical considerations surrounding emerging technologies often vary from country to country, influenced by cultural, political, and historical contexts. For example, attitudes toward privacy and surveillance differ significantly between countries like the United States and China. In the United States, individual privacy is enshrined as a fundamental right, while in China, state surveillance is more widely accepted as a means of maintaining social order. These differences in values and legal frameworks can make it difficult to create universal ethical standards for technologies such as facial recognition or data-driven surveillance systems (Zuboff, 2019; UNESCO, 2021).

Given these differences, it is essential to navigate the tension between global ethical standards and local cultural values. While international cooperation is necessary, it must be approached in a way that respects cultural diversity and accommodates varying ethical perspectives. This may involve finding common ground on key ethical principles—such as human dignity, justice, and equity—while allowing for flexibility in how those principles are applied in different cultural contexts. For example, while all countries may agree that AI should not be used to perpetuate discrimination, how each country defines and addresses discrimination may differ based on its social, political, and historical context (UNESCO, 2021; Council of Europe, 2024).

International organizations play a crucial role in establishing universal ethical standards for emerging technologies. Institutions such as the United Nations (UN), the World Health Organization (WHO), and the European Union (EU) can provide a platform for dialogue and negotiation between countries, helping to set global norms and guidelines for ethical technology development. These organizations can create frameworks that encourage the responsible use of emerging technologies while promoting shared values such as human rights, sustainability, and social justice (UNESCO, 2021; WHO, 2023; European Parliament & Council, 2024).

For example, the UN has established principles for the responsible use of AI, focusing on the need for AI to be transparent, accountable, and aligned with human rights. The EU has developed its own set of ethical guidelines for AI, emphasizing the importance of protecting fundamental rights and ensuring that AI technologies are used in ways that benefit society as a whole. These international initiatives are vital in setting a global agenda for technology ethics, encouraging countries to adopt common ethical standards and to hold companies and organizations accountable for the social impact of their innovations (UN General Assembly, 2024; European Parliament & Council, 2024).

In addition to setting ethical standards, international organizations also have a role in fostering global collaboration on research and development, ensuring that emerging technologies are developed in ways that benefit the global community. For example, the WHO has been instrumental in coordinating efforts to develop and distribute vaccines for diseases like COVID-19, ensuring that global health technologies are accessible and equitable. Similar cooperation is needed in areas such as AI, biotechnology, and environmental technologies, where cross-border collaboration can help ensure that technological advancements address global challenges such as poverty, climate change, and public health (WHO, 2023; UNESCO, 2021).

However, the establishment of universal ethical standards is not without its challenges. One of the most significant challenges is the diversity of cultural perspectives on ethics and technology. Different cultures have varying views on what is morally acceptable, and these differences can create tensions when trying to establish global norms. For instance, while some cultures may prioritize individual rights and freedoms, others may place a greater emphasis on collective welfare and social harmony. These divergent values can shape how countries approach ethical issues related to emerging technologies, such as genetic modification, data privacy, or the use of AI in decision-making (UNESCO, 2021; Zuboff, 2019).

To reconcile these diverse cultural perspectives, it is essential to engage in inclusive and respectful dialogue that allows for the exchange of ideas and perspectives. Multilateral forums, conferences, and dialogues can provide platforms for countries to discuss their ethical views on emerging technologies and to find common ground. These discussions should prioritize mutual understanding and the recognition of cultural differences, while also seeking to identify universal ethical principles that can guide the development and use of technology on a global scale (UNESCO, 2021; Council of Europe, 2024).

Another challenge in reconciling cultural differences is the uneven pace of technological development across the world. Some countries have advanced technological infrastructures and regulatory frameworks, while others are still grappling with basic issues such as access to technology and digital literacy. In this context, ethical guidelines must consider not only the needs of technologically advanced countries but also the challenges faced by less developed regions. For example, when creating ethical standards for AI, policymakers must consider how AI technologies can be used to address global challenges such as poverty, education, and healthcare, while also ensuring that these technologies do not exacerbate existing inequalities (UNESCO, 2021; OECD, 2019).

Ethical guidelines must also take into account the potential for emerging technologies to reinforce global power imbalances. As new technologies are developed, there is a risk that the countries and corporations that control these technologies will gain disproportionate power, leaving less developed nations at a disadvantage. This dynamic is particularly evident in the global data economy, where major tech companies based in the United States, China, and Europe dominate the collection and use of personal data. Ethical frameworks must address these power imbalances, ensuring that emerging technologies are developed in ways that promote global equity and inclusivity (Zuboff, 2019; UNESCO, 2021).

As part of reconciling cultural perspectives, it is also important to consider the role of education and capacity-building in promoting ethical awareness of emerging technologies. Countries with different levels of technological development may require different approaches to education and training in ethics and technology. For example, in regions where technology adoption is still in its infancy, there may be a greater need for basic education on the ethical implications of technology. In contrast, in more advanced technological settings, there may be a greater emphasis on developing sophisticated ethical frameworks to address the complex dilemmas posed by emerging technologies (National Academies, 2022; UNESCO, 2021).

In addition to formal education, public awareness campaigns can play a significant role in promoting ethical literacy and fostering informed debate about the implications of new technologies. By raising awareness of the ethical challenges posed by emerging technologies, governments and organizations can empower citizens to engage with these issues and participate in the policymaking process. This can help ensure that ethical considerations are not only addressed by experts but also by the broader public, whose lives will ultimately be affected by these technologies (Rowe & Frewer, 2000; UNESCO, 2021).

The ethical implications of emerging technologies are also influenced by the rapidly changing nature of these technologies. As technology evolves, so too do the ethical questions and challenges that arise. This dynamic nature of technology requires ethical guidelines to be adaptable and responsive to new developments. While international organizations can provide a framework for ethical standards, these standards must be flexible enough to accommodate future advancements and unforeseen consequences (OECD, 2019; UNESCO, 2021).

In conclusion, the global nature of emerging technologies necessitates international cooperation in addressing cross-border ethical dilemmas. By working together, governments, international organizations, and stakeholders can establish universal ethical standards that promote the responsible use of technology while respecting cultural diversity. However, reconciling cultural perspectives on ethics and technology is not without its challenges. Through inclusive dialogue, capacity-building, and a commitment to equity, these challenges can be overcome, allowing for the development of ethical guidelines that benefit the global community (UN General Assembly, 2024; Council of Europe, 2024).

ETHICAL LEADERSHIP IN TECHNOLOGY DEVELOPMENT

Ethical frameworks are only as effective as the individuals and institutions applying them. While establishing robust ethical guidelines is critical, it is the responsibility of leaders and practitioners within the technology sector to ensure these guidelines are adhered to and effectively integrated into their work. Ethical leadership plays a crucial role in shaping the direction of technological development, influencing how technologies are designed, implemented, and regulated. Without ethical leadership, technological innovations can veer off track, prioritizing profit or expedience over societal well-being. This section explores the importance of ethical leadership in the technology sectors, focusing on fostering a culture of ethical awareness among developers and engineers, training leaders to make decisions guided by ethical principles, and providing case studies of organizations that have successfully integrated ethics into their technological innovations (ACM, 2018; IEEE, 2021; NIST, 2023).

Fostering a culture of ethical awareness among developers and engineers is essential to ensuring that ethical considerations are embedded in every stage of technology development. Developers, engineers, and technical experts are often at the forefront of innovation, creating the algorithms, applications, and systems that power modern technologies. These professionals are not just responsible for the functionality of their creations but also for the broader implications of their work on society. By cultivating a deep understanding of the ethical consequences of their actions, organizations can empower their employees to consider the long-term impact of their designs and the social, environmental, and economic challenges that may arise (Friedman & Hendry, 2019; National Academies, 2022).

To foster this culture of ethical awareness, it is important to integrate ethics into the daily practices and values of technology development teams. This can be achieved by including ethics as a core component of the organizational culture, ensuring that ethical concerns are prioritized alongside technical specifications and performance metrics. One approach is to encourage continuous education on ethics, providing engineers with access to training and workshops that cover a wide range of ethical issues related to technology, such as privacy, data security, and algorithmic bias. Furthermore, creating spaces for open dialogue around these issues—such as regular ethics review meetings or ethical advisory boards—can help developers and engineers reflect on the moral implications of their work and encourage them to engage with ethical challenges proactively (Stilgoe et al., 2013; Rowe & Frewer, 2000).

Leadership plays a crucial role in setting the tone for ethical behavior within an organization. When leaders prioritize ethics, their teams are more likely to follow suit. Ethical leadership involves leading by example, consistently making decisions that reflect moral values and considering the wider implications of those decisions. This type of leadership requires a commitment to transparency, accountability, and a willingness to confront difficult ethical dilemmas head-on. Leaders in technology companies need to establish clear ethical guidelines and provide support for those who raise concerns about unethical practices. This leadership style not only promotes a positive ethical culture within the organization but also ensures that employees feel empowered to voice their concerns without fear of retaliation (ACM, 2018; IEEE, 2021).

In addition to fostering ethical awareness among employees, it is essential to train leaders to make decisions guided by ethical principles. Ethical decision-making requires leaders to consider not only the immediate business outcomes but also the broader social and environmental consequences of their decisions. Training leaders in ethical decision-making involves equipping them with the tools to evaluate complex dilemmas, weigh competing values, and navigate moral gray areas. This training can take many forms, including formal education on ethical theories, practical case studies, and mentoring from experienced leaders in the field. By developing a strong ethical foundation, leaders are better prepared to steer their organizations through the challenges of emerging technologies, ensuring that their innovations align with both corporate values and societal needs (Hursthouse & Pettigrove, 2020; Held, 2020).

Effective ethical leadership also requires an ongoing commitment to self-reflection and continuous improvement. Leaders must regularly evaluate their own values, decision-making processes, and the impact of their actions on their teams, their customers, and society at large. This process of self-examination ensures that leaders remain mindful of the ethical challenges they face and encourages them to make decisions that reflect their commitment to the greater good. As technology evolves, so too must the ethical leadership practices that guide it. Therefore, leaders must be willing to adapt and evolve, staying informed about new ethical challenges and the tools available to address them (Sinnott-Armstrong, 2023; Alexander & Moore, 2021).

One of the most significant ways organizations can demonstrate ethical leadership is by creating case studies that highlight their successes in integrating ethics into technological innovation. These case studies serve as valuable examples for other organizations, illustrating how ethical principles can be effectively applied in real-world scenarios. For example, a company that successfully implements privacy measures in its AI products by designing systems that prioritize user consent and data security demonstrates how ethical considerations can be incorporated into the development process. Similarly, organizations that prioritize diversity and inclusion in the development of algorithms, ensuring that their technologies do not perpetuate discrimination, set an example of how to apply ethical principles in practice (Mitchell et al., 2019; Gebru et al., 2021; Raji et al., 2020).

One prominent case study of ethical leadership in technology is Google's development of its AI principles, which emphasize fairness, accountability, and transparency. These principles were developed in response to concerns about the potential biases and discriminatory effects of AI algorithms. Google's commitment to ethical leadership in this case involved creating a set of guidelines that could inform the company's approach to AI development, ensuring that AI systems are designed with ethical considerations at their core. Additionally, Google has established an AI ethics board to review the development and deployment of AI technologies, providing oversight and ensuring that the company's AI practices align with its ethical commitments (Google, 2018; Mitchell et al., 2019).

Another example is Microsoft's AI and Ethics in Engineering and Research (AETHER) Committee, which was created to ensure that AI technologies are developed with a strong ethical framework. The AETHER Committee advises Microsoft's product teams on issues such as privacy, fairness, and transparency, ensuring that AI technologies meet ethical standards before they are released to the public. By establishing this dedicated committee, Microsoft has demonstrated its commitment to ethical leadership, providing a model for other organizations to follow in addressing the ethical challenges posed by emerging technologies (Microsoft, 2022; Microsoft, 2023).

Ethical leadership is also critical when it comes to the responsible development of autonomous systems, such as self-driving cars. Companies like Tesla and Waymo have made significant advancements in autonomous driving technology, but they have also faced scrutiny regarding the safety and ethical implications of these systems. Ethical leadership in this area requires a commitment to transparency in reporting accidents, as well as a proactive approach to addressing public concerns about safety and accountability. By establishing clear ethical guidelines for the development and deployment of autonomous vehicles, these companies can demonstrate that they are prioritizing the well-being of society over profit or expedience (SAE International, 2021; National Academies, 2022).

The importance of ethical leadership extends beyond large tech companies. Startups and smaller organizations can also have a significant impact on the ethical landscape of emerging technologies. These organizations may not have the same resources as larger companies, but they often have the advantage of being more agile and adaptable in their approach to ethics. By embedding ethical considerations into their core values from the outset, startups can build ethical practices into the foundation of their operations. This early commitment to ethics can set them apart in a competitive market and build trust with consumers, who are increasingly concerned about the ethical implications of the technologies they use (Friedman & Hendry, 2019; O'Neil, 2016).

Another critical aspect of ethical leadership is ensuring that ethical principles are integrated into the corporate governance structure of technology companies. This can be achieved by establishing ethics committees or advisory boards that provide oversight and guidance on ethical issues related to technology. These committees can consist of experts from various fields, including ethics, law, and technology, and their role is to evaluate the potential ethical risks of new technologies and advise leadership on how to mitigate those risks. By formalizing the ethical decision-making process, companies can ensure that ethical considerations are incorporated into every stage of product development, from conception to deployment (ACM, 2018; IEEE, 2021).

Moreover, ethical leadership also requires accountability. Leaders must be willing to take responsibility for the ethical decisions made within their organizations, even when those decisions result in negative outcomes. This means that when ethical lapses occur, organizations must be transparent about their mistakes and take steps to rectify them. Holding leaders accountable for their ethical decisions sends a clear message to employees and stakeholders that ethical behavior is a priority, fostering trust and credibility both within and outside the organization (ACM, 2018; NIST, 2023).

Ultimately, the role of ethical leadership in technology is to guide organizations toward creating technologies that benefit society while minimizing harm. Ethical leadership ensures that technology development is not driven solely by profit motives but by a commitment to the greater good. By fostering a culture of ethical awareness, training leaders to make decisions guided by ethical principles, and highlighting successful case studies, organizations can help shape a future in which technology is developed and deployed in ways that are aligned with human values and societal well-being. This approach to leadership is not just about preventing harm; it is about proactively using technology to enhance lives, promote equity, and contribute to a more just and sustainable world (UNESCO, 2021; OECD, 2019).

As the technology landscape continues to evolve, so too will the ethical challenges that arise. Ethical leadership in the tech sector will remain a critical factor in ensuring that emerging technologies are developed responsibly. By prioritizing ethics at all levels of technology organizations, companies can help create a future where innovation and morality go hand in hand, ensuring that technology serves the interests of society as a whole rather than a select few. Ethical leadership, when practiced effectively, has the power to transform the technology sector and the world at large, helping to realize the true potential of technology to improve human lives (National Academies, 2022; UNESCO, 2021).

THE FUTURE OF ETHICAL DECISION-MAKING IN TECHNOLOGY

As technological advancements continue to reshape every aspect of society, the need for robust ethical frameworks becomes ever more critical. This concluding section envisions the future of ethical frameworks in a world of accelerating technological change, emphasizing the importance of preparing future technologists to navigate these challenges. It explores the role of education in equipping emerging professionals with the necessary ethical decision-making skills, examines some of the most pressing ethical dilemmas on the horizon, such as brain-computer interfaces and digital immortality, and underscores the need for continuous dialogue and adaptation to address unforeseen challenges. By grounding technological advancements in ethical frameworks, society can better navigate the complexities of innovation while upholding principles of justice, equity, and respect for human dignity (UNESCO, 2021; NIST, 2023; National Academies, 2022).

The role of education in equipping future technologists with ethical decision-making skills cannot be overstated. As technology becomes increasingly integral to every facet of life, technologists are not only tasked with creating solutions but also with considering the broader social and ethical implications of their innovations. To ensure that these professionals are prepared to address the challenges posed by emerging technologies, it is crucial to embed ethics into educational curricula for future technologists. This includes offering courses that cover ethical theory, moral reasoning, and real-world applications of ethics in technology development (ABET, 2022; ACM, 2018).

Educational institutions must also emphasize the importance of interdisciplinary learning. The ethical challenges posed by new technologies are often complex, requiring a deep understanding of not only technical aspects but also social, legal, and philosophical considerations. By fostering collaboration between technology experts, ethicists, and social scientists, educational programs can prepare students to approach ethical dilemmas from multiple perspectives. This interdisciplinary approach will allow future technologists to anticipate and address potential risks and consequences before they arise (Stilgoe et al., 2013; Friedman & Hendry, 2019).

Furthermore, ethical training should not be limited to just a few specialized courses but should be woven throughout the entire educational experience of future technologists. From introductory courses to advanced technical training, students should be encouraged to engage with ethical questions at every stage of their education. This will help cultivate a mindset of ethical reflection that becomes ingrained in their professional identities, ensuring that they prioritize ethical considerations throughout their careers (UNESCO, 2021; National Academies, 2022).

As new technologies continue to emerge, so too will new ethical dilemmas. One of the most pressing issues on the horizon is the development of brain-computer interfaces (BCIs). BCIs hold enormous potential for enhancing human capabilities, from restoring lost sensory functions to enabling direct communication between the brain and machines. However, these technologies also raise significant ethical concerns. For instance, how can we ensure that BCIs are used safely and without coercion? What privacy protections are necessary to prevent unauthorized access to users' thoughts and memories? How do we address the potential for inequality, where only certain segments of society have access to these life-altering technologies? (Yuste et al., 2017; Ienca & Andorno, 2017; Royal Society, 2019).

The ethics of brain-computer interfaces will require careful consideration of issues such as consent, autonomy, and the potential for misuse. Developers of BCIs will need to navigate the tension between innovation and safeguarding individuals' rights. They will also need to engage with ethicists, regulators, and the public to develop guidelines that protect users from potential harm while fostering the responsible development of these technologies. As BCIs become more advanced, it will be essential for society to continuously reevaluate the ethical implications of their use and to remain vigilant in protecting human dignity and autonomy (OECD, 2019; Nuffield Council on Bioethics, 2013).

Another emerging ethical dilemma is the concept of digital immortality, where individuals' consciousness or personalities may be uploaded into digital systems, potentially allowing for a form of life after death. While this technology could offer profound benefits, such as the preservation of knowledge and the continuation of personal legacies, it also raises difficult questions about identity, personhood, and the very nature of life and death. What happens when a person's consciousness is digitized? Is this a true continuation of life, or is it simply an artificial replication? How do we handle issues of consent, especially for those who may be unable to make decisions about such matters in the future? (Sandberg & Bostrom, 2008; Öhman & Floridi, 2018).

The concept of digital immortality challenges our traditional understanding of ethics, particularly in the realms of personhood, rights, and dignity. As this technology develops, society will need to grapple with questions about who controls digital identities and how these identities are treated in legal and moral contexts. Issues of consent, ownership, and the preservation of human dignity will be central to discussions about digital immortality. Moreover, the potential for exploitation or the creation of "digital slaves" must be carefully considered, with strong ethical guidelines in place to protect individuals from harm (Burr & Cristianini, 2020; Öhman & Floridi, 2018).

These emerging ethical dilemmas underscore the need for continuous dialogue and adaptation of ethical frameworks. As new technologies develop, unforeseen challenges will inevitably arise, and society must remain flexible in its approach to ethical decision-making. This means creating spaces for ongoing conversations between technologists, ethicists, lawmakers, and the public to ensure that ethical considerations keep pace with innovation. Regularly updating ethical guidelines and frameworks will be essential to addressing the new risks and challenges posed by rapidly evolving technologies (UNESCO, 2021; OECD, 2019).

Incorporating feedback from diverse stakeholders, including marginalized groups who may be disproportionately affected by new technologies, will be crucial. Ethics in technology cannot be a one-size-fits-all approach; it must take into account the values, needs, and rights of all people, especially those who are vulnerable or disadvantaged. This requires a commitment to inclusivity in both the development and implementation of ethical frameworks, ensuring that all voices are heard and considered in the decision-making process (Benjamin, 2019; Eubanks, 2018).

The future of ethical frameworks in technology must also emphasize accountability and transparency. As technological systems become more complex, it will be essential to establish clear mechanisms for holding developers, corporations, and governments accountable for the ethical implications of their work. This might involve creating independent ethics boards or oversight committees to review new technologies before they are deployed, as well as establishing stronger regulations around the use of emerging technologies (Reisman et al., 2018; European Parliament & Council, 2024).

Education will continue to be a key tool in ensuring that the next generation of technologists is prepared to meet these ethical challenges. Schools, universities, and professional organizations must take a proactive approach to teaching ethics in technology, ensuring that students understand not only the technical skills required to innovate but also the moral responsibility they carry as creators of new technologies. By embedding ethics into technical education at all levels, we can prepare future technologists to make informed, responsible decisions that prioritize human dignity and societal well-being (ABET, 2022; ACM, 2018).

Additionally, companies in the technology sector must foster an internal culture that values ethical responsibility. This can be achieved by creating ethical codes of conduct that guide decision-making, implementing ethics training for all employees, and promoting a company-wide commitment to social responsibility. Leaders in the tech industry must set the tone by making ethical considerations a priority in all aspects of their work, from product development to corporate governance (IEEE, 2021; NIST, 2023).

In the future, the intersection of technology and ethics will continue to be a dynamic and evolving field. The rapid pace of technological innovation means that the ethical challenges of today may quickly become obsolete, while new challenges emerge almost overnight. This requires that ethical frameworks remain flexible, adaptable, and open to revision. Technologists, ethicists, and policymakers must work together to stay ahead of these changes, continuously updating ethical guidelines to address new risks and opportunities (UNESCO, 2021; OECD, 2019).

Ultimately, the role of ethics in technology is to ensure that innovation serves the greater good, rather than exacerbating inequality or harm. By grounding technological advancements in ethical frameworks, society can ensure that new technologies contribute to human flourishing, promote justice, and respect the dignity of all individuals. As the world becomes increasingly reliant on technology, the importance of ethical decision-making will only grow, shaping the future of both technology and society. By preparing future technologists to address the ethical challenges that lie ahead, we can help build a future where technology serves humanity, not the other way around (National Academies, 2022; UNESCO, 2021).

The integration of ethics into the fabric of technological development is essential for creating a future that is just, equitable, and sustainable. Ethical frameworks provide the moral compass that guides technological innovation, ensuring that advancements in fields such as AI, biotechnology, and digital interfaces are developed with the well-being of all in mind. As we look to the future, it is crucial that society, guided by these ethical principles, navigates the complexities of technological progress in ways that respect human dignity and promote the common good (ACM, 2018; UNESCO, 2021).

CONCLUSION

In conclusion, the rapid pace of technological innovation presents both immense opportunities and profound ethical challenges. As we continue to develop new technologies, from artificial intelligence to brain-computer interfaces, it is imperative that we ground these advancements in ethical frameworks that prioritize justice, equity, and human dignity. This chapter has explored various ethical theories—deontological ethics, consequentialism, virtue ethics, and care ethics—and demonstrated how they can guide decision-making in the face of emerging technological dilemmas. Each framework offers valuable perspectives on how to balance innovation with responsibility, ensuring that technology serves the broader good while minimizing harm, promoting fairness, and safeguarding fundamental rights (Alexander & Moore, 2021; Hursthouse & Pettigrove, 2020; Held, 2020; Sinnott-Armstrong, 2023).

The case studies presented throughout this chapter underscore the complexity of applying ethics to real-world technological challenges. From AI bias to autonomous weapons systems, each scenario reveals the need for careful consideration of not only technical feasibility but also the social, moral, and humanitarian implications. As we look ahead, the development of ethical guidelines tailored to emerging technologies will require collaboration across disciplines, as well as engagement with diverse stakeholders to ensure that the values of justice and equity are upheld (Reisman et al., 2018; OECD, 2019).

Moreover, the global nature of technological development demands international cooperation and the establishment of universal ethical standards. However, this must be done in a way that respects cultural differences and values, creating frameworks that can be adapted to different contexts while maintaining core ethical principles. Ethical leadership in technology sectors will be crucial to the successful integration of these frameworks, ensuring that technologists and innovators are equipped to make decisions guided by both technical expertise and ethical responsibility (UNESCO, 2021; Council of Europe, 2024).

Looking forward, education will play a pivotal role in shaping the next generation of technologists who will be responsible for navigating these ethical challenges. By embedding ethical reasoning into technological education, we can cultivate a generation of innovators who understand the importance of making decisions that prioritize the common good. In addition, the continuous dialogue between ethicists, technologists, policymakers, and the public will be essential to addressing emerging issues and adapting ethical frameworks to new realities (ABET, 2022; National Academies, 2022).

Ultimately, the future of technology depends not only on the technical solutions we create but on the ethical decisions we make as we develop and deploy them. By grounding technological advancements in thoughtful, well-established ethical frameworks, we can ensure that innovation continues to benefit humanity, promote social good, and protect the rights and dignity of all individuals. The responsibility to shape a technology-driven future that is just, equitable, and sustainable lies with all of us, and by committing to ethical principles, we can build a world where technology serves the betterment of society as a whole (ACM, 2018; UNESCO, 2021).

CHAPTER 5 ESSAY PROMPTS

Ethical Frameworks for Emerging Technologies

1. Discuss the key ethical frameworks (deontology, consequentialism, virtue ethics, and care ethics) and explain how each can be applied to address the ethical challenges presented by emerging technologies. Provide examples.
2. How does the ethical dilemma of AI bias challenge current technological development, and what role do ethical frameworks play in mitigating this issue? Discuss with examples from real-world applications.
3. Examine the ethical implications of autonomous weapons systems. How can different ethical theories guide decision-making regarding their development and use in modern warfare?
4. In the context of emerging technologies such as brain-computer interfaces, what are the potential ethical concerns? How can ethical frameworks help ensure that these technologies are developed responsibly?
5. Evaluate the role of ethical leadership in the technology sector. Why is it important for leaders in tech industries to apply ethical frameworks, and what might happen if these frameworks are neglected?
6. What are the challenges and opportunities of creating universal ethical standards for emerging technologies in a globalized world? Discuss how cultural diversity influences the application of ethics in technology development.
7. Consider the ethical issues surrounding data privacy and security in the age of big data and artificial intelligence. How can the principles of deontological ethics guide the development of more secure and transparent technologies?
8. How can care ethics be applied to the design of assistive technologies for marginalized or vulnerable populations? Provide examples of technologies where care ethics can improve outcomes.
9. What role does public engagement play in shaping ethical guidelines for new technologies? Discuss the importance of involving diverse stakeholders in the ethical policymaking process for emerging technologies.
10. Looking to the future, how can educational institutions contribute to shaping the ethical decision-making skills of future technologists? What ethical dilemmas should be prioritized in the curriculum of technology and engineering programs?

REFERENCES

ABET. (2022). *Criteria for accrediting engineering programs (2022–2023)*. ABET.

ACM. (2018). *ACM Code of Ethics and Professional Conduct*. Association for Computing Machinery.

AI Seoul Summit Leaders. (2024, May). *Leaders' statement of the AI Seoul Summit*. Government of the Republic of Korea & Government of the United Kingdom.

Alexander, L., & Moore, M. (2021). Deontological ethics. In E. N. Zalta (Ed.), *The Stanford Encyclopedia of Philosophy* (Fall 2021 ed.). (Original work published 2016.)

Amnesty International. (2024). *Ban the scan: A call to prohibit the use of facial recognition technology for mass surveillance*. Amnesty International.

Baldé, C. P., Kuehr, R., Yamamoto, T., McDonald, R., D'Angelo, E., Althaf, S., … Wagner, M. (2024). *Global e-waste monitor 2024*. International Telecommunication Union & United Nations Institute for Training and Research.

Benjamin, R. (2019). *Race after technology: Abolitionist tools for the new Jim code*. Polity.

Buolamwini, J., & Gebru, T. (2018). Gender shades: Intersectional accuracy disparities in commercial gender classification. *Proceedings of Machine Learning Research*, 81, 1–15.

Burr, C., & Cristianini, N. (2020). Can machines read our minds? *Minds and Machines, 30*(3), 461–480.

Council of Europe. (2024). *Framework Convention on Artificial Intelligence and human rights, democracy and the rule of law*. Council of Europe.

Department for Science, Innovation & Technology. (2023). *The Bletchley Declaration* (AI Safety Summit, 1 November 2023). GOV.UK.

Ensign, D., Friedler, S. A., Neville, S., Scheidegger, C., & Venkatasubramanian, S. (2018). Runaway feedback loops in predictive policing. In *Proceedings of the Conference on Fairness, Accountability, and Transparency (FAT*)* (pp. 160–171).

Eubanks, V. (2018). *Automating inequality: How high-tech tools profile, police, and punish the poor*. St. Martin's Press.

European Parliament. (2018). *European Parliament resolution of 12 September 2018 on autonomous weapon systems* (P8_TA(2018)0341).

European Parliament & Council. (2016). *Regulation (EU) 2016/679 (General Data Protection Regulation)*. *Official Journal of the European Union*, L 119.

European Parliament & Council. (2024). *Regulation (EU) 2024/1689 laying down harmonised rules on artificial intelligence (AI Act)*. *Official Journal of the European Union*.

Floridi, L., Cowls, J., Beltrametti, M., Chatila, R., Chazerand, P., Dignum, V., … Vayena, E. (2018). AI4People—An ethical framework for a good AI society. *Minds and Machines, 28*, 689–707.

Freedom House. (2024). *Freedom on the Net 2024: The struggle for trust online*. Freedom House.

Friedman, B., & Hendry, D. G. (2019). *Value sensitive design: Shaping technology with moral imagination*. MIT Press.

Gebru, T., Morgenstern, J., Vecchione, B., Vaughan, J. W., Wallach, H., Daumé III, H., & Crawford, K. (2021). Datasheets for datasets. *Communications of the ACM, 64*(12), 86–92.

Google. (2018). *AI at Google: Our principles*. Google.

Grother, P., Ngan, M., & Hanaoka, K. (2019). *Face Recognition Vendor Test (FRVT) Part 3: Demographic effects* (NISTIR 8280). National Institute of Standards and Technology.

Held, V. (2020). The ethics of care. In E. N. Zalta (Ed.), *The Stanford Encyclopedia of Philosophy* (Fall 2020 ed.). (Original work published 2017.)

Hursthouse, R., & Pettigrove, G. (2020). Virtue ethics. In E. N. Zalta (Ed.), *The Stanford Encyclopedia of Philosophy* (Fall 2020 ed.). (Original work published 2018.)

IEEE Standards Association. (2021). *IEEE Std 7000-2021: IEEE model process for addressing ethical concerns during system design*. IEEE.

International Energy Agency. (2024). *Data centres and data transmission networks*. International Energy Agency.

International Organization for Standardization & International Electrotechnical Commission. (2023). *ISO/IEC 23894:2023 — Information technology — Artificial intelligence — Risk management*. ISO/IEC.

International Telecommunication Union. (2024). *Measuring digital development: Facts and figures 2024*. ITU.

Ienca, M., & Andorno, R. (2017). Towards new human rights in the age of neuroscience and neurotechnology. *Life Sciences, Society and Policy, 13*, 5.

Lum, K., & Isaac, W. (2016). To predict and serve? *Significance, 13*(5), 14–19.

Microsoft. (2022). *Microsoft Responsible AI Standard (v2)*. Microsoft.

Microsoft. (2023). *AI and Ethics in Engineering and Research (AETHER): Overview*. Microsoft.

Mitchell, M., Wu, S., Zaldivar, A., Barnes, P., Vasserman, L., Hutchinson, B., … Gebru, T. (2019). Model cards for model reporting. In *Proceedings of the Conference on Fairness, Accountability, and Transparency (FAT*)** (pp. 220–229).

NIST. (2023). *AI Risk Management Framework 1.0 (NIST AI 100-1)*. National Institute of Standards and Technology.

Noble, S. U. (2018). *Algorithms of oppression: How search engines reinforce racism*. NYU Press.

Nuffield Council on Bioethics. (2013). *Novel neurotechnologies: Intervening in the brain*. Nuffield Council on Bioethics.

OECD. (2019). *Recommendation of the Council on Artificial Intelligence*. OECD Publishing.

OECD. (2019). *Recommendation of the Council on Responsible Innovation in Neurotechnology*. OECD Publishing.

OECD. (2022). *OECD guidelines on the protection of privacy and transborder flows of personal data* (Updated 2022). OECD Publishing.

OECD. (2024). *OECD employment outlook 2024: The net-zero transition and the labour market*. OECD Publishing.

Öhman, C., & Floridi, L. (2018). The political economy of death in the age of information. *Minds and Machines, 28*(4), 639–662.

O'Neil, C. (2016). *Weapons of math destruction: How big data increases inequality and threatens democracy*. Crown.

Obermeyer, Z., Powers, B., Vogeli, C., & Mullainathan, S. (2019). Dissecting racial bias in an algorithm used to manage the health of populations. *Science, 366*(6464), 447–453.

Raji, I. D., Smart, A., White, R. N., Mitchell, M., Gebru, T., Hutchinson, B., … Barnes, P. (2020). Closing the AI accountability gap. In *Proceedings of the Conference on Fairness, Accountability, and Transparency (FAccT)* (pp. 33–44).

Reisman, D., Schultz, J., Crawford, K., & Whittaker, M. (2018). *Algorithmic impact assessments: A practical framework for public agency accountability*. AI Now Institute.

Roff, H. M. (2019). The strategic robot problem. *Journal of Military Ethics, 18*(2–3), 85–103.

Rowe, G., & Frewer, L. J. (2000). Public participation methods: A framework for evaluation. *Science, Technology, & Human Values, 25*(1), 3–29.

Royal Society. (2019). *iHuman: Blurring lines between mind and machine*. The Royal Society.

SAE International. (2021). *Taxonomy and definitions for terms related to driving automation systems for on-road motor vehicles* (SAE J3016_202104). SAE.

Sandberg, A., & Bostrom, N. (2008). *Whole brain emulation: A roadmap* (Technical Report 2008-3). Future of Humanity Institute, University of Oxford.

Sharkey, N. (2016). Staying in the loop: Human supervisory control of weapons. In R. Alston (Ed.), *The ethics of autonomous weapons systems*. International Committee of the Red Cross.

Sinnott-Armstrong, W. (2023). Consequentialism. In E. N. Zalta (Ed.), *The Stanford Encyclopedia of Philosophy* (Fall 2023 ed.). (Original work published 2019.)

Sparrow, R. (2007). Killer robots. *Journal of Applied Philosophy, 24*(1), 62–77.

Stilgoe, J., Owen, R., & Macnaghten, P. (2013). Developing a framework for responsible innovation. *Research Policy, 42*(9), 1568–1580.

UNESCO. (2021). *Recommendation on the ethics of artificial intelligence*. UNESCO.

UN General Assembly. (2024). *Seizing the opportunities of safe, secure and trustworthy artificial intelligence systems for sustainable development* (A/RES/78/265). United Nations.

UN Secretary-General & ICRC. (2023). *Joint appeal on autonomous weapon systems*. United Nations / International Committee of the Red Cross.

WHO. (2023). *Ethics and governance of artificial intelligence for health: Guidance on large multi-modal models (LMMs)*. World Health Organization.

World Economic Forum. (2025). *The Future of Jobs Report 2025*. World Economic Forum.

Privacy, Surveillance, and Security in the Digital Age

6

ABSTRACT

Chapter 6 explores the complex intersections of privacy, surveillance, and security in the digital age, examining the ethical, legal, and practical challenges posed by emerging technologies. It delves into the balance between technological innovation and the protection of individual rights, highlighting the growing concerns surrounding data collection, surveillance practices, and the implementation of security measures. The chapter critically analyzes how advancements such as facial recognition, big data analytics, and cybersecurity protocols affect personal freedoms, public safety, and societal norms. It also considers the global implications of these issues, emphasizing the need for ethical frameworks, transparency, and international cooperation to ensure privacy and security are safeguarded in an increasingly interconnected world.

INTRODUCTION

In the digital age, privacy, surveillance, and security have become central concerns as technology continues to evolve at an unprecedented rate. The growth of the internet, mobile devices, and sophisticated data collection techniques has introduced new possibilities for both individuals and organizations, but these advances have also raised significant ethical, legal, and social challenges. While technology has the potential to enhance the quality of life, improve public safety, and foster innovation, it also poses risks to privacy, civil liberties, and the fundamental rights of individuals. As technology becomes more embedded in daily life, understanding the interaction between these domains is essential for ensuring that technological progress does not come at the expense of personal freedoms and societal norms (Solove, 2008; Lyon, 2018; Zuboff, 2019).

The intersection of privacy, surveillance, and security in the digital era is multifaceted and complex. On one hand, the ability to collect vast amounts of personal data and track individuals' activities can help improve security, prevent crime, and offer services tailored to user needs. However, this data collection raises profound questions about who owns personal information and how it is used. Individuals are increasingly aware of how their data is harvested, but the full scope of this data collection often remains hidden from the public eye. The extent to which personal information is shared and analyzed by governments, corporations, and other entities has made privacy a critical issue in the modern world (Richards & Hartzog, 2015; Nissenbaum, 2010).

Surveillance, a concept closely linked to privacy, has evolved dramatically with advances in technology. The introduction of facial recognition systems, geolocation tracking, and the proliferation of cameras in public spaces has reshaped the way individuals are monitored. Governments and private companies have adopted these tools for various purposes, ranging from improving public safety to targeting advertisements. While such technologies can offer security benefits, they also raise concerns about the potential for overreach. For example, the widespread use of surveillance technologies by governments to monitor citizens can erode the boundaries between necessary public security measures and excessive control over individuals' lives (Lyon, 2018; Zuboff, 2019).

At the same time, security in the digital world has become a matter of increasing importance. With the rise of cyberattacks, data breaches, and identity theft, securing personal and organizational data has become a critical focus for both public and private sectors. The methods used to safeguard information—such as encryption, firewalls, and intrusion detection systems—have advanced, but they also present new challenges to privacy. Strong security measures are essential for protecting sensitive data, but they can also enable intrusive surveillance practices, as the same technologies used to protect information can be used to monitor individuals. Striking the right balance between ensuring security and protecting privacy remains a critical issue (Schneier, 2015; OECD, 2022).

As the digital landscape continues to expand, so do the risks associated with personal privacy and security. The internet of things (IoT) has introduced millions of interconnected devices that collect data, often without users' full awareness of the implications. Devices like smart speakers, wearable fitness trackers, and connected home appliances can track user behavior, health, and preferences, which can be exploited for various purposes, including targeted advertising. This mass collection of data raises questions about consent and transparency, particularly when individuals do not fully understand the extent to which their personal information is being gathered or how it may be used in the future (Cavoukian, 2011; Kitchin, 2014).

Privacy concerns are not confined to personal data collected by corporations or governments. The increasing integration of surveillance technologies into everyday life has made it more difficult for individuals to maintain anonymity. Public spaces, once regarded as private areas where individuals could move freely without scrutiny, have become sites of constant surveillance. From security cameras to automatic license plate readers, the ability to monitor and track individuals in real time is expanding rapidly. This raises important ethical questions about the right to privacy in public spaces and whether constant monitoring creates a "chilling effect" that discourages free speech, activism, or other forms of expression (Penney, 2016; Lyon, 2018).

The role of governments in regulating surveillance and ensuring privacy protection is also a crucial part of the conversation. Many nations have enacted laws aimed at protecting citizens' personal data, such as the General Data Protection Regulation (GDPR) in the European Union, which gives individuals more control over their data and requires organizations to be more transparent about their data collection practices. However, these laws are not universally adopted, and the enforcement of privacy protections varies significantly between countries. While some governments have embraced stringent privacy regulations, others have prioritized national security and surveillance over individual privacy rights, complicating efforts to protect personal freedoms globally (European Parliament & Council, 2016; Greenleaf, 2018).

In addition to the legal challenges, ethical considerations regarding the collection and use of data remain a key concern. Many individuals are unaware of the full extent of their data's use or the ways in which it is shared between organizations. The ethical implications of collecting data without explicit consent are numerous, from undermining trust between consumers and companies to the potential for abuse of this information in ways that harm vulnerable populations. Companies and governments alike must recognize their responsibility to use data ethically, balancing the benefits of innovation with respect for individual autonomy and privacy (Floridi et al., 2018; Richards & Hartzog, 2015).

The trade-offs between privacy, security, and innovation are especially evident in the wake of major events such as the September 11 attacks. The aftermath of these events led to the rapid expansion of surveillance measures aimed at preventing terrorism and protecting national security. While many of these measures were designed to enhance public safety, they have also sparked debates about civil liberties, surveillance overreach, and the extent to which governments should monitor their citizens. The balance between securing a nation and protecting individual rights remains a contentious issue, with ongoing debates about the acceptable level of surveillance in democratic societies (Lyon, 2003; Solove, 2008).

The emergence of new technologies, such as artificial intelligence, facial recognition, and biometric data collection, presents new challenges to privacy and security. AI-driven surveillance systems, for instance, have the potential to identify individuals in real time, track their movements, and even predict their behavior based on vast datasets. While such technologies could improve security, they also present significant privacy risks, particularly in terms of potential misuse, surveillance bias, and the erosion of civil liberties. The development and deployment of these technologies must be accompanied by careful consideration of their ethical implications and the potential harms they could inflict on society (Buolamwini & Gebru, 2018; Grother, Ngan, & Hanaoka, 2019; Selbst et al., 2019).

Furthermore, the increasingly digital nature of society means that personal information is no longer contained within physical borders. The rise of global digital platforms and international data exchanges has made it difficult to regulate how personal data is collected, stored, and shared across borders. Different countries have varying levels of privacy protections and different attitudes toward surveillance, making international cooperation on privacy and security issues a complex and ongoing challenge. As such, establishing global standards and frameworks for privacy protection is essential to ensure that individual rights are upheld worldwide (Greenleaf, 2018; OECD, 2022).

The need for transparency and accountability in how data is collected and used is more critical than ever. Individuals should have the right to understand how their data is being used, who is accessing it, and for what purpose. Transparency in data practices not only fosters trust but also ensures that organizations are held accountable for any misuse or breach of privacy. Without transparency, individuals may be unaware of how their personal information is being exploited, leading to potential abuses of power and violations of privacy rights (OECD, 2022; NIST, 2023).

Finally, as technology continues to evolve, the ethical implications of privacy, surveillance, and security must be revisited and reconsidered regularly. New innovations will undoubtedly bring new challenges that require careful evaluation of their impact on society. Ongoing discussions and debates about these issues will be essential in shaping the future of technology in a way that respects individual rights while promoting the public good. Only through a balanced approach that considers both the benefits and risks of emerging technologies can we hope to create a digital landscape that serves society in a fair and ethical manner (Floridi et al., 2018; Zuboff, 2019).

PRIVACY IN THE DIGITAL AGE

Privacy is a fundamental human right that is deeply intertwined with individual freedom and autonomy. It provides individuals with control over their personal lives, ensuring they can make choices without fear of unwanted surveillance or intrusion. Privacy enables individuals to express themselves, seek information, and interact with others without the constant threat of being monitored or judged. In this way, privacy is essential to maintaining dignity, personal liberty, and democracy. However, the digital age has transformed how personal data is collected, shared, and

Privacy, surveillance, and security: balancing innovation and safety with rights, consent, and accountability in a data-saturated world

Read through a rights-and-risk lens, the image frames privacy, surveillance, and security as a three-way tension: we want safety from harms, but mass data collection and opaque monitoring can chill speech, shift power to unaccountable actors, and erode trust. The practical takeaway is that strong security doesn't require maximal data—only the right data used the right way: data minimization and purpose limits, consent that's meaningful, encryption and least-privilege access, audit trails and independent oversight, clear redress for misuse, and transparency about who sees what, when, and why. Design choices like privacy-by-default and accountability-by-design let institutions deter threats while protecting dignity and democratic participation.

used, raising complex challenges for how privacy is protected. The rapid expansion of technology, particularly in the areas of data collection and digital services, has fundamentally altered the way individuals' private information is handled (Solove, 2008; Richards & Hartzog, 2015).

The proliferation of social media platforms, mobile apps, and other digital services has led to widespread data collection practices by governments, corporations, and other entities. These platforms collect vast amounts of information, including users' preferences, browsing habits, location data, and even biometric information. The goal is often to create more personalized experiences, enhance targeted advertising, or improve user engagement. However, this collection is not always transparent or well-understood by users, and it frequently occurs without explicit informed consent. While these data practices are generally framed as beneficial for enhancing services and security, they also contribute to the growing concern over privacy violations (boyd & Crawford, 2012; Zuboff, 2019).

One of the most concerning aspects of this data collection is how much of it occurs without the knowledge or understanding of the individuals involved. Many people unknowingly agree to share their personal data when they sign up for digital services, often accepting lengthy privacy policies without fully comprehending the scope of data collection. Even when individuals take steps to control their data, the digital ecosystem is often so vast and fragmented that their efforts are ineffective. This lack of transparency and control makes it difficult for individuals to understand what data is being collected, who is using it, and for what purposes (Richards & Hartzog, 2015; Nissenbaum, 2010).

The line between beneficial innovation and privacy infringement is often thin. Many of the technologies that collect personal data—such as smart devices, social media platforms, and mobile apps—are designed to enhance user experience and offer more tailored services. For example, data about individuals' preferences and behaviors can improve recommendations on platforms like streaming services or e-commerce sites, and location tracking can provide personalized recommendations or alerts. However, the same data collection can also be used for more intrusive purposes, such as surveillance, manipulation, or the creation of detailed profiles for targeting advertisements. In the wrong hands, personal data can be exploited in ways that undermine privacy and violate individual rights (Zuboff, 2019; boyd & Crawford, 2012).

In some cases, the benefits of data collection are clear: personalized experiences and enhanced convenience are undeniable advantages of technology. Services can be customized based on the unique needs and preferences of users, making everyday tasks easier and more efficient. For instance, digital assistants can help users manage their daily schedules, while location-based apps help with navigation. Additionally, data collection can improve safety by providing authorities with information to prevent crime or respond more quickly to emergencies. But these benefits are not without their trade-offs, as the very technologies that enhance our lives can also compromise our privacy (Cavoukian, 2011; Schneier, 2015).

One significant challenge in the digital age is the pervasive nature of data collection. Unlike traditional forms of surveillance, which often required physical presence or active involvement, digital surveillance can occur passively, without individuals' awareness or consent. Internet browsers track users' browsing habits, apps collect information about their location and activities, and even devices like smart appliances or fitness trackers gather health data. This constant surveillance creates a digital trail that can be exploited by a range of entities, from corporations seeking to sell products to governments monitoring citizens. With the rise of artificial intelligence and machine learning, the ability to analyze and act on this data has become more powerful, amplifying concerns about its potential misuse (Kitchin, 2014; Narayanan & Shmatikov, 2008).

Another issue is the global nature of digital data. Personal data no longer resides solely within national borders. It flows across countries and continents, subject to various legal systems and privacy standards. While some regions, such as the European Union, have enacted strict privacy regulations like the GDPR, other countries have far more lax protections or none at all. This creates challenges for individuals whose data is collected by companies based in different jurisdictions, as well as for governments seeking to regulate cross-border data flows. The international nature of digital surveillance and data collection underscores the need for global cooperation and consistent standards to protect privacy (European Parliament & Council, 2016; Greenleaf, 2018).

The evolving relationship between privacy and security is another key concern. On one hand, the collection of personal data can be essential for enhancing security. For example, biometric data like fingerprints or facial recognition can improve authentication and prevent fraud, while data about an individual's behavior can be used to detect fraudulent activity or identify threats. In the realm of national security, governments collect data to monitor potential terrorist activities or prevent cyberattacks. While these efforts are designed to protect citizens, they often lead to the infringement of privacy rights. The more data that is collected and stored, the greater the risk that it could be misused or fall into the wrong hands, whether through hacking or government overreach (Schneier, 2015; Richards & Hartzog, 2015).

The question of consent is central to the privacy debate. In the digital age, consent is often obtained through complex and opaque privacy policies that users are required to accept in order to use digital services. These policies are typically long, written in legal jargon, and rarely read by users. This creates an imbalance of power between individuals and organizations, as users often have no choice but to accept terms if they wish to access services. Furthermore, consent is often not truly voluntary—users may feel compelled to give up their personal data in exchange for access to free or discounted services, leading to situations where people unknowingly sacrifice their privacy in exchange for convenience (Nissenbaum, 2010; Richards & Hartzog, 2015).

Despite these challenges, efforts are underway to give individuals more control over their personal information. Laws like the GDPR in the EU are a step in the right direction, requiring companies to provide clear explanations about how data is used and giving individuals the right to access, correct, or delete their data. However, such regulations are not universal, and enforcement can be inconsistent. In the absence of strong, comprehensive protections, there are significant gaps in privacy protection, particularly in industries that handle sensitive information, such as healthcare and finance (European Parliament & Council, 2016; Greenleaf, 2018).

Another challenge to privacy in the digital age is the collection of data without meaningful consent. Often, data is gathered passively, without individuals being fully aware of the extent of surveillance.

Many apps and services track location data, browsing history, and even conversations, all without explicit user knowledge. Even seemingly benign technologies, such as fitness trackers, can record intimate details about an individual's health, sleep patterns, and daily habits. The use of this data is not always clear, and individuals may not be informed about how it is being shared or used (boyd & Crawford, 2012; Kitchin, 2014).

Data breaches and cyberattacks further complicate the issue of privacy. With the vast amounts of personal data stored online, individuals are vulnerable to identity theft, financial fraud, and other forms of exploitation. Cybercriminals have increasingly sophisticated tools to access and steal sensitive information, while hackers may target organizations to access personal data for malicious purposes. These breaches can have devastating consequences for individuals, often leading to financial losses, reputational damage, and the loss of personal security. As digital platforms become more integral to everyday life, the need for robust security measures to protect personal data becomes even more critical (Verizon, 2024; Schneier, 2015).

While the need for privacy protection is undeniable, there is also a growing need for security. The digital age has brought with it new threats, including cybercrime, online harassment, and terrorism, all of which demand a level of surveillance and security measures that can sometimes infringe on privacy. Governments and corporations alike have a responsibility to protect their citizens and customers from these dangers, but the methods they use to do so must be carefully considered. Overreach in the name of security can result in the erosion of personal freedoms and the creation of a surveillance state. The challenge, then, is to balance the need for security with the right to privacy (Solove, 2008; Lyon, 2018).

A key aspect of this balance is ensuring that data collection is done ethically. Ethical data practices should prioritize transparency, user control, and the protection of sensitive information. Individuals should be informed about how their data is being used and have the ability to opt out or limit the data they share. Companies that collect and store data must adopt clear data privacy policies, implement strong security measures, and be held accountable for misuse. Ethical considerations must be embedded in the design of digital technologies, ensuring that privacy is respected from the outset (Cavoukian, 2011; OECD, 2022).

As more data is collected and stored, questions about data ownership and control become increasingly important. Who owns the data collected by digital services? Is it the individual, the company that collects it, or the platform on which it is stored? These questions have legal and ethical implications that must be addressed in order to protect individuals' rights. Without clear ownership and control, individuals have little recourse if their data is misused or compromised (Richards & Hartzog, 2015; Nissenbaum, 2010).

One of the most profound concerns about privacy in the digital age is the erosion of anonymity. In the past, individuals could engage in activities without the constant fear of being watched. Today, however, everything from online shopping to browsing history is tracked and recorded. The loss of anonymity raises questions about how much of our lives should be shared with others, particularly when individuals may be vulnerable to exploitation or manipulation. The ability to move through the world without being constantly monitored is a crucial component of personal freedom, yet this ability is rapidly diminishing as surveillance technologies become more pervasive (Penney, 2016; Lyon, 2018).

Another significant issue is the role of governments in protecting privacy. While privacy is a right, it is also a shared societal concern that must be protected at the national level. Governments must ensure that laws are in place to safeguard individuals' personal information and prevent abuse. However, governments themselves often engage in extensive surveillance, raising questions about the balance between national security and individual rights. In many cases, governments have expanded their surveillance capabilities in the name of security, sometimes without the necessary oversight or accountability. This poses a threat to democracy, as unchecked government surveillance can lead to abuses of power and the suppression of dissent (Solove, 2008; Lyon, 2018).

Ultimately, the question of privacy in the digital age is not just a technical or legal issue—it is a moral and philosophical one. It challenges our understanding of autonomy, freedom, and the role of technology in society. As we move further into an era where digital technologies are central to every aspect of our lives, we must carefully consider how to protect privacy while fostering innovation. Striking the right balance is a complex task, but it is essential for preserving the values that underpin individual freedom and human dignity. The future of privacy will depend on the choices we make today about how to navigate the ever-evolving landscape of data collection, digital innovation, and personal freedom (Nissenbaum, 2010; Zuboff, 2019).

THE RISE OF SURVEILLANCE TECHNOLOGIES

Surveillance has become an integral part of modern society, driven by advancements in technology. Technologies such as facial recognition, geolocation tracking, and big data analytics have revolutionized the way surveillance is conducted. These tools, once the realm of science fiction, are now embedded in daily life, influencing everything from public safety to consumer behavior. Governments and private companies alike utilize these technologies to monitor individuals, ostensibly to improve security, prevent crime, or mitigate threats. In urban spaces, for example, surveillance cameras are increasingly common, with their reach extending into public spaces and even private establishments. While surveillance can serve legitimate purposes, such as identifying threats or preventing criminal activities, it also raises concerns about the erosion of privacy and the potential for overreach (Lyon, 2018; Zuboff, 2019).

The question, therefore, is not whether surveillance should exist, but how much surveillance is necessary to ensure security without infringing upon personal freedoms. Surveillance has the potential to protect citizens from harm, improve law enforcement efficiency, and detect criminal behavior before it escalates. However, as surveillance technologies become more pervasive, they diminish the ability of individuals to remain anonymous. The widespread use of facial recognition, for instance, can identify individuals in real time in public spaces, erasing the distinction between public and private life. This lack of anonymity can have chilling effects on free speech, as people may be reluctant to express opinions or engage in behaviors that could attract unwanted attention, whether from the government or from others (Buolamwini & Gebru, 2018; Penney, 2016).

Moreover, the more integrated surveillance becomes into the fabric of society, the greater the risk of misuse. In the hands of governments, the power to monitor individuals raises significant concerns about civil liberties. Governments may use surveillance tools to monitor political opponents, suppress dissent, or curtail activism. Similarly, private companies, driven by profit motives, may use surveillance data to manipulate consumer behavior or engage in invasive marketing practices. This commercial surveillance, while not always as immediately threatening as state surveillance, can be equally insidious, as it often operates without individuals' full awareness or consent. The continuous collection of data can lead to a false sense of security, where privacy is sacrificed in the name of convenience or safety (Zuboff, 2019; Richards & Hartzog, 2015).

The erosion of privacy is perhaps one of the most troubling consequences of increased surveillance. As individuals' behaviors, preferences, and interactions are constantly monitored, the line between public and private life blurs. People may find themselves self-censoring, knowing that their every move is being watched, whether by the government, private companies, or even other citizens. This can have a profound impact on personal autonomy, limiting individuals' ability to act freely without fear of surveillance. The more ubiquitous surveillance becomes, the less control individuals have over how their data is collected, stored, and used. Surveillance, when unchecked, becomes a tool for control and conformity, rather than protection and safety (Lyon, 2018; Penney, 2016).

In addition to concerns about privacy, surveillance also raises questions about accountability and oversight. When governments or corporations monitor individuals, who is responsible for ensuring that these surveillance practices do not overstep legal and ethical boundaries? Without proper oversight, surveillance systems can be abused, leading to unjust practices, such as racial profiling or the disproportionate targeting of marginalized communities. In some countries, surveillance technologies have been used to track political dissidents or suppress opposition, raising fears of authoritarian control and state-sponsored surveillance programs. The need for robust checks and balances is paramount to ensure that surveillance systems are used responsibly and do not infringe upon fundamental rights (Richards & Hartzog, 2015; Buolamwini & Gebru, 2018).

The rapid advancement of surveillance technology also outpaces the development of laws and regulations that govern its use. Many countries have struggled to keep up with the technological innovations that allow for mass surveillance, leading to legal and ethical gaps. In some cases, surveillance practices have been implemented without adequate public debate or legal frameworks to protect citizens' rights. For example, the use of facial recognition technology in public spaces often occurs with little transparency, and individuals may not even be aware that their image is being captured and analyzed. Similarly, the use of big data analytics to monitor online behavior can raise questions about consent and transparency, as users may not fully understand how their data is being used or who has access to it (Grother et al., 2019; boyd & Crawford, 2012).

As surveillance technologies continue to evolve, the potential for overreach increases. Governments may justify the expansion of surveillance practices on the grounds of national security or crime prevention. However, when surveillance becomes too widespread, it can result in a surveillance state, where individuals are constantly monitored and tracked. The risk is that, in the name of security, societies could sacrifice essential freedoms, such as freedom of expression, the right to protest, and the right to privacy. The challenge is to determine where to draw the line between maintaining security and protecting personal freedoms. It is essential to ensure that surveillance practices are proportionate to the risks they aim to address and that they do not unnecessarily infringe on individual rights (Lyon, 2018; Solove, 2008).

In a democratic society, the need for surveillance must be carefully balanced with the protection of personal freedoms. A key aspect of this balance is ensuring that surveillance practices are transparent, accountable, and subject to public scrutiny. Citizens must have a say in how surveillance technologies are used and should have the ability to hold governments and corporations accountable for any abuses. Moreover, surveillance technologies should be implemented in ways that are respectful of privacy rights, with individuals' consent sought wherever possible. This is especially important when it comes to sensitive data, such as health information or financial records, which can be used to track individuals' movements, behaviors, and activities (OECD, 2022; Richards & Hartzog, 2015).

One area where surveillance and privacy intersect is in the realm of digital platforms. Social media, for example, is a major player in the collection of personal data, often using surveillance to target users with ads or content recommendations. While users may willingly share information on these platforms, the extent to which their data is collected, analyzed, and used is often opaque. Furthermore, social media platforms are increasingly being used by governments and private companies to monitor users' activities and influence their behavior. This has led to concerns about the manipulation of public opinion, as well as the erosion of privacy as individuals' online lives are tracked and analyzed in real time (Zuboff, 2019; boyd & Crawford, 2012).

Geolocation tracking is another example of how surveillance is becoming more pervasive. With the widespread use of smartphones and other mobile devices, individuals are constantly being tracked through GPS data. This data can be used to monitor movements, predict behaviors, and target users with location-based advertisements. While this may provide some convenience, such as helping users navigate or find nearby businesses, it also raises concerns about how this information is being used and who has access to it. The more precise the tracking, the greater the potential for abuse, especially if this data is shared with third parties without users' knowledge or consent (Kitchin, 2014; Richards & Hartzog, 2015).

The use of surveillance technologies in the workplace also raises important ethical considerations. Employers may use surveillance tools to monitor employees' activities, track productivity, or ensure compliance with company policies. While employers may have legitimate reasons for monitoring employees, such as maintaining security or protecting proprietary information, excessive surveillance can create a hostile work environment. Employees may feel that they are constantly under scrutiny, leading to anxiety and a reduction in morale. Moreover, the line between legitimate workplace surveillance and privacy infringement can be difficult to draw, particularly when it comes to monitoring employees' personal devices or activities outside of work (Ball, 2010; Moore & Robinson, 2016).

The advent of artificial intelligence (AI) and machine learning has further complicated the issue of surveillance. AI algorithms can analyze vast amounts of data and detect patterns that humans might miss. In the context of surveillance, this means that AI can be used to monitor individuals on an unprecedented scale, analyzing everything from facial expressions to social media posts to predict behavior. While this technology has the potential to improve security and efficiency, it also raises concerns about bias, discrimination, and the erosion of privacy. AI-powered surveillance systems may disproportionately target certain individuals or communities, exacerbating existing inequalities and leading to unjust outcomes (Buolamwini & Gebru, 2018; Selbst et al., 2019).

As surveillance continues to evolve, so too does the ethical and legal landscape that governs its use. Privacy advocates argue that individuals should have more control over how their data is used and should be given the opportunity to opt out of certain surveillance practices. Governments must enact laws and regulations that protect citizens' privacy rights while allowing for reasonable surveillance in the name of security. At the same time, corporations that rely on surveillance for marketing or other purposes must be held accountable for how they collect, store, and use personal data. Stronger privacy protections, better transparency, and increased public awareness are essential to ensuring that surveillance does not unduly infringe upon individual rights (European Parliament & Council, 2016; OECD, 2022).

In conclusion, the rise of surveillance technologies presents both opportunities and challenges. On the one hand, surveillance can be used to improve security, prevent crime, and protect citizens from harm. On the other hand, the pervasive nature of modern surveillance raises significant concerns about the erosion of privacy, the potential for overreach, and the risks of misuse. As surveillance becomes more widespread, it is crucial to strike a balance between maintaining security and protecting individual freedoms. The future of surveillance in society will depend on how governments, corporations, and individuals navigate this complex and evolving landscape (Lyon, 2018; Solove, 2008; Zuboff, 2019).

SECURITY AND ITS IMPACT ON PRIVACY

Security, in the context of digital technologies, refers to the measures and protocols that are put in place to protect sensitive information, systems, and networks from unauthorized access, manipulation, or destruction. With the increasing dependence on digital systems for everything from communication to financial transactions, the threat of cyberattacks, data breaches, and identity theft has become a growing concern. The rapid digitalization of personal, public, and corporate data has made it a prime target for malicious actors seeking to exploit vulnerabilities for financial gain, political purposes, or personal agendas. As a result, security practices and technologies have had to evolve, ranging from the implementation of basic encryption and firewalls to more sophisticated, AI-driven monitoring systems. All of these measures are designed with the ultimate goal of safeguarding sensitive data and maintaining public trust in the digital ecosystem. However, as security measures become more advanced and pervasive, they often come at the cost of privacy, raising difficult ethical questions about the balance between protection and individual rights. (NIST, 2024; Verizon, 2025; European Union Agency for Cybersecurity, 2024.)

The advent of mass digitalization has opened up new avenues for malicious actors to exploit vulnerabilities in digital infrastructure. From corporate databases to government systems, personal data is increasingly stored in centralized digital repositories, making it an attractive target for hackers and cybercriminals. As threats to security grow more sophisticated, so too must the technologies and strategies to protect against them. Security practices today range from basic encryption of communications and data storage to more complex mechanisms such as multi-factor authentication, AI-powered intrusion detection systems, and blockchain technologies for transaction verification. These measures are vital in preventing unauthorized access, manipulation, or theft of sensitive information. However, the very nature of these security technologies can also become a double-edged sword, as the techniques designed to protect information may, in turn, compromise privacy. (NIST, 2024.)

Encryption, for example, is one of the most widely used security tools, ensuring that data transmitted online is scrambled so that only authorized parties can access it. While encryption is essential for safeguarding privacy, it also poses a dilemma in the context of law enforcement and national security. Governments, in their efforts to combat terrorism or organized crime, have demanded access to encrypted data under the argument that encryption can be a barrier to national security. This has led to conflicts between security agencies and tech companies, particularly when government requests for access to encrypted communications are perceived as encroachments on personal privacy and civil liberties. The tug-of-war between law enforcement needs and individual rights to privacy is a critical issue in the ongoing debate over digital security. (Privacy and Civil Liberties Oversight Board [PCLOB], 2023.)

Similarly, firewalls, which serve to protect networks from unauthorized intrusions, and intrusion detection systems (IDS), which analyze network traffic to detect malicious activity, are important safeguards in the digital age. These technologies protect organizations from cyberattacks and prevent unauthorized access to sensitive data. However, the use of IDS and firewalls can be seen as a form of surveillance in itself, as these systems often monitor and analyze all data that enters or leaves a network, including the activities of employees or users. While such monitoring may be necessary for security purposes, it raises concerns about the extent to which individuals' online activities are being scrutinized, even when they are not engaged in any illegal activity. (NIST, 2024.)

The increasing reliance on AI in cybersecurity is another area where the balance between security and privacy becomes particularly complicated. AI-driven systems are being used to monitor network traffic in real time, identify anomalies, and predict potential threats before they happen. While these AI systems are capable of quickly identifying and responding to cyberattacks, they also collect vast amounts of data on user behavior, communications, and interactions. This data can be used to fine-tune security measures and predict future attacks, but it also raises the possibility of overreach, where the line between legitimate security monitoring and invasive surveillance becomes blurred. Furthermore, the use of AI in security can create new vulnerabilities. If these systems are compromised, they could potentially be used against the very individuals and organizations they are meant to protect. (European Union Agency for Cybersecurity, 2024; NIST, 2024.)

Data breaches are a constant threat in the digital age. High-profile incidents, such as the breach of credit card information at major retailers or the exposure of personal data through online platforms, highlight the risks that individuals face when their personal information is stored digitally. The financial and reputational damage from such breaches can be significant, affecting millions of people worldwide. Organizations must therefore invest heavily in cybersecurity to safeguard their systems from external attacks and insider threats. However, securing data often means collecting even more data, further expanding the amount of personal information that is available to be potentially compromised. This creates a vicious cycle: the more information we collect to protect our data, the greater the risk of exposure. (Verizon, 2025; European Union Agency for Cybersecurity, 2024.)

Governments have a responsibility to protect their citizens from cyber threats and maintain national security, but this mandate often leads to practices that blur the lines between security and privacy. In the name of protecting national interests, governments may engage in mass surveillance or data collection programs. For instance, intelligence agencies may intercept communications or monitor online activity in an effort to detect potential threats before they escalate. While these efforts may be effective in combating terrorism or cyberattacks, they also pose significant threats to individual privacy. Programs such as the U.S. National Security Agency's PRISM surveillance program have sparked debates over the ethical implications of mass surveillance and the need for greater transparency and oversight. The question remains: how much surveillance is justifiable in the name of security, and at what point does it violate the privacy rights of citizens? (PCLOB, 2023.)

Corporate data collection practices also play a significant role in the tension between security and privacy. In an era where data is considered a valuable commodity, companies are incentivized to collect, store, and analyze vast amounts of personal information from their customers. This data is often used to improve services, personalize experiences, or enhance security features. However, it also exposes individuals to risks of data misuse, whether through breaches, unauthorized sharing, or the sale of personal information to third parties. While consumers may benefit from more personalized services or enhanced security features, they often do so without fully understanding the extent to which their data is being collected or how it might be used. For example, data collected from online shopping habits, social media activity, and even wearable devices may be analyzed to create detailed profiles, which could be exploited for targeted advertising or, in some cases, sold to third-party organizations without the individual's knowledge or consent. (European Parliament & Council, 2016; Federal Trade Commission, 2019, 2024.)

One of the critical challenges in ensuring digital security is the vulnerability of personal data to cyberattacks. Personal information such as credit card details, health records, and social security numbers is highly valuable to cybercriminals, and its theft can lead to devastating financial and social consequences. As more personal and sensitive data is digitized, it becomes a prime target for hackers who can exploit vulnerabilities in online systems. A data breach may not only lead to identity theft but can also have long-term effects on an individual's financial security and reputation. For organizations, the reputational damage caused by a breach can result in a loss of consumer trust and loyalty, affecting their bottom line. Security measures such as encryption, multi-factor authentication, and secure coding practices are essential to protecting data from breaches, but even these measures are not foolproof. (Verizon, 2025; NIST, 2024.)

As the use of digital systems continues to expand, security must be prioritized, but it should not come at the cost of privacy. While cybersecurity measures are essential for preventing data breaches, they must be carefully implemented in ways that respect individuals' rights to privacy. This means ensuring that the data collected for security purposes is used responsibly, with clear guidelines on how it can be accessed and shared. Data retention policies must also be carefully crafted to prevent the long-term storage of unnecessary or outdated information. Additionally, individuals should have greater control over their personal data, including the ability to access, correct, and delete information that has been collected about them. (European Parliament & Council, 2016.)

The role of technology companies in securing personal data is also crucial. Tech companies, especially those that provide cloud storage, social media platforms, and online services, hold vast amounts of personal data and are responsible for ensuring that it is protected. However, there is growing concern that some companies prioritize profit over security and privacy. For instance, a company may choose to collect vast amounts of user data to improve its algorithms or advertising model, but it may not invest sufficient resources in securing that data against potential breaches. This trade-off between profitability and security is one of the fundamental ethical issues surrounding data privacy. Additionally, companies must be transparent about their data practices and give consumers the information they need to make informed decisions about whether to share their personal data. (FTC, 2019.)

The increasing integration of the Internet of Things (IoT) into daily life further complicates the balance between security and privacy. IoT devices, such as smart home systems, wearables, and connected cars, generate vast amounts of data, often without the user's full understanding of what is being collected. These devices can provide valuable services, such as enhanced convenience or better health tracking, but they also create new entry points for cyberattacks. For example, a smart thermostat or home security camera can be hacked, giving cybercriminals access to personal data or even physical spaces. The more interconnected our devices become, the more vulnerable we are to cyberattacks, making it essential to ensure that these devices are secured and that consumers are aware of the potential risks. (National Institute of Standards and Technology, 2020, 2023, 2024; NIST, 2025.)

The ethical considerations surrounding the collection and use of personal data are central to the ongoing debate over digital security. As more personal information is collected, analyzed, and stored, individuals must be assured that their privacy is protected and that security measures are implemented in a transparent and responsible manner. Governments and organizations must work together to develop laws and regulations that balance the need for security with the protection of privacy. Moreover, individuals must be empowered to take control over their own data, ensuring that they are informed about the risks and benefits of sharing personal information in the digital age. (European Parliament & Council, 2016; FTC, 2024a.)

In conclusion, the pursuit of digital security must be balanced with respect for individual privacy. While robust security measures are essential to protecting against cyberattacks and data breaches, they should not infringe upon personal freedoms or lead to unnecessary surveillance. The tension between security and privacy is an ongoing challenge in the digital age, requiring careful consideration of the ethical, legal, and social implications of data collection and monitoring practices. By ensuring that security measures are transparent, accountable, and respectful of privacy, society can maintain trust in digital systems and continue to benefit from the advancements that technology offers. (NIST, 2024.)

THE TRADE-OFFS BETWEEN PRIVACY, SURVEILLANCE, AND SECURITY

The most significant trade-offs between privacy, surveillance, and security become evident in government surveillance programs. In the wake of events like the 9/11 attacks, many governments worldwide escalated their surveillance efforts in order to combat terrorism, cyberattacks, and other national security threats. This shift marked a fundamental change in how governments viewed surveillance, with a greater focus on monitoring individuals and groups that could pose a risk to national security. While these measures are often justified by the need to protect citizens from immediate threats, they have sparked significant concerns about the erosion of civil liberties, particularly regarding the invasion of privacy. The question then arises: how much surveillance is too much, and at what point does the pursuit of security violate individual rights? (PCLOB, 2023.)

Governments argue that expanded surveillance programs are essential for national security, allowing them to identify and neutralize potential threats before they manifest. In this view, the trade-off between privacy and security is seen as necessary, as protecting citizens from terrorism and cyberattacks requires a proactive approach. Programs like the National Security Agency's (NSA) mass surveillance initiatives, which monitor phone records, internet activity, and communication patterns, were justified as crucial tools for identifying terrorist cells and foreign intelligence threats. However, as these programs expanded, so did concerns over the scope of their reach. Critics argue that these programs often infringe upon the privacy of individuals who have no connection to criminal activity, thereby violating civil liberties in the name of security. (PCLOB, 2023.)

The balance between national security and individual privacy is a contentious issue, and it is essential that governments be held accountable for any misuse of power or violations of privacy. This is especially important when surveillance programs operate with minimal oversight, as was the case with the NSA's metadata collection. In some cases, individuals' personal information is gathered without their knowledge or consent, raising significant questions about the ethics of government surveillance. While the stated goal is to protect citizens, the unchecked power of surveillance agencies can easily lead to abuses. In democratic societies, it is crucial that there are strong legal frameworks and independent oversight bodies to ensure that surveillance powers are used only when absolutely necessary and that the privacy rights of citizens are respected. (PCLOB, 2023.)

One of the most publicized examples of the tension between privacy and surveillance is the collection of personal data by tech companies. Social media platforms, search engines, and e-commerce websites collect vast amounts of data on users, which is often used for targeted advertising and commercial purposes. These companies track users' online behaviors, preferences, and activities, compiling detailed profiles to tailor ads, content, and services to individual users. While this personalized approach can lead to a more convenient and engaging online experience, it raises serious concerns about the extent to which personal information is being harvested, stored, and potentially shared with third parties without full transparency or user consent. (European Parliament & Council, 2016.)

The practice of collecting personal data by tech companies raises important questions about consent and control. In many cases, users unknowingly agree to data collection practices by accepting terms and conditions they may not fully understand. This lack of informed consent can undermine individuals' ability to maintain control over their own personal information. Furthermore, there are concerns about how this data is being used. While some companies collect data for commercial purposes, such as improving services or targeting ads, others may share or sell this information to third parties. This practice raises the risk that personal data could be exploited for profit or used in ways that individuals may not have anticipated when they first shared their information. (European Parliament & Council, 2016.)

The Cambridge Analytica scandal is a glaring example of how personal data can be exploited for political gain. In this case, millions of Facebook users' personal data was harvested without their consent and used to target them with political advertisements. This targeted campaigning was designed to influence voting behaviors during the 2016 U.S. presidential election. The scandal revealed how data can be weaponized, not only undermining trust in the companies involved but also raising alarm about the potential for foreign interference in democratic processes. The use of personal data for political manipulation undermines the core principles of democratic engagement, where citizens should be free to make decisions without undue influence. (UK House of Commons Digital, Culture, Media and Sport Committee, 2019.)

In response to growing concerns about the abuse of personal data, regulations like the European Union's General Data Protection Regulation (GDPR) have been introduced. The GDPR aims to give individuals more control over their personal data by requiring companies to obtain explicit consent before collecting data, ensuring transparency in data processing, and granting individuals the right to access, correct, and delete their personal information. The GDPR also imposes strict penalties on companies that fail to comply with these regulations, ensuring that they are held accountable for mishandling user data. While the GDPR is a significant step forward in protecting personal privacy, the rapid pace of technological innovation and the global nature of the internet pose ongoing challenges to enforcement and compliance. (European Parliament & Council, 2016.)

The trade-off between privacy and security also plays out in the context of surveillance in public spaces. Many governments and private entities have increasingly turned to surveillance cameras, facial recognition software, and geolocation tracking to monitor public spaces and gather data on individuals' movements and activities. These measures are often justified as necessary for preventing crime, enhancing public safety, or even monitoring the spread of infectious diseases. However, the widespread use of such surveillance raises significant concerns about the loss of anonymity in public life. With the proliferation of surveillance technologies, individuals may feel that they are constantly being watched, which can have a chilling effect on free speech and personal expression. (Grother, Ngan, & Hanaoka, 2019.)

The introduction of facial recognition technology is particularly controversial, as it can be used to identify individuals in real-time without their consent. While facial recognition has been touted as a valuable tool for law enforcement and security agencies, it also poses serious risks to privacy. In some cases, this technology has been deployed in public spaces, such as airports or shopping malls, to track individuals' movements and behaviors. However, critics argue that this technology can lead to mass surveillance, with individuals being monitored and profiled without their knowledge or approval. The use of facial recognition technology also raises concerns about its accuracy, especially when it comes to identifying people of color, as studies have shown that these systems are more prone to misidentifying certain demographic groups. (Buolamwini & Gebru, 2018; Grother et al., 2019.)

While facial recognition technology may provide security benefits, such as identifying criminal suspects or preventing fraud, it also poses significant risks to personal privacy. The ability to track individuals without their consent undermines the principle of anonymity in public spaces, potentially subjecting people to unwanted scrutiny. This raises the question of how much surveillance is necessary to achieve security, and whether the loss of privacy is a fair trade-off. In democratic societies, it is essential that surveillance technologies be deployed with strict oversight and regulation to ensure they are used proportionately and in accordance with fundamental rights. (European Parliament & Council, 2016.)

In addition to government and corporate surveillance, the proliferation of digital tracking technologies further complicates the balance between privacy and security. Devices such as smartphones, wearables, and smart home systems continuously collect data on users' activities, locations, and preferences. While these devices can enhance convenience and safety, they also create new vulnerabilities to privacy. For instance, location tracking features on smartphones can reveal users' movements, potentially exposing them to unwanted surveillance or even physical danger. In some cases, this data can be accessed by third parties, including marketers, advertisers, and even hackers. The widespread collection of personal data through digital devices poses a challenge to privacy protections, as it becomes increasingly difficult for individuals to maintain control over their own information. (FTC, 2024a.)

As more data is collected and stored by governments and corporations, the risk of data breaches and misuse grows. High-profile incidents, such as the Equifax data breach, which exposed the personal information of millions of individuals, highlight the vulnerabilities of data storage systems. Data breaches can have serious consequences for individuals, including identity theft, financial loss, and reputational damage. These incidents also raise questions about the responsibility of companies and governments to protect the personal information they collect. If organizations are going to collect data for security or commercial purposes, they must take the necessary precautions to safeguard that data and ensure it is not misused. (FTC, 2019.)

The trade-off between privacy and security extends to the realm of cybersecurity as well. As cyberattacks become more sophisticated, governments and private companies are increasingly adopting aggressive cybersecurity measures to protect critical infrastructure and sensitive data. However, these measures often come at the expense of privacy. For example, some cybersecurity tools, such as malware scanners and intrusion detection systems, require access to users' personal data in order to identify potential threats. While these tools are essential for protecting against cyberattacks, they can also compromise privacy if they are not properly secured or if they collect unnecessary personal information. (NIST, 2024.)

In light of these challenges, it is essential to develop frameworks that balance the need for privacy, security, and surveillance. Governments, tech companies, and individuals must work together to ensure that surveillance technologies are used responsibly and ethically. Transparency, accountability, and oversight are key to maintaining public trust and ensuring that privacy rights are not unduly infringed upon in the pursuit of security. This includes the implementation of robust data protection regulations, the use of privacy-enhancing technologies, and the promotion of a culture of responsible data use. (European Parliament & Council, 2016; NIST, 2024.)

Ultimately, the trade-offs between privacy, surveillance, and security are complex and multifaceted. While security measures are essential for protecting individuals and societies from harm, they must be implemented in ways that respect the fundamental right to privacy. Striking this balance requires careful consideration of ethical, legal, and social implications, and it is an ongoing challenge that will require constant vigilance and adaptation as technology continues to evolve.

THE ETHICS OF SURVEILLANCE: RACIAL PROFILING AND DISCRIMINATION

The ethical implications of surveillance and data collection are deeply intertwined with issues of equity and justice. Surveillance technologies and security measures are not distributed equally across society, and this disparity often leads to significant disadvantages for marginalized communities. As surveillance becomes more embedded in public and private life, it disproportionately impacts specific groups, particularly those who are already vulnerable to systemic discrimination and inequality. This creates an ethical dilemma: while surveillance can improve security and public safety, it also has the potential to deepen existing social divides.

For example, law enforcement agencies are more likely to deploy intensive surveillance on minority populations. This has been particularly evident in the use of facial recognition technologies and predictive policing tools. Research has shown that these technologies are more likely to misidentify people of color, especially Black and Latino individuals, leading to higher rates of surveillance and suspicion. This kind of over-policing, driven by biased data and algorithms, can contribute to a cycle of racial profiling, criminalization, and disproportionate punishment. The ethical concerns around such surveillance practices include the risk of perpetuating systemic racism and the erosion of trust between marginalized communities and law enforcement. (Buolamwini & Gebru, 2018; Grother et al., 2019.)

Beyond the criminal justice system, surveillance can also be used by corporations to monitor and exploit marginalized groups. For example, targeted advertising can reinforce social inequalities by profiling individuals based on their race, gender, or socioeconomic status. These companies collect vast amounts of data on consumers, often without explicit consent or transparency, to create highly personalized marketing campaigns. While these practices may seem harmless or even beneficial from a consumer standpoint, they can exacerbate existing inequalities. For instance, individuals from lower-income backgrounds or underrepresented communities may face manipulation through targeted ads that promote products they cannot afford or services that exploit their vulnerabilities.

Similarly, the collection and analysis of personal data can exacerbate social inequalities by disproportionately affecting those who are already at a disadvantage in terms of access to technology or resources. Individuals without the latest devices or reliable internet connections may find themselves excluded from digital platforms that offer essential services, such as online banking, healthcare, or education. This digital divide can create a barrier to equitable access to opportunities and further entrench social stratification. At the same time, those who have access to technology may be more vulnerable to data exploitation, as they often have less control over how their personal data is collected or used. The more data an individual generates, the greater the potential for exploitation, particularly when data collection practices are opaque or unregulated.

The lack of control over personal data raises significant ethical questions about autonomy and consent. Many individuals are unaware of the extent to which their data is collected, stored, and analyzed. In many cases, consent is implied through terms of service agreements that users often do not read or fully understand. This lack of informed consent is particularly concerning when it comes to marginalized groups, who may already be socially or economically disadvantaged. These groups may have fewer options for opting out of data collection or surveillance programs, leaving them vulnerable to exploitation and manipulation by corporations, governments, or other entities.

The ethical considerations surrounding surveillance and data collection are further complicated by the role of algorithms and artificial intelligence. Algorithms are often used to process and analyze vast amounts of data, making decisions about everything from credit scores to hiring practices. However, these algorithms are not neutral; they are shaped by the biases inherent in the data they process. If the data collected for surveillance or security purposes is biased, the algorithms that analyze this data may perpetuate those biases, leading to unfair outcomes. This is particularly problematic when algorithms are used in critical areas like law enforcement, healthcare, and employment, where biased decisions can have life-altering consequences for marginalized individuals.

In the context of surveillance, algorithmic bias can manifest in a variety of ways. For example, predictive policing algorithms, which rely on historical crime data, often reflect patterns of racial bias in law enforcement. These algorithms may predict higher crime rates in predominantly Black or Latino neighborhoods, leading to more frequent surveillance and policing in these areas. Similarly, the use of facial recognition technology has been shown to have higher error rates for people with darker skin tones, increasing the likelihood of false identification and misidentification in minority populations. These biases, whether intentional or unintentional, have serious implications for equity and justice, as they exacerbate the marginalization of already vulnerable communities. (Buolamwini & Gebru, 2018; Grother et al., 2019.)

Moreover, the ethical concerns about surveillance and data collection extend to the global stage. Many multinational corporations and governments collect data on individuals across national borders, often without regard for local privacy laws or cultural norms. In some cases, personal data is shared or sold to third parties, raising concerns about the exploitation of individuals' information on a global scale. This transnational nature of data collection and surveillance makes it difficult to protect the privacy rights of individuals, particularly in countries where legal frameworks for data protection are weak or nonexistent. The global nature of surveillance also raises questions about sovereignty and control over personal data, as individuals may have little recourse if their data is misused by foreign governments or corporations.

The trade-off between security and privacy becomes even more complicated when considering the potential for mass surveillance. Governments may justify mass surveillance programs by arguing that they are necessary to combat terrorism, cyberattacks, or other threats to national security. While these programs may help prevent crimes and protect public safety, they also risk violating the privacy rights of entire populations, especially marginalized groups. Mass surveillance often relies on the collection of vast amounts of personal data, including communications, locations, and behavioral patterns. The ethical dilemma arises when such programs disproportionately target specific communities, raising concerns about discrimination, overreach, and the violation of civil liberties.

To address the ethical implications of surveillance and data collection, it is essential to develop privacy and security policies that protect all individuals, particularly those from marginalized groups. These policies should be designed to ensure that no group is unfairly targeted or discriminated against. For example, data collection practices should be transparent, and individuals should be given clear and accessible options to control their personal information. Additionally, surveillance programs should be subject to independent oversight to ensure they are used responsibly and fairly. This oversight can help mitigate the risk of bias, discrimination, and abuse, ensuring that surveillance and data collection are conducted in ways that promote equity and justice.

Furthermore, it is important to include marginalized communities in discussions about surveillance and data collection. These communities often bear the brunt of surveillance practices, yet they are frequently excluded from decision-making processes that affect their lives. Involving marginalized groups in policy development and regulatory processes ensures that their voices are heard and that their concerns are addressed. This can help create more inclusive and equitable privacy and security policies that consider the unique challenges faced by vulnerable populations.

In addition to legal and policy reforms, there is also a need for technological innovations that prioritize privacy and equity. Privacy-enhancing technologies, such as end-to-end encryption and decentralized data storage, can help protect individuals' personal information while still allowing for secure communication and data sharing. These technologies can empower individuals to retain control over their own data, reducing the risks of exploitation and surveillance. Moreover, ethical considerations should be integrated into the design and development of surveillance technologies, ensuring that they do not reinforce existing biases or perpetuate discrimination.

Ultimately, addressing the ethical implications of surveillance and data collection requires a multi-faceted approach that takes into account the intersection of privacy, security, equity, and justice. It is not enough to focus solely on security or privacy in isolation; both must be considered in tandem, with a particular emphasis on protecting vulnerable communities. As surveillance technologies continue to evolve, it is essential to prioritize the rights and dignity of individuals, ensuring that security measures do not come at the expense of justice and equality.

THE GLOBAL IMPACT OF PRIVACY AND SECURITY

Technological innovation, particularly in the digital realm, is inherently global. The rise of the internet and the proliferation of digital technologies have created a landscape where data flows freely across borders, and digital platforms are accessible worldwide. This interconnectedness has transformed how we communicate, do business, and interact with one another. However, it has also raised significant privacy and security concerns that are no longer confined to individual countries or regions. The consequences of cyberattacks, data breaches, or identity theft can reverberate globally, affecting individuals, companies, and even governments in multiple countries. As such, the challenges surrounding privacy and security have taken on an international dimension (ENISA, 2024; Verizon, 2025).

One of the primary challenges in addressing these issues on a global scale is the lack of a unified approach to privacy laws. Countries around the world differ widely in their attitudes toward privacy, with some placing a high value on individual privacy rights, while others prioritize national security or economic interests. This divergence in perspectives leads to a patchwork of privacy regulations, making it difficult to establish universal standards for privacy protection and data security. For example, the European Union has implemented the General Data Protection Regulation (GDPR), a comprehensive set of rules designed to protect individuals' data and privacy. However, such regulations are not universally adopted, and countries outside the EU may not have similar laws in place, creating discrepancies in how personal data is handled across borders (European Parliament & Council, 2016; OECD, 2022).

Moreover, the cultural differences between nations play a significant role in shaping privacy policies and attitudes. In some countries, there is a strong cultural emphasis on personal privacy, with individuals expecting their data to be protected by law. In others, privacy is less prioritized, with a greater focus on collective security or economic development. These cultural differences complicate efforts to reach a global consensus on privacy standards and regulations. For example, some countries may view government surveillance as a necessary tool for maintaining national security, while others may see it as an infringement on individual freedoms. These contrasting views make it difficult to craft international agreements that respect both security needs and personal privacy rights (Council of Europe, 2024).

The lack of global alignment on privacy regulations creates significant challenges for multinational corporations and international organizations. Companies operating in multiple countries must navigate a complex web of privacy laws and regulations, which can be costly and time-consuming. For instance, a tech company based in the United States may have to comply with the GDPR when operating in the European Union, while also adhering to different privacy laws in other regions. This can lead to confusion and inconsistencies in how personal data is handled, particularly when it comes to cross-border data transfers. The absence of a unified global approach also makes it harder to hold companies accountable for privacy violations, as enforcement mechanisms may differ from one jurisdiction to another (European Commission, 2023; European Parliament & Council, 2016).

As technological innovation continues to advance, the challenges surrounding privacy and security will only intensify. Emerging technologies such as the Internet of Things (IoT), brain-computer interfaces, and artificial intelligence (AI) are creating new opportunities for data collection and analysis. While these technologies have the potential to improve lives in countless ways, they also raise significant privacy and security concerns. The IoT, for example, connects everyday devices—such as home appliances, wearables, and vehicles—to the internet, creating vast networks of interconnected devices that constantly collect and exchange data. This creates new opportunities for hackers to exploit vulnerabilities and for corporations to gather personal information without users' explicit consent (ENISA, 2024).

Brain-computer interfaces (BCIs) are another emerging technology that poses new privacy and security risks. BCIs allow for direct communication between the brain and external devices, enabling individuals to control technology with their thoughts. While this technology has the potential to revolutionize healthcare and communication, it also raises concerns about the potential for unauthorized access to individuals' thoughts and emotions. The ethical and security implications of BCIs are still being explored, but it is clear that new technologies like these will require innovative solutions to ensure that individuals' privacy is protected (Royal Society, 2019; OECD, 2019; Yuste et al., 2017).

Artificial intelligence (AI) presents yet another layer of complexity in the privacy and security landscape. AI systems are increasingly being used to process and analyze large amounts of personal data, from facial recognition to behavior tracking. While AI can enhance security and improve services, it also raises concerns about the potential for biased decision-making, surveillance, and data misuse. For example, AI-powered surveillance systems have been used in some countries to monitor public spaces and track individuals, leading to concerns about privacy violations and state-sponsored surveillance. Additionally, AI systems that analyze personal data may perpetuate existing biases, leading to unfair outcomes in areas such as hiring, credit scoring, and law enforcement (UNESCO, 2021; NIST, 2019; Amnesty International, 2024).

The rapid pace of technological innovation also presents challenges for policymakers and regulators. Privacy laws that were designed for earlier generations of technology may no longer be adequate to address the complexities of new technologies. For example, data protection laws that focus primarily on personal information may not fully address the risks posed by the IoT, BCIs, or AI. As a result, there is an urgent need for international cooperation to update existing privacy regulations and create new frameworks that account for emerging technologies. This will require collaboration among governments, international organizations, technology companies, and privacy advocates to develop regulations that balance the benefits of innovation with the protection of individuals' privacy and security (OECD, 2022; Council of Europe, 2024).

At the same time, the global nature of the digital landscape means that no single country can effectively address privacy and security concerns on its own. Cyberattacks and data breaches are often carried out by actors operating across borders, making it difficult for any one government to prevent or mitigate these threats. For example, a cyberattack on a U.S. company may be perpetrated by hackers in another country, and the stolen data may be sold or used to target individuals in multiple regions. This highlights the need for international cooperation and coordination in responding to cyber threats. Countries must work together to establish common standards for cybersecurity, share information about emerging threats, and develop joint strategies for responding to cyberattacks (WEF, 2025; ENISA, 2024).

Despite the challenges, there are some positive developments in international efforts to address privacy and security concerns. The European Union's GDPR has set a strong precedent for data protection, and other countries are beginning to adopt similar laws. In addition, international organizations such as the United Nations and the Organization for Economic Cooperation and Development (OECD) have begun to address privacy and security in the context of the digital economy. These efforts are a step in the right direction, but much more work remains to be done. For example, there is still no global treaty on privacy, and countries continue to struggle with issues such as data localization, cross-border data flows, and surveillance (European Parliament & Council, 2016; OECD, 2022).

As the digital landscape continues to evolve, it is essential that privacy and security concerns remain at the forefront of international discussions. The interconnectedness of the global digital economy means that privacy and security are not issues that can be resolved within individual countries or regions. Instead, there must be a collective effort to create a unified, global approach to privacy and data protection. This will require balancing the competing interests of privacy, security, and innovation while ensuring that the rights of individuals are respected. As new technologies continue to emerge, it will be critical for international cooperation to keep pace with these changes and ensure that privacy and security remain protected in the digital age (European Commission, 2023; UNESCO, 2021).

EDUCATION AND ADVOCACY IN PRIVACY AND SECURITY

Given the complexity and rapid pace of technological change, education plays a critical role in addressing privacy and security concerns. As technology continues to evolve and permeate every aspect of our lives, the need for individuals to understand the implications of their digital actions has never been greater. Privacy and security issues are no longer confined to specific sectors or industries but affect nearly every facet of society, from personal data collection by corporations to the use of surveillance technologies by governments. Educational institutions are uniquely positioned to equip students with the knowledge and skills necessary to navigate this landscape (UNESCO, 2021).

One of the primary ways education can address these concerns is by raising awareness about the importance of privacy rights. In an increasingly interconnected world, where personal information is often shared, stored, and analyzed in real time, understanding one's privacy rights is essential. Educational programs that emphasize the significance of data ownership, consent, and the potential risks of sharing personal information can help individuals make informed choices. By teaching students about the legal protections available to them and the potential consequences of breaches, educators can empower individuals to take control over their own privacy (European Parliament & Council, 2016; OECD, 2022).

In addition to privacy rights, data protection is another critical area of focus in today's digital landscape. The vast amounts of personal and professional data collected by online platforms, businesses, and public institutions raise questions about how that information is secured and who has access to it. Schools and universities can integrate lessons on encryption, secure data storage, and safe online practices into curricula, helping students understand how to protect sensitive information. Educating students about the potential dangers of cyberattacks and identity theft, as well as the technological solutions designed to counter these threats, can foster a culture of digital responsibility and proactive security measures (NIST, 2020; ENISA, 2024).

Furthermore, the ethical implications of surveillance in the digital age are another important topic that must be addressed in education. Surveillance technologies, such as facial recognition and location tracking, have raised significant concerns regarding individual freedoms and human rights. Students must learn about the balance between security and privacy, as well as the ethical dilemmas that arise when surveillance tools are used by governments, corporations, or other entities. By teaching future generations to critically analyze surveillance practices and their broader societal impacts, educators can help students develop a deeper understanding of the ethical considerations surrounding technology (NIST, 2019; Amnesty International, 2024).

Incorporating discussions about the legal aspects of technology into education also plays a vital role in preparing students for the future. Laws and regulations related to data privacy, intellectual property, and digital communications are evolving at a rapid pace. Students need to understand the current legal framework that governs technology and be aware of potential changes on the horizon. Educating individuals about their rights under these laws, as well as the responsibilities of businesses and governments, is key to ensuring that they can advocate for policies that prioritize individual freedoms and fairness in the digital realm (European Parliament & Council, 2016; European Commission, 2023).

The social dimensions of technology, including its impact on inequality and access, should also be a core component of educational curricula. While technology offers tremendous opportunities for advancement, it can also exacerbate existing disparities in society. Students need to explore how access to technology and the internet can either bridge or widen gaps in education, healthcare, employment, and other key areas. Teaching students about the social consequences of technological inequities will help them understand the broader societal implications of digital divides and empower them to advocate for more equitable access to technology (UNESCO, 2021).

Another vital aspect of preparing students for the digital world is fostering critical thinking and ethical decision-making. In a world where technology often outpaces the development of laws and ethical guidelines, individuals must be able to make sound decisions based on an understanding of the ethical, legal, and social implications of their actions. Education that emphasizes critical thinking can help students assess the consequences of their online behaviors, whether it involves the ethical use of data, the impact of social media on mental health, or the responsible development of emerging technologies. By equipping students with the tools to think critically about technology, educators can prepare them to be responsible digital citizens (UNESCO, 2021).

Ultimately, the role of education in addressing privacy and security concerns is essential for shaping a future where individuals can thrive in the digital world without sacrificing their rights and freedoms. As technology continues to evolve, educational institutions must remain at the forefront of this conversation, ensuring that students are not only knowledgeable about the technical aspects of technology but also equipped to navigate its ethical and social challenges. By teaching future generations about the ethical, legal, and social dimensions of technology, we can help create a society where technology serves the greater good while safeguarding individual rights (UNESCO, 2021; NIST, 2020).

CONCLUSION

The challenges of privacy, surveillance, and security in the digital age are multifaceted and complex, driven by the rapid evolution of technology. As digital platforms and surveillance systems become more integrated into everyday life, there is a growing concern about how much personal information is being collected and how it is being used. Technological innovations, such as artificial intelligence, biometrics, and ubiquitous connectivity, have created unprecedented opportunities for monitoring and analyzing behavior, raising significant questions about how to balance the advantages of these tools with the need to protect personal freedoms. While technology has the potential to improve safety, health, and convenience, it also carries risks related to the erosion of privacy and autonomy (UNESCO, 2021; NIST, 2019).

To address these concerns, it is essential to strike a balance between the potential benefits of new technologies and the protection of individual rights. Surveillance systems, for example, can help prevent crime and terrorism, but they must be designed and implemented in a way that respects citizens' privacy and freedom from unwarranted intrusion. Security measures, such as encryption and data protection, are necessary to safeguard sensitive personal information from cyber threats, but they must be carefully balanced with law enforcement's need to access data for legitimate security purposes. Transparent policies that clearly define the scope of surveillance and security practices, as well as their limitations, are crucial to ensuring that technology serves the public interest without overstepping ethical boundaries (European Parliament & Council, 2016; OECD, 2022).

Ethical considerations play a vital role in navigating the complexities of privacy, surveillance, and security in the digital age. Policymakers and technology developers must engage in continuous dialogue to ensure that new innovations are not only technologically feasible but also morally justifiable. This includes addressing concerns about discrimination, bias, and unintended consequences that may arise from the widespread use of surveillance technologies. Additionally, ethical frameworks must guide decisions on data collection, usage, and storage to prevent exploitation or misuse. Public engagement is also essential, as individuals need to understand the implications of surveillance and security measures and have a say in how their personal information is managed (UNESCO, 2021).

Ultimately, fostering a culture of ethical awareness and encouraging ongoing dialogue among policymakers, technology developers, and the public is crucial to creating a future where technology supports both individual autonomy and societal security. By considering privacy as a fundamental right and security as a collective responsibility, we can work towards a balanced approach where technological advancements enhance public safety without sacrificing privacy. This vision requires collaboration across sectors, transparent governance, and a commitment to protecting human dignity in the face of rapidly advancing technologies (Council of Europe, 2024; European Commission, 2023).

TEN ESSAY QUESTIONS FOR CHAPTER 6

Privacy, Surveillance, and Security in the Digital Age

1. How has the rapid advancement of technology in the digital age impacted individual privacy, and what are the main risks associated with this shift?
2. Discuss the ethical dilemmas that arise from the widespread use of surveillance technologies in modern society. How can these technologies be balanced with the protection of civil liberties?
3. In what ways can government surveillance programs enhance national security, and what are the potential consequences for personal freedoms and privacy?
4. Analyze the role of big data and artificial intelligence in surveillance systems. How do these technologies raise concerns about privacy, bias, and discrimination?
5. What are the implications of the "right to be forgotten" in the digital age, and how should this right be balanced against freedom of expression and public interest?
6. How can transparent policies and ethical considerations guide the development of surveillance and security technologies to ensure that they serve the greater good without infringing on individual rights?
7. What role does public engagement and awareness play in shaping the ethical use of surveillance and security measures? Should citizens have a say in the policies that govern their privacy rights?
8. Evaluate the impact of social media platforms on privacy. To what extent should these platforms be responsible for safeguarding user data, and how can users protect their own privacy online?
9. How do international differences in privacy laws and surveillance practices affect global citizens, particularly in terms of cross-border data sharing and security concerns?
10. With the rise of biometric surveillance, such as facial recognition technology, what are the potential risks to personal privacy, and how can legislation address these concerns without stifling technological innovation?

REFERENCES

Amnesty International. (2024). *Ban the scan: A call to prohibit the use of facial recognition technology for mass surveillance.* Amnesty International.

Ball, K. (2010). Workplace surveillance: An overview. *Labor History*, 51(1), 87–106.

boyd, d., & Crawford, K. (2012). Critical questions for big data. *Information, Communication & Society*, 15(5), 662–679.

Buolamwini, J., & Gebru, T. (2018). Gender shades: Intersectional accuracy disparities in commercial gender classification. *Proceedings of Machine Learning Research*, 81, 1–15.

Cavoukian, A. (2011). *Privacy by design: The 7 foundational principles.* Information and Privacy Commissioner of Ontario.

Council of Europe. (2024). *Framework Convention on Artificial Intelligence and human rights, democracy and the rule of law.* Council of Europe.

European Commission. (2023). *Commission Implementing Decision (EU) 2023/1795 on the adequate level of protection of personal data under the EU–U.S. Data Privacy Framework.* European Commission.

European Union Agency for Cybersecurity. (2024). *ENISA threat landscape 2024.* European Union Agency for Cybersecurity.

Federal Trade Commission. (2019, July 22). *Equifax to pay $575 million as part of settlement with FTC, CFPB, and states.* Federal Trade Commission.

Federal Trade Commission. (2024, December 18). *In the matter of Mobilewalla, Inc. (FTC Docket C-4805) [Decision and Order].* Federal Trade Commission.

Floridi, L., Cowls, J., Beltrametti, M., Chatila, R., Chazerand, P., Dignum, V., … Vayena, E. (2018). AI4People—An ethical framework for a good AI society. *Minds and Machines*, 28, 689–707.

Greenleaf, G. (2018). Global data privacy laws 2017: 120 national data privacy laws including Indonesia and Turkey. *Privacy Laws & Business International Report*, 145, 10–13.

Grother, P., Ngan, M., & Hanaoka, K. (2019). *Face Recognition Vendor Test (FRVT) Part 3: Demographic effects* (NISTIR 8280). National Institute of Standards and Technology.

Kitchin, R. (2014). *The data revolution: Big data, open data, data infrastructures & their consequences.* Sage.

Lyon, D. (2018). *The culture of surveillance: Watching as a way of life.* Polity.

Narayanan, A., & Shmatikov, V. (2008). Robust de-anonymization of large sparse datasets. In *2008 IEEE Symposium on Security and Privacy* (pp. 111–125). IEEE.

National Institute of Standards and Technology. (2020). *Profile of the IoT core baseline for consumer IoT products* (NISTIR 8259A). National Institute of Standards and Technology.

National Institute of Standards and Technology. (2023). *AI risk management framework 1.0* (NIST AI 100-1). National Institute of Standards and Technology.

National Institute of Standards and Technology. (2024). *Cybersecurity framework 2.0.* National Institute of Standards and Technology.

National Institute of Standards and Technology. (2025, October 2). *NIST initiates effort to update core IoT device cybersecurity baseline* [Blog post]. National Institute of Standards and Technology.

Nissenbaum, H. (2010). *Privacy in context: Technology, policy, and the integrity of social life.* Stanford University Press.

OECD. (2019). *Recommendation of the Council on responsible innovation in neurotechnology.* Organisation for Economic Co-operation and Development.

OECD. (2022). *OECD guidelines on the protection of privacy and transborder flows of personal data* (updated 2022). Organisation for Economic Co-operation and Development.

Penney, J. (2016). Chilling effects: Online surveillance and Wikipedia use. *Berkeley Technology Law Journal*, 31(1), 117–182.

Privacy and Civil Liberties Oversight Board. (2023). *Report on the surveillance program operated pursuant to Section 702 of the Foreign Intelligence Surveillance Act* (Unclassified Public Report). Privacy and Civil Liberties Oversight Board.

Richards, N. M., & Hartzog, W. (2015). *Privacy's blueprint: The battle to control the design of new technologies.* Harvard University Press.

Royal Society. (2019). *iHuman: Blurring lines between mind and machine.* The Royal Society.

Schneier, B. (2015). *Data and Goliath: The hidden battles to collect your data and control your world.* W. W. Norton.

Selbst, A. D., Boyd, D., Friedler, S. A., Venkatasubramanian, S., & Vertesi, J. (2019). Fairness and abstraction in sociotechnical systems. In *Proceedings of the Conference on Fairness, Accountability, and Transparency (FAT '19)* (pp. 59–68). ACM.

Solove, D. J. (2008). *Understanding privacy.* Harvard University Press.

UK House of Commons Digital, Culture, Media and Sport Committee. (2019). *Disinformation and "fake news": Final report* (HC 1791). House of Commons.

UNESCO. (2021). *Recommendation on the ethics of artificial intelligence.* United Nations Educational, Scientific and Cultural Organization.

Verizon. (2024). *Data Breach Investigations Report.* Verizon.

Verizon. (2025). *Data Breach Investigations Report.* Verizon.

WEF. (2025). *Global cybersecurity outlook 2025.* World Economic Forum.

Yuste, R., Goering, S., Bi, G., Carmena, J. M., Carter, A., Fins, J. J., Huguenard, J., et al. (2017). Four ethical priorities for neurotechnologies and AI. *Nature*, 551(7679), 159–163.

Zuboff, S. (2019). *The age of surveillance capitalism.* PublicAffairs.

Artificial Intelligence: Ethics and Decision-Making

7

ABSTRACT

Artificial Intelligence (AI) is reshaping society by offering transformative opportunities while posing significant ethical challenges. Its integration into decision-making processes across sectors such as healthcare, education, and criminal justice highlights the immense potential of AI to enhance efficiency, reduce human error, and address global challenges like climate change. However, the rapid adoption of AI also raises concerns about bias, transparency, accountability, and privacy. Ethical dilemmas emerge as AI systems influence critical decisions, often blurring lines of human responsibility and agency. Addressing these challenges requires a balanced approach, emphasizing fairness, inclusivity, and human oversight. Comprehensive regulatory frameworks, ethical design principles, and collaborative governance are essential to mitigate risks and ensure AI aligns with societal values. By fostering innovation responsibly, society can harness AI as a force for good, tackling global issues while safeguarding individual rights and moral integrity.

INTRODUCTION

Artificial Intelligence (AI) is rapidly redefining the landscape of modern society, impacting diverse domains and fundamentally altering the way individuals and organizations approach problem-solving and decision-making (UNESCO, 2021; European Parliament & Council of the European Union, 2024). Its potential to revolutionize industries such as healthcare, finance, education, and criminal justice is unparalleled, yet this transformative power brings with it a myriad of ethical dilemmas (Executive Office of the President, 2023; ISO/IEC, 2023). The dual-edged nature of AI demands a nuanced exploration of its capabilities, risks, and broader societal implications.

At its core, AI is about harnessing the power of data to create systems that can learn, reason, and act with a level of intelligence that mimics or even surpasses human capabilities (National Academies of Sciences, Engineering, and Medicine, 2022). Its applications range from mundane tasks like personalized shopping recommendations to critical functions such as diagnosing diseases or predicting criminal activity (WHO, 2023). While these innovations promise efficiency and precision, they also raise questions about the fairness, equity, and transparency of decisions made by machines (Mitchell et al., 2019; NIST, 2023).

One of the most profound impacts of AI is in decision-making. Traditional decision-making processes rely heavily on human judgment, which is often limited by biases, cognitive overload, and emotional influences (National Academies of Sciences, Engineering, and Medicine, 2022). AI systems, by contrast, can analyze massive datasets and identify patterns that humans might overlook. However, this apparent objectivity is not without flaws. The data used to train AI systems often reflect historical inequities, which can result in biased or discriminatory outcomes when left unchecked (Buolamwini & Gebru, 2018; Obermeyer et al., 2019; NIST, 2023).

In healthcare, for instance, AI tools are being used to diagnose diseases, optimize treatment plans, and even predict patient outcomes (WHO, 2023). While these tools have the potential to save lives and reduce medical errors, they also introduce ethical concerns. Who is responsible if an AI system makes an incorrect diagnosis? How do we ensure that these systems are accessible to underserved populations? These questions highlight the need for accountability and inclusivity in AI development (UNESCO, 2021; ISO/IEC, 2023).

In the criminal justice system, predictive algorithms are increasingly used to inform decisions about bail, sentencing, and parole. While these tools aim to make the system more efficient, they risk perpetuating systemic biases against marginalized communities (Selbst et al., 2019). For example, if an AI tool is trained on data that reflects historical over-policing of certain demographics, it may unfairly label individuals from those groups as higher-risk (Buolamwini & Gebru, 2018).

Beyond specific applications, the sheer scale of AI's integration into society raises broader ethical concerns. AI-powered surveillance technologies, for instance, have the potential to enhance public safety but at the cost of individual privacy (Amnesty International, 2024; PCLOB, 2023). The balance between security and personal freedoms is a contentious issue, with implications for democratic values and human rights.

The "black box" nature of many AI systems compounds these challenges. Advanced algorithms, particularly those based on deep learning, often operate in ways that are not fully understood even by their creators (Doshi-Velez & Kim, 2017). This lack of transparency makes it difficult to explain or justify AI-driven decisions, eroding public trust and complicating regulatory oversight (Rudin, 2019).

AI also poses significant risks to employment and economic stability. As automation becomes more sophisticated, jobs traditionally performed by humans are increasingly at risk of displacement. While AI creates opportunities for innovation and new industries, it also necessitates workforce reskilling and rethinking economic models to address the resulting disruptions (International Labour Organization, 2023; World Economic Forum, 2025).

Despite these challenges, the opportunities presented by AI are too significant to ignore. AI has the potential to address some of the world's most pressing problems, from climate change to global health crises. For example, AI models can analyze climate data to predict and mitigate the impact of natural disasters or optimize energy consumption to reduce carbon footprints (National Academies of Sciences, Engineering, and Medicine, 2022).

To fully harness the benefits of AI while mitigating its risks, ethical considerations must be at the forefront of its development and deployment. This involves addressing questions of fairness, ensuring that AI systems do not reinforce existing inequalities, and promoting inclusivity by involving diverse voices in AI design and governance (UNESCO, 2021; European Parliament & Council of the European Union, 2024).

The global nature of AI development adds another layer of complexity. Technologies developed in one country often have far-reaching impacts, requiring international collaboration to establish ethical standards and regulatory frameworks. The European Union's Artificial Intelligence Act, for example, represents a significant step toward creating guidelines for the responsible use of AI, but it also highlights the challenges of harmonizing regulations across jurisdictions (European Parliament & Council of the European Union, 2024).

Governments, industry leaders, and civil society all have roles to play in shaping the future of AI. Policymakers must establish regulations that protect individual rights without stifling innovation, while developers must adhere to ethical guidelines that prioritize transparency, accountability, and inclusivity (Executive Office of the President, 2023; ISO/IEC, 2023).

Education and public engagement are also critical. As AI becomes increasingly integrated into daily life, fostering AI literacy can empower individuals to make informed decisions about its use and implications. This includes understanding how AI works, recognizing its limitations, and advocating for ethical practices in its application (National Academies of Sciences, Engineering, and Medicine, 2022).

AI at the decision point: balancing efficiency with fairness, transparency, privacy, and human oversight

Read through a governance lens, the image frames AI-assisted decisions as threshold choices—who gets a loan, a job interview, a benefit, a warning, or a false flag—where speed and consistency can also amplify bias, opacity, and automation complacency. The takeaway is practical: pair model accuracy with fairness testing across groups, purpose-fit explanations, human oversight with real authority to override, documented data lineage, audit logs and independent review, clear appeal and redress pathways, and continuous monitoring for drift and error costs so efficiency doesn't eclipse equity or accountability.

Ultimately, the integration of AI into society is not just a technological issue but a deeply human one. The ethical challenges it presents force us to confront fundamental questions about responsibility, agency, and the kind of future we want to build. By engaging in thoughtful dialogue and collaboration, we can ensure that AI serves as a force for good, enhancing human potential while safeguarding our shared values (UNESCO, 2021).

This chapter delves into the complex interplay between AI's opportunities and ethical considerations, emphasizing the need for a balanced and proactive approach. As we navigate this transformative era, the focus must remain on creating systems that align with human dignity, fairness, and sustainability, ensuring that AI's benefits are equitably shared and its risks responsibly managed (UNESCO, 2021; ISO/IEC, 2023).

OPPORTUNITIES OF AI IN DECISION-MAKING

AI's capabilities offer transformative potential, making it a cornerstone of modern innovation. By processing vast datasets at speeds far beyond human capacity, AI systems can uncover patterns, predict outcomes, and provide actionable insights with unparalleled precision (Mitchell et al., 2019). In healthcare, AI-powered tools are revolutionizing medical diagnostics, enabling early detection of conditions like cancer and cardiovascular diseases. These technologies not only optimize treatment plans but also significantly improve patient outcomes by personalizing interventions based on an individual's unique data (WHO, 2023; Obermeyer et al., 2019).

In the financial sector, AI enhances operational efficiency and security. Advanced algorithms detect fraudulent activities in real time, safeguarding transactions and preventing financial crimes (NIST, 2024). Personalized investment strategies tailored to individual risk appetites and goals are another innovation, offering ordinary investors access to high-level financial planning once reserved for the affluent.

Education is another domain where AI is making significant strides. Adaptive learning platforms powered by AI cater to diverse learner needs by tailoring content delivery to an individual's pace and comprehension level. This personalized approach helps students achieve better outcomes, reduces dropout rates, and ensures inclusivity in education. AI systems can also provide educators with valuable insights into student performance, allowing for timely interventions (National Academies of Sciences, Engineering, and Medicine, 2022).

Beyond sector-specific applications, AI has the potential to address complex global challenges. In climate science, AI models analyze vast datasets to predict weather patterns, monitor deforestation, and optimize renewable energy systems (National Academies of Sciences, Engineering, and Medicine, 2022). These capabilities are critical in mitigating climate change and developing sustainable strategies. During pandemics, AI-driven analytics facilitate the rapid development of vaccines, monitor disease spread, and optimize resource allocation in healthcare systems (WHO, 2023).

AI's ability to democratize access to services is particularly noteworthy. By reducing costs and eliminating traditional barriers, AI enables underserved populations to access critical resources like healthcare, education, and financial tools. For example, telemedicine platforms powered by AI provide remote consultations, while language translation tools bridge communication gaps in global education initiatives (WHO, 2023).

However, the transformative potential of AI is not without challenges. One significant issue is the potential for bias in AI systems. Because these systems learn from historical data, they may inadvertently replicate or amplify existing societal inequalities (Buolamwini & Gebru, 2018; Obermeyer et al., 2019). For instance, biased algorithms in hiring processes have excluded qualified candidates based on gender or ethnicity, perpetuating discriminatory practices (Selbst et al., 2019).

The "black box" problem adds another layer of complexity. Many AI systems, particularly those using deep learning, operate in ways that are difficult to interpret or explain (Doshi-Velez & Kim, 2017). This lack of transparency poses significant risks in high-stakes applications such as autonomous vehicles or medical diagnostics, where accountability and trust are paramount (Rudin, 2019).

Ethical concerns also arise regarding the concentration of power in AI development. A handful of corporations and nations dominate AI innovation, raising concerns about monopolistic practices and geopolitical imbalances (World Economic Forum, 2025). These disparities risk widening the digital divide, leaving underrepresented communities further marginalized (International Labour Organization, 2023).

AI's integration into decision-making processes also raises profound philosophical questions. As machines take on roles traditionally reserved for humans, such as judging legal cases or making hiring decisions, society must grapple with issues of accountability. If an AI system makes an error, who is responsible—the developer, the user, or the machine itself? (ISO/IEC, 2023).

Privacy and security are additional areas of concern. AI systems rely on vast amounts of data, often collected without explicit user consent. This creates vulnerabilities, as personal information can be misused or compromised (Amnesty International, 2024; NIST, 2023). Invasive surveillance technologies powered by AI threaten individual freedoms, forcing societies to confront the delicate balance between security and privacy (PCLOB, 2023).

Despite these challenges, AI holds immense promise for societal advancement. In disaster response, AI systems can analyze satellite imagery to identify affected areas and prioritize relief efforts. In agriculture, AI-powered tools optimize crop yields, monitor soil health, and reduce resource waste, contributing to food security (National Academies of Sciences, Engineering, and Medicine, 2022).

To maximize AI's benefits while mitigating its risks, ethical frameworks and regulatory measures are essential. Policymakers must establish clear guidelines to ensure AI's responsible development and use. These regulations should address issues like bias, transparency, and accountability, while fostering innovation and economic growth (European Parliament & Council of the European Union, 2024; Executive Office of the President, 2023).

Collaboration between governments, academia, and the private sector is crucial for creating inclusive and effective AI policies. International cooperation is particularly important, as AI's impact transcends borders. Frameworks such as the European Union's Artificial Intelligence Act provide a valuable starting point for addressing these global challenges (European Parliament & Council of the European Union, 2024).

AI developers also have a critical role to play. By adhering to ethical principles such as fairness, transparency, and accountability, developers can ensure that their innovations align with societal values (ISO/IEC, 2023; IEEE Standards Association, 2021). Incorporating diverse perspectives in AI design teams can help mitigate biases and create systems that reflect a broader range of human experiences (Gebru et al., 2021).

Education and public awareness are equally important. As AI becomes more pervasive, fostering AI literacy among the general population is essential. This includes understanding how AI systems work, recognizing their limitations, and advocating for ethical practices in their deployment (National Academies of Sciences, Engineering, and Medicine, 2022).

The integration of AI into the workforce presents both challenges and opportunities. While automation may displace certain jobs, it also creates new roles and industries. Preparing for this transition requires proactive strategies, including reskilling programs and policies that promote equitable economic adjustments (International Labour Organization, 2023; World Economic Forum, 2025).

AI's impact on cultural and social dynamics is another area worth exploring. As AI systems become more capable of mimicking human behavior, they blur the lines between human and machine. This raises philosophical questions about identity, agency, and what it means to be human in an increasingly automated world (UNESCO, 2021).

The potential for AI to exacerbate existing inequalities cannot be ignored. Without deliberate efforts to address these disparities, AI risks reinforcing systemic inequities (Selbst et al., 2019). For example, access to advanced AI tools and technologies remains limited in developing regions, further entrenching the global digital divide (World Economic Forum, 2025).

In the healthcare domain, ethical AI practices must ensure equitable access to innovations. This includes developing affordable AI-driven solutions for low-income populations and addressing language and cultural barriers in healthcare delivery (WHO, 2023).

AI's influence on governance and democracy also warrants attention. While AI has the potential to enhance public policy through data-driven insights, it can also be weaponized to manipulate public opinion or suppress dissent (PCLOB, 2023). Establishing safeguards to protect democratic processes is imperative (UNESCO, 2021).

The societal implications of AI extend beyond its immediate applications. By reshaping economic structures, cultural norms, and global power dynamics, AI challenges traditional notions of governance, identity, and ethics (UNESCO, 2021). Addressing these broader impacts requires a multidisciplinary approach that incorporates insights from philosophy, sociology, and economics.

Despite its challenges, AI offers unprecedented opportunities to address global challenges and improve quality of life. By fostering collaboration, promoting ethical practices, and ensuring inclusivity, society can harness AI's transformative potential while safeguarding human values and rights (UNESCO, 2021; ISO/IEC, 2023). The road ahead requires vigilance, innovation, and a shared commitment to building an equitable and sustainable future shaped by AI.

RISKS ASSOCIATED WITH AI INTEGRATION

Despite its transformative potential, AI introduces significant risks that demand careful scrutiny and thoughtful consideration. One of the most pressing concerns is the potential for bias in AI systems (Buolamwini & Gebru, 2018; Obermeyer et al., 2019). AI models are designed to learn from vast datasets, often drawn from historical data that reflect existing societal patterns. Unfortunately, these datasets can carry inherent biases, which AI systems may replicate or even amplify. For instance, in hiring algorithms, AI systems have been found to favor candidates from certain demographics, typically male or white, while excluding others based on gender or ethnicity. This not only perpetuates long-standing inequalities in the workforce but also raises questions about fairness and representation (Selbst et al., 2019). Similarly, predictive policing tools, which rely on historical crime data, have been shown to disproportionately target minority communities, reinforcing systemic racial biases in law enforcement (Buolamwini & Gebru, 2018).

The risks associated with bias are not confined to law enforcement or employment. In healthcare, biased AI algorithms can lead to inaccurate diagnoses or unequal access to care, especially for individuals from underrepresented communities (Obermeyer et al., 2019; WHO, 2023). For example, facial recognition technology, which is often used in security and healthcare applications, may perform poorly for people of color, leading to misidentification or unequal treatment (Grother et al., 2019). Moreover, AI-powered tools for medical diagnostics can fail to account for the unique health needs of diverse populations, potentially exacerbating health disparities. Addressing these biases requires a concerted effort from developers, regulators, and stakeholders to ensure that AI systems are fair, inclusive, and equitable. This includes diversifying the data used to train AI models, regularly auditing algorithms for fairness, and implementing transparent accountability mechanisms (Mitchell et al., 2019; NIST, 2023).

Another critical risk posed by AI is the lack of transparency, often referred to as the "black box" problem. Many AI systems, especially those based on deep learning and neural networks, operate in ways that are complex and opaque, making it difficult for humans to understand how they arrive at certain decisions (Doshi-Velez & Kim, 2017). This lack of interpretability is particularly concerning in high-stakes domains such as medical diagnoses, legal judgments, and criminal sentencing. When an AI system makes a decision that affects an individual's health, freedom, or livelihood, it is crucial that the reasoning behind that decision is transparent and understandable (Rudin, 2019). The inability to explain how a decision was made undermines accountability and trust, leaving individuals and society at large unsure of the fairness and validity of AI-driven outcomes. In healthcare, for instance, an AI system that recommends a particular treatment without clear justification could lead to mistrust among patients and healthcare providers, ultimately hindering the adoption of AI technologies (WHO, 2023).

The lack of transparency also poses challenges in the realm of law and justice. In the criminal justice system, for example, AI-driven risk assessment tools are used to predict the likelihood of reoffending, influencing parole and sentencing decisions. However, when these tools cannot explain their reasoning, it becomes difficult for judges, lawyers, and defendants to challenge potentially harmful or inaccurate outcomes (Selbst et al., 2019). Without interpretability, AI systems can perpetuate injustices and reduce confidence in the fairness of legal processes. As AI continues to play a larger role in decision-making, it is essential to prioritize transparency in its design and implementation. This includes developing methods for explaining AI decision-making processes in a way that is understandable to non-experts and establishing clear standards for accountability (ISO/IEC, 2023; IEEE Standards Association, 2021).

In addition to bias and lack of transparency, AI poses significant risks to privacy and security. AI systems often rely on vast amounts of data, including sensitive personal information, to function effectively. The collection, analysis, and storage of this data raise significant privacy concerns, particularly when individuals' consent is not obtained or their data is used for purposes beyond their original intent (NIST, 2023). AI-driven data collection tools, such as surveillance cameras, can infringe upon personal privacy by tracking individuals' movements and behaviors without their knowledge or consent (Amnesty International, 2024). This is particularly concerning in the context of facial recognition technology, which can be used for mass surveillance by governments or private companies, leading to the potential erosion of civil liberties (PCLOB, 2023; Grother et al., 2019). Unauthorized profiling, where AI systems infer personal details such as political beliefs or health conditions from seemingly innocuous data, is another privacy concern. The ability of AI to aggregate data from multiple sources also increases the risk of unauthorized access and exploitation, making individuals vulnerable to identity theft, discrimination, and other forms of harm (NIST, 2023).

The widespread use of AI also raises cybersecurity concerns. As AI systems become more integrated into critical infrastructure, such as power grids, transportation networks, and healthcare systems, they become prime targets for cyberattacks (ENISA, 2024; Verizon, 2024, 2025). Hackers could exploit vulnerabilities in AI systems to disrupt services, steal sensitive information, or even manipulate decision-making processes. For instance, an AI system controlling a power grid could be hacked to cause widespread blackouts, or an AI-driven medical device could be tampered with to deliver incorrect treatments. These potential consequences highlight the need for robust cybersecurity measures to protect AI systems from malicious attacks (NIST, 2024; WEF, 2025). Additionally, AI systems themselves are susceptible to adversarial attacks, where small, often imperceptible changes to input data can cause AI systems to make incorrect predictions or decisions. Ensuring the security of AI systems is critical to maintaining trust in their reliability and safety, particularly as they take on more responsibilities in society (ISO/IEC, 2023).

The risks associated with AI also extend to the manipulation of decision-making processes. AI systems are increasingly being used to make decisions that affect individuals' lives, such as credit scoring, job applications, and even legal rulings. However, these systems can be manipulated by malicious actors to serve specific agendas. For example, AI-powered algorithms used in social media platforms can be exploited to spread misinformation or manipulate public opinion, potentially influencing elections or fueling societal divisions (PCLOB, 2023). Similarly, AI-driven content recommendation algorithms may prioritize sensational or biased content, reinforcing echo chambers and polarizing public discourse. This manipulation of information has serious implications for democracy, public trust, and social cohesion. To mitigate these risks, it is essential to establish ethical guidelines and regulatory frameworks that govern the use of AI in decision-making processes, ensuring that these systems are used transparently and responsibly (UNESCO, 2021; European Parliament & Council of the European Union, 2024).

Furthermore, the risk of AI systems being used to reinforce or exacerbate existing power imbalances is a significant concern. In many sectors, AI is being developed and deployed by large corporations, which have access to vast amounts of data and the resources to build sophisticated AI systems. This concentration of power in the hands of a few entities could lead to monopolistic practices and the erosion of privacy rights (World Economic Forum, 2025). AI systems could be used to surveil populations, control access to information, and manipulate consumer behavior in ways that benefit powerful corporations at the expense of individuals' rights. In the absence of adequate regulations, the deployment of AI could further entrench existing societal inequalities, disproportionately affecting marginalized groups (International Labour Organization, 2023). To prevent this, it is crucial to ensure that AI development is guided by ethical principles that prioritize human rights, privacy, and equity (UNESCO, 2021; ISO/IEC, 2023).

As AI continues to evolve, it will be essential to address these risks through comprehensive regulation, ethical frameworks, and robust accountability mechanisms. Policymakers must develop laws and regulations that address the challenges posed by AI, including its potential to perpetuate bias, undermine transparency, infringe on privacy, and threaten security (European Parliament & Council of the European Union, 2024; Executive Office of the President, 2023). These regulations should be designed to promote fairness, accountability, and transparency in AI systems while fostering innovation and ensuring the responsible use of AI technologies. Additionally, international cooperation is necessary to address the global implications of AI, as its impact transcends national borders and affects individuals around the world (UNESCO, 2021; ISO/IEC, 2023).

In conclusion, while AI holds immense promise, its risks cannot be ignored. Bias, lack of transparency, privacy violations, security threats, and manipulation of decision-making processes are all significant concerns that must be addressed to ensure that AI benefits society as a whole. By adopting a proactive approach to regulation, accountability, and ethical development, we can mitigate these risks and harness the transformative potential of AI in a way that promotes fairness, equity, and respect for human rights (UNESCO, 2021; European Parliament & Council of the European Union, 2024). As AI continues to play an increasingly prominent role in our lives, it is essential to remain vigilant and prioritize the ethical implications of its use to ensure that its benefits are realized while minimizing its potential harms (NIST, 2023; ISO/IEC, 2023).

ETHICAL CONSIDERATIONS IN AI DEVELOPMENT AND USE

The development and deployment of AI systems raise profound ethical questions that society must urgently address. As AI continues to permeate every aspect of life—from healthcare and education to criminal justice and warfare—the ethical implications of these technologies are becoming more complex and far-reaching (UNESCO, 2021). At the heart of these considerations is the question of responsibility: who is accountable when AI systems make decisions? This issue becomes particularly pressing in applications where AI has the potential to make life-or-death choices, such as autonomous vehicles and lethal autonomous weapons (ISO/IEC, 2023). In these cases, the lack of a clear chain of accountability is unsettling. If an autonomous car causes an accident or a drone strike results in civilian casualties, who is responsible—the developer, the manufacturer, the operator, or the AI system itself? Determining liability and accountability is a major challenge, as AI systems are often opaque and operate on algorithms that are difficult to interpret. The lack of clear legal frameworks to address such issues underscores the urgency of establishing clear policies and ethical guidelines to ensure accountability in AI decision-making (European Parliament & Council of the European Union, 2024).

One of the most difficult aspects of AI accountability is the question of how to assign responsibility when decisions are made by machines without human intervention. For example, autonomous cars are equipped with algorithms that allow them to make real-time decisions, such as braking to avoid an accident (ISO/IEC, 2023). However, when an accident occurs, it may be unclear whether the fault lies with the car's programming, the design of the AI, or external factors that the system could not predict. In the case of lethal autonomous weapons, which can make decisions about targeting and engaging without human oversight, the consequences are even more severe (UNESCO, 2021). How can we ensure that such systems do not violate international humanitarian law or commit war crimes? The deployment of autonomous weapons raises significant ethical concerns about the removal of human judgment from military decisions and the risk of unintended consequences. These concerns call for an international treaty or set of guidelines to govern the use of AI in warfare, ensuring that such technologies are used responsibly and ethically (UNESCO, 2021).

Beyond accountability, fairness is another critical ethical concern in the development and deployment of AI systems. AI has the potential to perpetuate and even exacerbate existing inequalities if not designed and implemented carefully (Selbst et al., 2019). Many AI systems rely on historical data to make predictions and decisions, and if that data is biased, the AI will reflect and magnify those biases. For example, biased hiring algorithms can favor certain demographics over others, reinforcing existing societal inequalities. Predictive policing tools that rely on past arrest data may disproportionately target minority communities, perpetuating racial profiling and systemic discrimination (Buolamwini & Gebru, 2018). The ethical principle of fairness calls for AI systems to be designed and tested in ways that eliminate discriminatory outcomes and promote equity. Developers must take steps to ensure that AI systems are not only efficient but also just and inclusive, addressing biases in data, algorithms, and outcomes (Mitchell et al., 2019; ISO/IEC, 2023).

Achieving fairness in AI is not a simple task, as it requires addressing biases that may be deeply embedded in historical data. For instance, criminal justice data may reflect societal prejudices, such as racial bias, that have influenced arrests and convictions over time (Selbst et al., 2019). If AI systems are trained on such data, they may reinforce these biases in their predictions and recommendations. To counteract this, AI developers must implement strategies to ensure that training data is diverse, representative, and free from discriminatory patterns. Additionally, AI systems should be designed with mechanisms for auditing and reviewing their decisions, allowing for ongoing evaluation and adjustment (Mitchell et al., 2019). The creation of independent oversight bodies could also help ensure that AI systems operate fairly and equitably. Ethical AI requires that organizations prioritize inclusivity and fairness in the design, implementation, and evaluation of AI systems, ensuring that marginalized groups are not disproportionately harmed by technological advancements (UNESCO, 2021).

In the context of fairness, it is also important to consider the role of algorithms in shaping social outcomes. Algorithms often determine who gets access to opportunities, services, and resources, from credit scoring and loan approvals to job applications and healthcare services. If these algorithms are biased or lack transparency, they can lead to unequal treatment and exacerbate existing disparities (Rudin, 2019; Doshi-Velez & Kim, 2017). For example, an AI-driven loan approval process may disproportionately reject applications from low-income individuals or people of color, even if they meet the required criteria. This can further entrench inequality and limit social mobility for disadvantaged groups. Ensuring fairness in AI requires careful design, testing, and regulation to prevent these types of discriminatory outcomes. Ethical AI must promote inclusivity, transparency, and accountability, ensuring that all individuals have equal access to the opportunities and services provided by AI systems (ISO/IEC, 2023; IEEE Standards Association, 2021).

At the same time, ethical AI must also balance innovation with the preservation of human agency. While AI systems have the potential to enhance decision-making and improve efficiency, they should not undermine the autonomy and critical judgment of individuals and organizations. One of the key ethical principles in AI is ensuring that humans remain "in the loop" in decision-making processes, especially in high-stakes situations (UNESCO, 2021; ISO/IEC, 2023). For example, in healthcare, AI systems can assist in diagnosing diseases and recommending treatments, but the final decision should always involve human oversight. Medical professionals must remain responsible for the care they provide, ensuring that AI tools are used as aids rather than replacements for human judgment. This principle extends beyond healthcare to other areas, such as finance, criminal justice, and education, where AI systems should support rather than replace human decision-making (National Academies of Sciences, Engineering, and Medicine, 2022).

The preservation of human agency is particularly important in contexts where individuals may be vulnerable to the influence of AI. For instance, AI-powered recommendation systems on social media platforms can influence people's beliefs, behaviors, and purchasing decisions. These algorithms often prioritize content that generates strong emotional reactions, which can manipulate users' perceptions and reinforce echo chambers (PCLOB, 2023). Ethical AI in this context means designing algorithms that respect users' autonomy and promote healthy, diverse, and informed interactions. Developers should ensure that AI systems do not exploit users' vulnerabilities or manipulate them for profit, but rather empower individuals to make informed choices (UNESCO, 2021). This requires a commitment to transparency, user control, and respect for privacy, as well as an ongoing dialogue between developers, users, and policymakers about the ethical implications of AI in everyday life.

Another essential aspect of maintaining human agency in AI systems is the design of user-friendly interfaces and tools that enable individuals to understand and interact with AI decisions. The complexity and opacity of many AI systems, particularly those based on deep learning, can make it difficult for users to understand how decisions are made (Doshi-Velez & Kim, 2017). This "black box" problem undermines trust in AI and can lead to a lack of accountability. To address this, developers should prioritize explainability in AI design, ensuring that users can access clear, understandable explanations of how AI systems work and how decisions are made (Rudin, 2019). This is particularly important in fields like healthcare and criminal justice, where individuals' lives can be significantly impacted by AI-driven decisions. By providing users with the tools and information they need to engage with AI systems responsibly, developers can foster greater transparency and trust (ISO/IEC, 2023).

In addition to transparency and explainability, it is essential to ensure that AI systems are designed with ethical values at their core. This includes prioritizing human dignity, privacy, and fairness in the design of AI technologies (UNESCO, 2021). Developers should be mindful of the broader societal implications of their work, considering how AI will impact various groups and individuals, especially those who may be marginalized or disadvantaged. Ethical AI design requires an interdisciplinary approach, drawing on expertise from fields like philosophy, law, sociology, and psychology, as well as computer science and engineering (National Academies of Sciences, Engineering, and Medicine, 2022). Collaboration between these diverse disciplines can help ensure that AI systems are developed in a way that reflects a wide range of perspectives and values, ultimately benefiting society as a whole.

To achieve ethical AI, it is crucial that organizations implement rigorous ethical frameworks and oversight mechanisms. These frameworks should include clear guidelines for the development, deployment, and evaluation of AI systems, ensuring that ethical considerations are integrated into every stage of the AI lifecycle (IEEE Standards Association, 2021; ISO/IEC, 2023). Independent ethics boards or review committees can play a key role in overseeing AI projects and holding organizations accountable for their ethical practices. Additionally, regulatory bodies at the national and international levels must establish standards for ethical AI development, providing a regulatory framework that guides AI innovation while protecting fundamental human rights (European Parliament & Council of the European Union, 2024; UNESCO, 2021).

Furthermore, public engagement is essential in shaping the ethical direction of AI. As AI systems become more integrated into society, it is important that the public has a voice in the decision-making process. Public consultations, open forums, and citizen engagement initiatives can help ensure that the development of AI reflects the values and needs of diverse communities (UNESCO, 2021). By involving the public in discussions about AI ethics, we can foster greater understanding, trust, and collaboration, ensuring that AI serves the broader public interest.

In conclusion, the ethical considerations surrounding AI are complex and multifaceted. As AI systems become more advanced and pervasive, it is essential that we address key issues related to accountability, fairness, and human agency (UNESCO, 2021; ISO/IEC, 2023). By prioritizing transparency, inclusivity, and ethical design in AI development, we can ensure that AI technologies contribute positively to society while minimizing potential harms. Ethical AI is not just about mitigating risks but about actively shaping AI systems that reflect our collective values and aspirations.

It is an ongoing process that requires collaboration, vigilance, and a commitment to the well-being of all individuals and communities.

THE ROLE OF REGULATION AND GOVERNANCE

The ethical challenges posed by AI demand the development of comprehensive regulatory frameworks and governance structures (European Parliament & Council of the European Union, 2024; National Institute of Standards and Technology, 2020, 2023, [NIST], 2024). As AI technologies advance, it becomes increasingly clear that relying solely on voluntary guidelines or self-regulation is insufficient to address the profound ethical concerns arising from their use. Governments, international bodies, and private sector stakeholders must work collaboratively to establish legal standards that ensure the safe, fair, and transparent deployment of AI while continuing to foster innovation. The sheer scale and complexity of AI applications across various sectors—ranging from healthcare and finance to criminal justice and autonomous vehicles—necessitate regulation that can adapt to the rapidly evolving landscape of AI (UNESCO, 2021). Policymakers must act swiftly, yet thoughtfully, to ensure that regulatory measures strike a balance between encouraging technological progress and safeguarding fundamental rights and values (OECD, 2022).

A major challenge in establishing regulatory frameworks for AI is the global nature of AI development and deployment. Unlike more localized technologies, AI operates across borders, transcending national boundaries and affecting people and societies worldwide. This reality underscores the need for international cooperation in the development of AI governance structures (Council of Europe, 2024). Global consensus on AI standards and best practices is essential to prevent fragmentation in regulations, which could result in inconsistency and confusion. Moreover, fragmented regulations could create an uneven playing field, where multinational companies may navigate through different sets of rules depending on where they operate, while smaller companies or governments may struggle to keep up. An international approach to AI regulation would ensure that AI technologies are developed in a manner that is consistent with universal human rights, ethical principles, and the long-term interests of society (European Commission, 2023; OECD, 2022).

International cooperation on AI regulation is critical not only to create harmonized standards but also to facilitate the sharing of knowledge and best practices. Countries and regions with more advanced AI capabilities have much to gain from collaborating with less developed regions, ensuring that AI benefits are equitably distributed. Such collaboration could help address concerns related to the digital divide and ensure that developing nations are not left behind in the global AI race. The European Union's Artificial Intelligence (AI) Act, for example, represents a significant step in the global regulation of AI. The Act aims to establish a legal framework for AI that addresses its risks and challenges while ensuring that AI technologies align with European values, such as human dignity and non-discrimination. By categorizing AI systems based on their potential risks, the Act introduces different levels of regulatory scrutiny, with high-risk AI applications subject to stricter requirements. This risk-based approach could serve as a model for other nations and regions to develop their own AI governance frameworks (European Parliament & Council of the European Union, 2024).

However, creating effective regulations is no easy task, and it involves navigating various complexities. One major challenge is the fast-paced evolution of AI technologies, which can outpace legislative processes. AI systems and their applications are constantly evolving, and regulators often find themselves struggling to keep up with new developments. This creates a dynamic tension between the need for timely regulation and the need for regulations that remain relevant and adaptable in the face of rapid technological change. Policymakers must ensure that AI regulations are flexible enough to accommodate future innovations while providing sufficient oversight and protection (NIST, 2023). This will likely require continuous collaboration between governments, the private sector, and academic institutions to monitor and assess the impact of AI on society and adjust regulatory frameworks as needed (UNESCO, 2021).

In addition to regulatory frameworks, organizations and developers also bear a responsibility to adopt ethical guidelines and best practices when designing and implementing AI systems. Ethical AI practices are crucial for ensuring that AI technologies are developed in ways that minimize harm and maximize societal benefits. Many organizations are already beginning to embrace AI ethics principles, which emphasize transparency, fairness, accountability, and respect for human rights (IEEE Global Initiative on Ethics of Autonomous and Intelligent Systems, 2019; UNESCO, 2021). These principles serve as a foundation for responsible AI development and provide guidance for organizations to follow as they create and deploy AI systems. By incorporating ethical considerations into the design and implementation of AI technologies, developers can help ensure that their systems align with societal values and contribute to positive outcomes.

Transparency is a key component of ethical AI, as it ensures that AI systems and their decisions are understandable and accessible to users, regulators, and other stakeholders. Transparent AI systems allow users to know how decisions are being made, enabling them to trust the technology and challenge decisions if necessary. Moreover, transparency is essential for accountability, as it allows for the identification of errors, biases, or unintended consequences in AI systems. Ethical guidelines for AI development should include clear standards for explainability, ensuring that AI systems provide understandable rationales for their decisions. This is particularly important in high-stakes domains such as healthcare, criminal justice, and finance, where AI systems can have significant impacts on individuals' lives (UNESCO, 2021; European Parliament & Council of the European Union, 2024).

Fairness is another essential principle in the development of ethical AI. AI systems must be designed and tested to avoid perpetuating or exacerbating existing biases in society. As mentioned earlier, AI systems that are trained on biased data can reinforce existing inequalities, such as discrimination based on race, gender, or socioeconomic status. Organizations and developers must take proactive steps to identify and mitigate bias in AI algorithms, ensuring that their systems operate fairly and do not inadvertently harm marginalized or vulnerable groups (Buolamwini & Gebru, 2018; Grother et al., 2019).

Accountability is also crucial in the ethical design and implementation of AI systems. Developers, organizations, and policymakers must ensure that AI systems are held accountable for their decisions and actions. This means establishing clear lines of responsibility for AI outcomes, particularly in high-risk applications where human lives or freedoms are at stake. One way to promote accountability is through independent oversight, such as the establishment of ethics review boards or third-party audits, which can evaluate AI systems and ensure they comply with ethical standards (Privacy and Civil Liberties Oversight Board, 2023). Additionally, developers should implement mechanisms for users to challenge AI decisions or seek redress when harm occurs. This ensures that AI systems do not operate in isolation but are subject to the scrutiny and oversight of human actors who can intervene when necessary (UNESCO, 2021).

To further support ethical AI development, organizations can also embrace frameworks such as the IEEE Global Initiative on Ethics of Autonomous and Intelligent Systems. This initiative aims to provide comprehensive guidelines for the ethical development and deployment of AI, focusing on key issues such as privacy, security, and transparency. By adhering to established ethical frameworks and principles, organizations can demonstrate their commitment to responsible AI development and help build public trust in AI technologies. Furthermore, organizations can engage in ongoing ethical training for their teams to ensure that AI ethics remains a core consideration throughout the entire development process (IEEE Global Initiative on Ethics of Autonomous and Intelligent Systems, 2019).

As part of their ethical responsibilities, AI developers must also ensure that AI systems do not undermine human agency or autonomy. AI should augment, not replace, human decision-making, particularly in critical domains like healthcare and criminal justice. The deployment of autonomous systems, such as self-driving cars or decision-making algorithms used in legal contexts, raises concerns about the potential for AI to make decisions that override human judgment (European Parliament & Council of the European Union, 2024). Ethical AI development should prioritize human oversight and control, ensuring that humans remain "in the loop" and retain the final say in important decisions.

Moreover, AI ethics must include a commitment to privacy protection and data security. As AI systems rely heavily on vast amounts of personal data to function effectively, there is a risk that this data could be misused, mismanaged, or exposed to breaches. Developers must prioritize the protection of individuals' privacy rights and implement robust security measures to safeguard personal information (OECD, 2022; NIST, 2020). Ethical guidelines should also ensure that individuals have control over their own data and are informed about how their data is being used, stored, and shared. This is especially important in sensitive areas such as healthcare, where AI systems may have access to individuals' medical histories and other personal information.

In addition to internal guidelines, external regulation plays a vital role in ensuring that AI is developed and deployed ethically. Governments and international bodies must collaborate to establish regulatory frameworks that protect human rights, ensure transparency, and foster innovation (Council of Europe, 2024; European Parliament & Council of the European Union, 2024). Such regulations should be flexible enough to accommodate new developments in AI while providing robust safeguards against potential risks. A well-designed regulatory system will not only protect individuals from harm but will also help build public trust in AI technologies, encouraging their widespread adoption and responsible use.

Public engagement is another important aspect of ethical AI governance. As AI becomes an integral part of daily life, it is essential that the public is informed about AI's capabilities, risks, and ethical considerations. Policymakers, developers, and researchers should engage with the public through consultations, discussions, and educational initiatives to foster a deeper understanding of AI. Public engagement can help shape the ethical and regulatory frameworks for AI, ensuring that they reflect the needs and values of society as a whole (UNESCO, 2021). This participatory approach will also help mitigate concerns about AI being developed in a top-down manner without consideration for the broader social impact.

In conclusion, the ethical challenges posed by AI are complex and multifaceted, requiring a concerted effort from governments, organizations, developers, and civil society to address them. Comprehensive regulatory frameworks, grounded in principles of fairness, transparency, and accountability, are essential for ensuring that AI technologies are developed and deployed in ways that benefit society while minimizing harm. As AI continues to evolve, it is crucial that ethical guidelines and best practices remain at the forefront of development efforts, ensuring that AI contributes positively to society and aligns with our collective values (NIST, 2024; ENISA, 2024; World Economic Forum, 2025).

THE INTERSECTION OF AI AND SOCIETY

The societal implications of AI are profound and far-reaching, extending beyond its immediate applications to influence cultural norms, economic structures, and power dynamics in ways that are only beginning to unfold (UNESCO, 2021). As AI becomes more integrated into daily life, it reshapes the way people interact with technology, altering societal expectations and behaviors. In many ways, AI is becoming an invisible thread that weaves itself into the fabric of daily activities, from shopping recommendations and social media algorithms to navigation systems and financial tools. As AI continues to evolve, it will undoubtedly shape the norms, values, and conventions that guide societal interactions.

One of the most pressing societal concerns related to AI is its impact on employment. Automation driven by AI technologies is rapidly reshaping industries, replacing human workers with machines capable of performing tasks with greater efficiency and accuracy. While AI has the potential to increase productivity and drive economic growth, it also raises concerns about job displacement, particularly in industries reliant on manual labor or routine tasks. The displacement of workers in sectors such as manufacturing, retail, and transportation is already occurring, and the pace of automation is expected to accelerate in the coming years (World Economic Forum, 2025). This shift could lead to significant disruptions in the labor market, with many workers potentially facing job loss without immediate prospects for new employment.

Addressing the impact of AI on employment requires proactive strategies at both the societal and governmental levels. One of the key strategies is the implementation of reskilling and upskilling programs designed to help workers adapt to the changing job market. As AI technologies advance, there will be an increasing demand for workers with skills in areas such as data science, machine learning, and AI system management. Governments and organizations must invest in education and training initiatives that equip workers with the skills needed to thrive in an AI-driven economy (World Economic Forum, 2025). This approach will help mitigate the negative effects of automation by preparing workers for new opportunities in emerging fields.

In addition to reskilling, policies that promote equitable economic transitions will be critical in addressing the societal impact of AI. As certain industries shrink or evolve, others will emerge, but it is essential that these transitions are managed in ways that do not exacerbate existing inequalities. Equitable economic policies must ensure that workers in sectors affected by AI-driven automation are not left behind. This could include targeted support for communities that are most vulnerable to job loss, such as providing retraining opportunities, job placement services, and economic assistance for those impacted by automation. Furthermore, policies should focus on fostering inclusive growth, ensuring that the benefits of AI-driven productivity gains are distributed across society and not concentrated in the hands of a few (OECD, 2022).

Another significant societal implication of AI is its potential to challenge traditional notions of identity and agency. As AI systems become more capable of mimicking human behavior, they blur the lines between human and machine, raising important philosophical and ethical questions. AI technologies, such as advanced chatbots, virtual assistants, and deep learning models, are increasingly able to simulate human conversation and decision-making processes. This ability to replicate human-like behavior leads to questions about what it means to be human, particularly as AI systems are integrated into roles traditionally filled by humans, such as customer service representatives, therapists, and even educators (Council of Europe, 2024).

The rise of AI in personal interactions also challenges traditional concepts of agency—the capacity to act independently and make informed choices. As AI systems become more integrated into everyday life, they begin to influence decision-making processes, subtly guiding individuals in their choices without explicit awareness of their involvement. For example, recommendation algorithms on social media platforms or streaming services not only suggest content based on users' past behavior but also shape their preferences by highlighting certain types of information and suppressing others. This creates a scenario in which AI systems indirectly influence individuals' decisions, potentially limiting the autonomy of users and raising concerns about the erosion of personal agency (UNESCO, 2021).

The relationship between humans and machines also brings up questions of trust and reliance. As AI systems become more sophisticated, they are increasingly capable of performing tasks that were once considered uniquely human, such as diagnosing medical conditions, driving vehicles, or even writing content. While this creates tremendous efficiencies, it also raises concerns about over-reliance on machines and the potential for individuals to lose critical thinking skills or the ability to make decisions without technological assistance. The risk is that humans may begin to defer too heavily to AI, reducing their ability to exercise judgment and critical reasoning on their own (UNESCO, 2021).

AI's growing presence also poses risks to personal autonomy, particularly in areas such as surveillance and privacy. With AI-powered technologies such as facial recognition and behavioral tracking, it has become easier to monitor individuals' actions and behaviors without their knowledge or consent. While such technologies can enhance security and improve services, they also raise concerns about the potential for abuse. The ability of AI systems to track and analyze personal data on a massive scale creates new opportunities for both government and private entities to invade personal privacy, leading to a growing tension between the convenience of personalized services and the need to protect individual rights (Grother et al., 2019; Buolamwini & Gebru, 2018).

The pervasive role of AI in shaping cultural norms is also evident in its impact on human relationships and social interactions. As AI systems, such as chatbots and virtual assistants, become more advanced, there is the potential for humans to form connections with machines in ways that resemble human relationships. This raises concerns about the emotional and social consequences of relying on AI for companionship or support. While AI can provide useful assistance, particularly for individuals who are socially isolated or need help with tasks, it is important to consider the implications of substituting human relationships with AI-based interactions. Over time, this could lead to changes in how people view human connection, potentially eroding social skills or leading to a devaluation of authentic human relationships (UNESCO, 2021).

Moreover, AI's impact on cultural norms extends to issues of social justice and equity. As AI systems are increasingly used to make decisions in areas such as hiring, lending, and criminal justice, there is a risk that they could perpetuate existing biases and reinforce systemic inequalities. AI algorithms are only as good as the data they are trained on, and if historical data contains biases or reflects inequities in society, these biases can be embedded in the AI systems themselves. This could result in unfair treatment for individuals from marginalized groups, exacerbating social divides and creating new forms of discrimination. Ensuring that AI systems are designed and implemented in ways that promote fairness and inclusivity is essential to avoid perpetuating these harmful patterns (OECD, 2022; European Parliament & Council of the European Union, 2024).

The economic implications of AI are also far-reaching. While AI has the potential to drive significant growth and innovation, it also has the potential to concentrate wealth and power in the hands of a few corporations and individuals. Companies that develop and control AI technologies can dominate industries, gaining a competitive edge over smaller players who lack the resources to invest in AI research and development. This concentration of power can lead to monopolistic behaviors and exacerbate income inequality, as those who control AI technologies are able to reap the majority of the financial benefits. The challenge, then, is to ensure that the economic benefits of AI are shared more equitably across society (World Economic Forum, 2025).

As AI continues to evolve, governments and policymakers must address these challenges through regulations that promote social and economic equity. This could involve measures such as taxing AI-driven automation to fund reskilling programs, ensuring fair competition in the AI industry, and providing support for communities that may be negatively affected by the widespread adoption of AI. Furthermore, governments must work to create frameworks that encourage the ethical use of AI while preventing the misuse of the technology for surveillance or control. These efforts will be critical in ensuring that AI technologies benefit society as a whole rather than exacerbating existing inequalities (OECD, 2022; Council of Europe, 2024).

The growing reliance on AI also raises questions about the future of human creativity and innovation. As AI systems become more capable of performing tasks traditionally associated with human creativity, such as generating music, art, and even writing, it is important to consider how these developments might impact creative industries. On one hand, AI could enhance creativity by offering new tools and capabilities for artists, musicians, and writers. On the other hand, there is the potential for AI to replace human creativity altogether, leading to a situation where machines dominate the creative process. As AI continues to blur the lines between human and machine-generated content, society must grapple with questions about authorship, originality, and the value of human artistic expression (UNESCO, 2021).

In conclusion, the societal implications of AI are vast and complex, encompassing everything from employment and economic transitions to cultural norms and personal identity. As AI technologies become increasingly embedded in society, they will continue to shape and redefine the way people live, work, and interact with one another. While AI holds great potential to drive progress and innovation, it also presents significant challenges that require thoughtful and proactive responses from governments, organizations, and individuals. Addressing the societal impact of AI will require a holistic approach that considers not only the technological aspects of AI but also its ethical, social, and economic dimensions. Only by engaging with these issues in a comprehensive and inclusive manner can society ensure that AI benefits everyone while minimizing its potential harms (European Parliament & Council of the European Union, 2024; UNESCO, 2021).

BUILDING AN ETHICAL AI FUTURE

To navigate the opportunities and risks of AI, society must adopt a proactive and collaborative approach. This approach involves not only leveraging the technological advancements that AI offers but also engaging in thoughtful discussions about its ethical implications. The rapid development of AI has the potential to create significant societal changes, but these changes must be guided by informed decisions and a commitment to ensuring that AI serves the broader public good. One critical element of this process is the investment in AI literacy, ensuring that individuals at all levels of society can understand the implications of AI, how it works, and its potential impact on various aspects of life (UNESCO, 2021). By promoting AI literacy, we can empower people to make informed choices, participate in meaningful conversations about the technology, and advocate for policies that prioritize fairness, accountability, and transparency.

AI literacy is not just about understanding how algorithms work or the technicalities of machine learning models; it is about fostering a broader understanding of how AI can shape society and how its benefits and risks must be balanced. People need to recognize that AI systems are not neutral tools; they are shaped by the values and decisions of those who design and implement them (IEEE Global Initiative on Ethics of Autonomous and Intelligent Systems, 2019). As AI becomes more pervasive, its influence on everything from healthcare and education to employment and governance will grow, making it essential for individuals to understand not only how to interact with AI but also how to critically evaluate its role in their lives.

Engaging diverse perspectives in AI design and governance is essential to creating systems that reflect a broad range of values, priorities, and needs. AI development has historically been dominated by certain groups—predominantly technologists, researchers, and corporations based in a few parts of the world. This lack of diversity can lead to systems that reflect a narrow set of cultural and socioeconomic norms. To avoid this, it is crucial to involve a wider array of voices in AI's development. This includes not only technologists and business leaders but also ethicists, sociologists, policymakers, and representatives from marginalized communities (UNESCO, 2021). These diverse perspectives ensure that AI systems are designed with inclusivity in mind, addressing the needs and concerns of various segments of society rather than reinforcing existing biases or inequalities.

Moreover, the governance of AI needs to be inclusive and participatory. AI policies and regulations should not be developed in isolation by a small group of experts or technocrats. Instead, they should involve input from multiple stakeholders, including civil society organizations, labor groups, and individuals who are directly affected by AI systems. This collaborative approach to governance can help ensure that AI policies reflect a more comprehensive understanding of its social, economic, and ethical implications. Such inclusive governance frameworks also help foster trust in AI systems, as people are more likely to support policies that they have had a role in shaping and that align with their values and interests (Council of Europe, 2024; UNESCO, 2021).

One of the most important goals in navigating AI's potential is ensuring that its development and deployment align with ethical principles. AI has the potential to bring about remarkable advancements, from revolutionizing healthcare and improving education to addressing climate change and poverty. However, without a commitment to ethical standards, these advancements could come at a high cost. The development of AI must prioritize human dignity, rights, and fairness. This means ensuring that AI systems do not discriminate against individuals or groups, that they respect privacy, and that they operate transparently (OECD, 2022; UNESCO, 2021). In addition, AI systems should be designed in a way that allows for accountability and redress when things go wrong. Without these ethical considerations, AI's transformative potential could be overshadowed by negative consequences, such as deepening social inequalities or violating fundamental rights.

At the same time, it is important to recognize that the ethical use of AI is not a static goal but an ongoing process. As AI systems evolve, new ethical challenges will emerge that may not have been anticipated at the outset. For example, the use of AI in surveillance, predictive policing, and military applications raises questions about privacy, freedom, and the potential for abuse (Privacy and Civil Liberties Oversight Board, 2023). Similarly, the growing use of AI in decision-making processes in areas such as hiring, lending, and criminal justice introduces risks of bias and discrimination (Buolamwini & Gebru, 2018; Grother et al., 2019). It is essential that the development of AI is guided by continuous ethical reflection, ensuring that new developments are carefully examined through the lens of fairness, equity, and respect for human rights.

Ultimately, the goal of AI development should be to harness its potential as a force for good, promoting innovation that can enhance human well-being while safeguarding ethical principles. AI has the power to address some of the most pressing challenges facing humanity, from tackling climate change to improving healthcare outcomes and advancing education. However, this potential can only be realized if AI is developed and deployed in ways that align with the broader goals of society. This means prioritizing human values, ensuring that AI systems are transparent, accountable, and fair, and addressing the risks associated with the technology before they become entrenched (NIST, 2023; NIST, 2024).

The ethical considerations of AI must also take into account the broader societal impact of the technology. This includes ensuring that the benefits of AI are distributed equitably and that marginalized groups are not left behind in the race to develop and implement AI systems. AI should be a tool for promoting social good, not just for maximizing profits or consolidating power. By adopting a responsible approach to AI, society can unlock its potential to enhance decision-making, improve quality of life, and tackle global challenges (World Economic Forum, 2025; OECD, 2022).

This chapter emphasizes the need for a nuanced and balanced approach to AI ethics. Recognizing both the transformative possibilities of AI and its profound responsibilities is essential for navigating the future of this technology. AI holds great promise, but it also presents significant ethical challenges that require careful consideration and action. By addressing these challenges head-on, society can ensure that AI serves as a tool for progress and positive change, rather than a force that exacerbates inequality or undermines fundamental rights (European Parliament & Council of the European Union, 2024; UNESCO, 2021).

The need for a balanced approach to AI ethics also highlights the importance of interdisciplinary collaboration. The development of AI is not just a technical challenge but a moral and societal one. Ethicists, policymakers, sociologists, and technologists must work together to understand the full implications of AI and develop frameworks that promote its ethical use. By bringing together expertise from a variety of fields, we can ensure that AI is developed in ways that are both innovative and responsible (IEEE Global Initiative on Ethics of Autonomous and Intelligent Systems, 2019).

One of the key components of ethical AI is transparency. As AI systems become more complex, it becomes increasingly difficult to understand how they arrive at decisions. This "black box" problem makes it challenging to ensure that AI systems are operating in ways that are fair and accountable. To address this, AI developers must prioritize transparency in the design and implementation of AI systems. This includes making the algorithms and decision-making processes behind AI systems more understandable and accessible to the public, regulators, and other stakeholders. Transparency can help build trust in AI systems and allow for more effective oversight and accountability (UNESCO, 2021; European Parliament & Council of the European Union, 2024).

Accountability is another critical aspect of ethical AI. When AI systems make decisions that affect people's lives, it is important that there are mechanisms in place to hold those systems accountable. This includes establishing clear lines of responsibility for AI outcomes and ensuring that individuals or organizations are held accountable when AI systems cause harm. It also means providing avenues for people to seek redress when they are negatively impacted by AI decisions. By fostering a culture of accountability, society can ensure that AI is used responsibly and that its benefits are distributed fairly (UNESCO, 2021).

Another important ethical consideration is the preservation of human agency. As AI systems become more capable of performing tasks traditionally carried out by humans, there is a risk that individuals may lose their ability to make independent decisions or exercise critical judgment. AI should not be used to replace human decision-making entirely but should instead augment and enhance human capabilities. Maintaining human oversight in AI-driven processes is essential to ensuring that technology serves the needs and values of society, rather than undermining them (European Parliament & Council of the European Union, 2024).

The ethical use of AI also requires a commitment to inclusivity. AI systems should be designed in ways that are accessible and beneficial to all members of society, regardless of their background or socioeconomic status. This means addressing biases in AI training data, ensuring that AI systems do not perpetuate discrimination, and providing access to AI-driven benefits for underserved communities. Inclusive AI design can help bridge digital divides and ensure that technology contributes to social equity (OECD, 2022; UNESCO, 2021).

Finally, the ethical challenges posed by AI require robust regulatory frameworks that can keep pace with the rapid development of the technology. Policymakers must establish clear standards for AI development and use, ensuring that AI systems are used responsibly and in ways that align with societal values. This includes regulating AI applications in areas such as healthcare, criminal justice, and finance to ensure that they are transparent, fair, and accountable (European Parliament & Council of the European Union, 2024; NIST, 2024). By creating strong regulatory frameworks, society can mitigate the risks associated with AI while fostering its potential for innovation and positive impact.

In conclusion, the ethical use of AI is not only about addressing the technical challenges of the technology but also about ensuring that it serves the broader goals of society. By promoting transparency, accountability, fairness, and inclusivity, we can harness the transformative potential of AI while safeguarding human rights and dignity. This balanced and collaborative approach is essential for ensuring that AI contributes to the well-being of all people and that its benefits are shared equitably across society. Only through responsible development and governance can we ensure that AI remains a force for good in the future (UNESCO, 2021; OECD, 2022).

CONCLUSION

In conclusion, the ethical challenges and opportunities posed by AI require a careful and deliberate approach to ensure that its development and deployment serve the broader good of society. As AI technologies continue to evolve and integrate into various aspects of daily life, it is essential to prioritize ethical principles such as transparency, accountability, fairness, and inclusivity (UNESCO, 2021; OECD, 2022). By addressing these issues proactively, society can unlock AI's potential to enhance decision-making, improve quality of life, and tackle pressing global challenges while safeguarding fundamental human rights and values.

The development of AI systems must be guided by thoughtful, interdisciplinary collaboration, involving technologists, ethicists, policymakers, and diverse stakeholders to ensure that AI is shaped by a wide range of perspectives. This collaboration should aim to mitigate biases, ensure that AI applications do not perpetuate inequality, and ensure that vulnerable groups are not marginalized in the process of technological advancement (Buolamwini & Gebru, 2018; Grother et al., 2019). Moreover, the design and use of AI must prioritize human agency and preserve autonomy, ensuring that AI serves to augment human decision-making rather than replacing it entirely (European Parliament & Council of the European Union, 2024).

Governance frameworks must also evolve to address the unique challenges AI presents, including the potential for misuse, the risk of discrimination, and the need for redress when AI systems cause harm. These frameworks should be transparent, adaptable, and rooted in principles that promote the equitable and responsible use of AI (Council of Europe, 2024; NIST, 2024). At the same time, AI literacy should be fostered across all levels of society, equipping individuals with the knowledge and understanding necessary to engage with AI in meaningful ways (UNESCO, 2021).

Ultimately, AI has the potential to transform society in profound ways, but this transformation must be carefully managed to ensure it aligns with the values of fairness, justice, and human dignity. By approaching AI development with a commitment to ethical principles and a focus on the well-being of all people, society can ensure that AI remains a tool for good, driving innovation while protecting the rights and freedoms of individuals (European Parliament & Council of the European Union, 2024). The journey toward ethical AI is ongoing, but with concerted effort, collaboration, and vigilance, AI can be a force that enhances human potential and contributes to the common good for generations to come (World Economic Forum, 2025; ENISA, 2024).

TEN ESSAY QUESTIONS FOR CHAPTER 7

Artificial Intelligence—Ethics and Decision-Making

1. What are the key ethical concerns surrounding the use of artificial intelligence in decision-making? Discuss how these concerns differ across various sectors such as healthcare, criminal justice, and finance.
2. In what ways can AI systems perpetuate or even exacerbate existing biases? Provide examples of biased AI applications and propose strategies for addressing bias in AI development and deployment.
3. How can transparency and accountability be ensured in AI decision-making processes, especially in high-stakes environments like medical diagnoses and legal judgments?
4. Discuss the moral implications of autonomous AI systems, such as self-driving cars and lethal autonomous weapons. Who should be held responsible for the decisions made by these systems?
5. Examine the ethical trade-offs between the efficiency of AI decision-making and the preservation of human agency. Should AI ever be allowed to make decisions that significantly impact human lives without human intervention?
6. How does AI challenge traditional notions of fairness and justice? Should AI decision-making systems be subject to the same ethical standards as human decision-making processes? Why or why not?
7. Analyze the potential risks of AI in terms of privacy and security. How can regulations be designed to protect individual privacy while fostering AI innovation?
8. What role should regulation and governance play in shaping the future of AI? Discuss existing efforts, such as the European Union's AI Act, and propose recommendations for enhancing AI oversight.
9. In what ways can AI contribute to societal good, and how can ethical guidelines be integrated into AI systems to ensure that these contributions are realized?
10. What are the long-term societal impacts of widespread AI integration? Consider issues such as employment displacement, shifts in power dynamics, and the evolving relationship between humans and machines. How should society prepare for these changes?

REFERENCES

Amnesty International. (2024). *Ban the scan: A call to prohibit the use of facial recognition technology for mass surveillance*. Amnesty International.

Buolamwini, J., & Gebru, T. (2018). Gender shades: Intersectional accuracy disparities in commercial gender classification. *Proceedings of Machine Learning Research*, 81, 1–15.

Council of Europe. (2024). *Framework Convention on Artificial Intelligence and human rights, democracy and the rule of law*. Council of Europe.

Doshi-Velez, F., & Kim, B. (2017). Towards a rigorous science of interpretable machine learning. *arXiv preprint*, arXiv:1702.08608.

ENISA. (2024). *ENISA threat landscape 2024*. European Union Agency for Cybersecurity.

European Commission. (2023). *Commission Implementing Decision (EU) 2023/1795 on the adequate level of protection of personal data under the EU–U.S. Data Privacy Framework*. European Commission.

European Parliament, & Council of the European Union. (2024). Regulation (EU) 2024/1689 laying down harmonised rules on artificial intelligence (Artificial Intelligence Act). *Official Journal of the European Union*.

Executive Office of the President. (2023). *Executive Order 14110—Safe, secure, and trustworthy development and use of artificial intelligence*. The White House.

Gebru, T., Morgenstern, J., Vecchione, B., Vaughan, J. W., Wallach, H., Daumé III, H., & Crawford, K. (2021). Datasheets for datasets. *Communications of the ACM*, 64(12), 86–92.

Grother, P., Ngan, M., & Hanaoka, K. (2019). *Face Recognition Vendor Test (FRVT) Part 3: Demographic effects (NISTIR 8280)*. National Institute of Standards and Technology.

IEEE Global Initiative on Ethics of Autonomous and Intelligent Systems. (2019). *Ethically aligned design: A vision for prioritizing human well-being with autonomous and intelligent systems* (1st ed.). IEEE.

IEEE Standards Association. (2021). *IEEE Std 7000-2021: IEEE model process for addressing ethical concerns during system design*. IEEE.

International Labour Organization. (2023). *Generative AI and jobs: A global analysis of potential effects on job quantity and quality*. International Labour Office.

ISO/IEC. (2023). *ISO/IEC 23894:2023—Information technology—Artificial intelligence—Risk management*. International Organization for Standardization and International Electrotechnical Commission.

Mitchell, M., Wu, S., Zaldivar, A., Barnes, P., Vasserman, L., Hutchinson, B., … Gebru, T. (2019). Model cards for model reporting. In *Proceedings of the 2019 Conference on Fairness, Accountability, and Transparency (FAT*)* (pp. 220–229).

National Academies of Sciences, Engineering, and Medicine. (2022). *Fostering responsible computing research: Foundations and practices*. National Academies Press.

National Institute of Standards and Technology. (2020). *Profile of the IoT core baseline for consumer IoT products (NISTIR 8259A)*. NIST.

National Institute of Standards and Technology. (2023). *AI Risk Management Framework 1.0 (NIST AI 100-1)*. NIST.

National Institute of Standards and Technology. (2024). *Cybersecurity Framework 2.0*. NIST.

Obermeyer, Z., Powers, B., Vogeli, C., & Mullainathan, S. (2019). Dissecting racial bias in an algorithm used to manage the health of populations. *Science*, 366(6464), 447–453.

OECD. (2022). *OECD guidelines on the protection of privacy and transborder flows of personal data (updated 2022)*. Organisation for Economic Co-operation and Development.

PCLOB (Privacy and Civil Liberties Oversight Board). (2023). *Report on the surveillance program operated pursuant to Section 702 of the Foreign Intelligence Surveillance Act (Unclassified Public Report)*. PCLOB.

Rudin, C. (2019). Stop explaining black box machine learning in high-stakes decisions and use interpretable models instead. *Nature Machine Intelligence*, 1(5), 206–215.

UNESCO. (2021). *Recommendation on the ethics of artificial intelligence*. United Nations Educational, Scientific and Cultural Organization.

Verizon. (2024). *Data Breach Investigations Report (DBIR)*. Verizon.

Verizon. (2025). *2025 Data Breach Investigations Report (DBIR)*. Verizon.

WEF (World Economic Forum). (2025). *Global cybersecurity outlook 2025*. World Economic Forum.

WHO (World Health Organization). (2023). *Ethics and governance of artificial intelligence for health: Guidance on large multimodal models (LMMs)*. World Health Organization.

World Economic Forum. (2025). *The Future of Jobs Report 2025*. World Economic Forum.

Bridging the Digital Divide: Equity and Inclusion in Technology Access

8

ABSTRACT

Chapter 8, *Bridging the Digital Divide: Equity and Inclusion in Technology Access*, explores the growing disparities in access to technology, particularly the internet and digital devices, and the urgent need to address these gaps to ensure equitable opportunities for all. The chapter examines how unequal access to technology exacerbates social, economic, and educational inequalities, particularly among rural, low-income, and minority communities. It discusses the importance of expanding broadband infrastructure, improving digital literacy, and providing targeted programs for underserved populations. By emphasizing the role of governments, businesses, and educational institutions in promoting digital inclusion, the chapter advocates for a collaborative approach to ensure that technology is accessible, affordable, and beneficial for everyone, fostering a more inclusive digital future.

INTRODUCTION: THE IMPORTANCE OF BRIDGING THE DIGITAL DIVIDE

In the 21st century, technology has become an indispensable part of daily life, shaping the way people interact, learn, work, and participate in society. It is a powerful tool for social, economic, and educational advancement, offering opportunities for personal growth, better access to information, and improved quality of life. From online education platforms to telemedicine, technology holds immense potential to address global challenges and transform industries (Organisation for Economic Co-operation and Development, 2024; World Economic Forum, 2025). However, despite its promise, the benefits of technology are not distributed equally among all populations. A significant barrier to equitable access is the "digital divide," which refers to the disparities in access to technology, particularly the internet and digital devices, between different social, economic, and demographic groups (International Telecommunication Union, 2024).

The digital divide is most evident in areas where access to technology is either limited or absent. For instance, rural communities often lack the necessary infrastructure to support reliable internet access, which creates a digital gap that hinders educational opportunities, access to healthcare, and employment prospects (Federal Communications Commission, 2024; U.S. Department of Agriculture, Rural Utilities Service, 2024). Without consistent and affordable internet access, individuals in these areas are left behind in an increasingly digital world. This issue is compounded by the rising costs of digital devices such as computers and smartphones, making it financially unfeasible for many low-income families to equip themselves with the tools needed to fully engage with technology (Pew Research Center, 2024).

In addition to financial barriers, the digital divide also affects various demographic groups differently. Minority populations, particularly African American, Hispanic, and Indigenous communities, experience lower rates of internet access and technology ownership (Pew Research Center, 2024; International Telecommunication Union, 2024). These groups often face systemic challenges, including lower household incomes, fewer educational opportunities, and geographic isolation, all of which contribute to their limited participation in the digital economy. As a result, members of these communities are disproportionately impacted by the digital divide, further exacerbating existing inequalities in education, healthcare, and employment (OECD, 2024).

One of the most significant consequences of the digital divide is its impact on education. In recent years, digital technology has become essential to modern learning environments, and the COVID-19 pandemic underscored how critical technology is for accessing education. Remote learning, in particular, revealed the stark differences between students who have access to devices and reliable internet connections at home and those who do not. The "homework gap" — the disparity between students who can complete assignments online and those who cannot — continues to widen educational inequalities, leaving disadvantaged students further behind (U.S. Department of Education, Office of Educational Technology, 2024). This divide in educational opportunities has long-term effects on students' academic performance and future prospects, perpetuating cycles of poverty and limited opportunity (OECD, 2024).

Digital literacy is another critical element in addressing the digital divide. While providing access to technology is important, it is not sufficient if individuals lack the skills to use it effectively.

Digital literacy includes the ability to navigate the internet, use digital tools, and engage safely and responsibly with technology. Many individuals, particularly older adults, may not have had the opportunity to develop these skills, leaving them vulnerable to exploitation or misinformation. Moreover, low-income households often lack the resources to acquire digital literacy skills, further deepening the divide between those who are technologically proficient and those who are not (National Digital Inclusion Alliance, 2024; OECD, 2024).

In addition to these practical concerns, the digital divide also poses broader societal challenges. The increasing reliance on technology for accessing essential services such as healthcare, government programs, and job opportunities means that those without access to technology are at risk of being excluded from these vital resources. For instance, telemedicine has become a key tool for delivering healthcare services, especially in rural or underserved areas. However, individuals without access to the internet or digital devices may be unable to access medical care, further exacerbating health disparities (Centers for Disease Control and Prevention, National Center for Health Statistics, 2024).

Bridging the digital divide requires a comprehensive approach that addresses both access and digital literacy. Expanding broadband infrastructure in rural and underserved urban areas is a critical first step (Federal Communications Commission, 2024; U.S. Department of Agriculture, Rural Utilities Service, 2024). Public policy initiatives, such as subsidies for low-income families or publicly funded Wi-Fi hotspots in underserved areas, can help ensure that internet access is available to all (National Telecommunications and Information Administration, 2024a). Additionally, educational institutions must invest in programs that teach digital literacy from an early age, integrating these skills into school curriculums so that future generations are equipped to navigate the digital world effectively (U.S. Department of Education, Office of Educational Technology, 2024).

Governments and educational institutions are not the only entities responsible for bridging the digital divide. Businesses, particularly those in the technology sector, have a crucial role to play in ensuring equitable access to technology. Tech companies can partner with local communities, schools, and nonprofits to provide access to affordable devices and offer training programs that help individuals develop the digital skills needed to thrive in the modern workforce (National Digital Inclusion Alliance, 2024; OECD, 2024). By investing in digital inclusion initiatives, businesses can contribute to building a more diverse and equitable workforce while helping to close the digital divide.

Workforce development programs are also essential in addressing the digital divide, especially as technology continues to reshape the job market. Many industries now require workers with digital skills, such as data analysis, software development, and digital marketing. However, individuals from disadvantaged backgrounds often lack the resources and opportunities to develop these skills.

Expanding workforce development programs that provide training and certification in high-demand digital fields can help create pathways for upward mobility and economic empowerment for communities that are otherwise at risk of being left behind (World Economic Forum, 2025).

The digital divide also has implications for civic participation. As more civic processes, including voting registration, government services, and community engagement, shift to online platforms, individuals without access to technology may be excluded from these essential democratic functions. Ensuring equitable access to technology is critical for preserving the principles of democracy and inclusivity, as marginalized groups are disproportionately impacted by digital exclusion (OECD, 2024).

Addressing the digital divide requires collaboration across sectors. Governments, businesses, non-profits, and educational institutions must work together to develop policies and programs that promote digital inclusion (National Telecommunications and Information Administration, 2024a; OECD, 2024). This includes providing financial support to families who cannot afford the technology, developing educational initiatives to teach digital skills, and ensuring that public services are accessible online to all, regardless of income or geography.

One promising solution to addressing the digital divide is the concept of community-driven initiatives. By empowering local communities to take an active role in designing and implementing digital inclusion programs, solutions can be tailored to the specific needs and preferences of those most affected by the divide. This approach ensures that interventions are culturally relevant, well-received, and sustainable (National Digital Inclusion Alliance, 2024).

Finally, bridging the digital divide is not just about providing access to technology; it is about creating an equitable society where everyone can benefit from the opportunities that technology offers. By prioritizing digital inclusion, governments and organizations can create a more level playing field in which all individuals, regardless of their socioeconomic status, race, or geographic location, can participate in the digital world. Achieving this goal requires a commitment to equity, collaboration, and a recognition of the transformative power of technology for all. Only by addressing the digital divide can we ensure that technology serves as a force for good and an equalizer for all members of society (International Telecommunication Union, 2024; OECD, 2024).

Bridging the digital divide: access, affordability, and digital skills for an inclusive future

Read through an inclusion lens, the image reframes the digital divide as a stack of barriers—availability of reliable broadband, affordability of service and devices, accessible design, digital skills and confidence, and the relevance and safety of online content. It suggests that gaps compound for rural, low-income, older, disabled, and linguistically diverse communities, turning "access" into an equity issue rather than a purely technical one. The practical prompt beneath the image is layered solutions: invest in last-mile infrastructure and community networks; pair subsidies with device access and repair; embed accessibility and multilingual support by default; provide trusted, local training; and measure success not by coverage maps but by adoption and outcomes that expand education, work, health, and civic participation.

MANIFESTATIONS OF THE DIGITAL DIVIDE

The digital divide manifests in several ways, primarily through unequal access to devices, broadband internet, and digital literacy. These disparities create significant barriers to full participation in today's increasingly digital society. While the divide is commonly understood as an issue of access to technology, it is important to recognize that it encompasses a complex array of challenges, from affordability to infrastructure limitations. In rural and underserved urban areas, the lack of adequate broadband infrastructure is particularly pronounced, leaving entire communities at a disadvantage in accessing critical online services, including education, healthcare, and employment opportunities (Federal Communications Commission, 2024; U.S. Department of Agriculture, Rural Utilities Service, 2024). In many of these regions, internet access is either too slow, too expensive, or simply unavailable, effectively cutting off individuals from the benefits of a connected world.

This lack of reliable internet access is not only an inconvenience but also an obstacle to participation in essential activities such as online education. The COVID-19 pandemic highlighted just how essential internet access is for students' success, as the shift to remote learning left many children without the necessary tools to fully engage with their studies. For students in low-income households, the barriers to educational access are compounded by the inability to afford the necessary digital devices, such as computers or tablets, required to participate in online learning. Without these devices, students fall further behind in their academic work, which contributes to the widening achievement gap between affluent and disadvantaged students (U.S. Department of Education, Office of Educational Technology, 2024; Pew Research Center, 2024).

While digital devices have become more affordable over time, the cost remains a significant barrier for many families. A study by the Pew Research Center found that about 25% of lower-income families do not have a computer at home, and many more lack access to high-speed internet (Pew Research Center, 2024). In households where financial constraints limit the purchase of a computer, children are forced to rely on smartphones or share devices with siblings, which makes it difficult to complete homework assignments or participate in virtual classrooms. This lack of access undermines the quality of education and the overall academic experience for these students, setting them back in their learning journey.

Furthermore, the digital divide is not solely about the availability of devices and internet access; it also involves the critical issue of digital literacy. Having access to technology is not enough if individuals do not have the skills necessary to use it effectively. Many people, especially older adults and those without formal education in technology, may lack the digital literacy skills required to navigate the internet, use software programs, or engage in online learning (OECD, 2024; National Digital Inclusion Alliance, 2024). This skills gap is particularly problematic in an era where digital proficiency is essential for success in both the job market and daily life.

The digital literacy gap is often compounded by inadequate training opportunities for those who lack access to technology. Many communities, particularly those in rural areas or lower-income urban neighborhoods, do not have affordable programs to teach people how to use digital tools. Without these resources, individuals are at a disadvantage when it comes to finding and applying for jobs, accessing healthcare services, or even completing basic tasks such as paying bills online. This digital exclusion creates a cycle of disadvantage, where those who are already marginalized by socioeconomic factors are further isolated by their inability to access or navigate the digital world (National Digital Inclusion Alliance, 2024).

In addition to these educational and personal challenges, the digital divide has broader societal implications, especially in the realm of economic opportunity. As more job applications and hiring processes shift online, individuals without access to technology are increasingly excluded from the workforce. Many employers now require applicants to submit resumes and conduct interviews through online platforms, making it nearly impossible for those without digital access to compete for jobs. This exclusion exacerbates economic disparities, leaving low-income and minority communities at a disadvantage in an increasingly digital job market (World Economic Forum, 2025; OECD, 2024).

Moreover, the digital divide is not limited to education and employment. It also has significant consequences for access to healthcare. Telemedicine, a rapidly growing field, has become an essential way for individuals to receive medical consultations, particularly in rural areas where healthcare providers are scarce. However, individuals without internet access or digital devices cannot benefit from telemedicine services, limiting their ability to receive medical care (Centers for Disease Control and Prevention, National Center for Health Statistics, 2024). This is particularly problematic for older adults and individuals with chronic health conditions, who are often more likely to require remote healthcare services. As a result, the digital divide exacerbates health disparities, with underserved populations lacking access to the same level of care as those in more affluent areas.

The digital divide also impacts civic participation. In recent years, many government services, including voting registration, have moved online. Citizens without internet access are therefore unable to fully participate in civic activities, such as registering to vote or accessing government benefits. This exclusion undermines the democratic process and further disenfranchises already marginalized groups, such as low-income individuals, people of color, and those with disabilities. As the government continues to move more services online, it becomes increasingly important to ensure that all citizens have access to the technology needed to engage in these activities (OECD, 2024).

As technology continues to shape every aspect of modern life, bridging the digital divide becomes a critical issue for ensuring equity and inclusion. One key approach to addressing this gap is expanding broadband infrastructure to underserved areas. By increasing investment in broadband expansion, particularly in rural and economically disadvantaged urban areas, governments can help ensure that more communities have access to high-speed internet (National Telecommunications and Information Administration, 2024a; Federal Communications Commission, 2024). Public-private partnerships can also play a key role in this effort, with technology companies and telecommunications providers working together to bring internet access to more people (OECD, 2024).

Another important solution to the digital divide is the provision of affordable devices. Programs that provide subsidized computers, tablets, and smartphones to low-income households can help ensure that more individuals have the tools they need to participate in the digital world. Such programs can be supported by both public and private sector funding, with tech companies offering discounts or donating devices to underserved communities (National Digital Inclusion Alliance, 2024). Additionally, educational institutions can work with nonprofits and tech companies to provide students with the devices and internet access they need to succeed in their studies (U.S. Department of Education, Office of Educational Technology, 2024).

Digital literacy programs are also essential for addressing the digital divide. In addition to providing access to technology, communities must also provide opportunities for individuals to develop the skills necessary to use that technology effectively. Public libraries, community centers, and schools can offer digital literacy courses, teaching individuals how to use the internet, navigate online resources, and stay safe in the digital environment (National Digital Inclusion Alliance, 2024; U.S. Department of Education, Office of Educational Technology, 2024). By providing these resources, communities can empower individuals to take full advantage of the opportunities that technology provides.

Workforce development programs are another important aspect of addressing the digital divide. By offering digital skills training programs, governments and businesses can help individuals gain the necessary skills to succeed in the modern workforce. Many of the fastest-growing industries, such as technology, healthcare, and finance, require workers with digital proficiency. Providing training and certification programs in these fields can help bridge the skills gap and create new opportunities for individuals in underserved communities (World Economic Forum, 2025).

Moreover, policymakers must ensure that the digital divide is not perpetuated by their own decisions. For example, policies that prioritize the development of affordable broadband infrastructure in underserved areas are essential. Policymakers must also ensure that technology companies are incentivized to create accessible and affordable devices for all consumers. This may include regulations that limit the cost of essential devices or that provide subsidies to families who cannot afford them (National Telecommunications and Information Administration, 2024a, 2024b, 2024c; OECD, 2024).

Community-driven solutions are also essential for addressing the digital divide. Local communities have unique insights into the specific challenges they face in accessing technology. By empowering communities to take the lead in designing and implementing digital inclusion programs, solutions can be tailored to meet the needs of those most affected by the divide (National Digital Inclusion Alliance, 2024). This approach ensures that interventions are culturally relevant, well-received, and sustainable, as local stakeholders are more likely to be invested in the success of these programs.

Finally, bridging the digital divide requires a long-term commitment to digital inclusion. The process of closing the divide will not happen overnight, but with continued investment in infrastructure, affordable devices, digital literacy, and workforce development, progress can be made. As technology continues to advance and permeate every aspect of life, it is critical that efforts to bridge the digital divide remain a top priority. Only by ensuring that all individuals have access to the tools, skills, and opportunities they need to succeed in the digital age can we create a more equitable and inclusive society (OECD, 2024; International Telecommunication Union, 2024).

DEMOGRAPHIC DISPARITIES IN ACCESS

The disparities in access to technology are not solely financial but also deeply demographic. Minority communities, particularly those in rural and low-income urban areas, face significant barriers to digital inclusion. These communities often experience compounded challenges that prevent them from fully participating in the digital world, making it harder for them to access opportunities and resources that are increasingly available online. Factors such as geography, socioeconomic status, and historical systemic inequalities contribute to the digital divide, exacerbating the barriers that these populations face (OECD, 2024).

African American, Hispanic, and Indigenous populations, for example, experience lower rates of internet access compared to their White counterparts. According to studies from the Pew Research Center, African American and Hispanic households are less likely to have high-speed broadband at home, which places them at a disadvantage when it comes to accessing critical online services (Pew Research Center, 2024). This digital exclusion can significantly hinder these communities' ability to access educational resources, healthcare information, job opportunities, and government services, further entrenching existing social and economic inequalities (International Telecommunication Union, 2024; OECD, 2024).

The reasons for these disparities are multifaceted. In many cases, the lack of infrastructure in rural areas means that these communities have limited access to high-speed internet, making it nearly impossible for individuals to engage in activities that require reliable online connectivity (U.S. Department of Agriculture, Rural Utilities Service, 2024; Federal Communications Commission, 2024). Even in urban areas, where broadband infrastructure may be more readily available, poverty and the high cost of internet services can still prevent individuals from accessing digital tools. In households where the internet is unaffordable, families may be forced to rely on slower or less reliable connections, such as mobile data or public Wi-Fi, which are inadequate for tasks like remote work or online learning (Pew Research Center, 2024).

For many low-income and minority families, purchasing a computer or other digital devices is simply not financially feasible. The cost of technology remains a significant barrier for these communities. Despite the decrease in the price of some devices over the years, a personal computer, tablet, or smartphone can still represent a major financial investment. Without access to affordable technology, these populations are excluded from the educational, social, and economic opportunities that digital tools make possible, perpetuating cycles of poverty and marginalization (Pew Research Center, 2024).

Beyond the lack of access to devices and the internet, systemic inequalities also play a role in the digital divide. Many minority communities face challenges rooted in historical and structural disadvantages, such as limited access to quality education and employment opportunities, which further exacerbate their difficulties in gaining digital inclusion. In some cases, the lack of educational resources in these communities means that individuals may not develop the digital literacy skills necessary to effectively navigate the internet and technology, even if they have access to the hardware. This creates a compounded disadvantage that makes it difficult for these groups to succeed in an increasingly digital world (OECD, 2024; U.S. Department of Education, Office of Educational Technology, 2024).

For African American, Hispanic, and Indigenous populations, the digital divide also reflects broader patterns of exclusion and discrimination. These groups often face biases that affect their access to financial resources, education, and employment, which in turn affect their ability to access technology. For example, minority communities are more likely to live in neighborhoods with limited access to high-quality broadband services or in homes where affordability is a constant challenge. These barriers are often reinforced by a lack of policies and investments that specifically address the needs of these communities, leaving them behind in the digital age (OECD, 2024; International Telecommunication Union, 2024).

The exclusion of these populations from the digital economy is a critical issue that has long-term implications for economic mobility and social equity. As more job opportunities, educational programs, and government services move online, those without access to digital resources are at risk of being excluded from vital opportunities. This creates a vicious cycle in which individuals from minority communities are unable to gain the skills or education necessary to participate in the digital economy, further entrenching social and economic disparities. Without concerted efforts to bridge the digital divide, these groups will continue to face systemic disadvantages in nearly every area of life (OECD, 2024).

In education, for example, students in minority communities may lack the devices and internet connections needed to fully participate in online learning. This problem was particularly apparent during the COVID-19 pandemic, when schools across the country shifted to remote learning. Minority students in low-income families faced a higher likelihood of being without the necessary technology, leading to increased disparities in educational outcomes. These students struggled to access course materials, complete assignments, and participate in virtual classrooms, putting them at a significant disadvantage compared to their peers who had reliable internet access and digital devices (U.S. Department of Education, Office of Educational Technology, 2024; Pew Research Center, 2024).

In healthcare, the digital divide also exacerbates health disparities. Telemedicine, which has become an essential tool for healthcare delivery, particularly in rural and underserved areas, requires reliable internet access and digital devices. Without these tools, individuals in minority communities are unable to benefit from remote healthcare services, which are critical for those living in areas with limited access to medical providers. This lack of access to telehealth services results in poorer health outcomes, as individuals are less likely to receive timely medical advice or treatment. In addition, individuals without access to digital tools may have difficulty obtaining important health information online, further preventing them from managing chronic conditions or making informed health decisions (Centers for Disease Control and Prevention, National Center for Health Statistics, 2024).

In the job market, digital exclusion means that individuals in minority communities are less likely to have the skills needed for today's high-demand digital jobs. The shift toward a digital economy, where many jobs require technological proficiency, places those without access to technology at a distinct disadvantage. Even jobs that do not directly require digital skills may still require the ability to apply online platforms for applications, interviews, and job search. Without reliable internet access or digital literacy, minority individuals may struggle to apply for positions, leading to lower employment rates and reduced economic mobility within these communities (World Economic Forum, 2025; OECD, 2024).

Moreover, these disparities in access to technology have a negative impact on civic participation. The digital divide affects the ability of individuals in minority communities to engage in democratic processes, such as voting. Increasingly, voter registration and other civic activities are moving online, leaving those without internet access excluded from full participation in the democratic process. This exclusion further perpetuates the marginalization of these communities, as they are unable to influence political decisions that affect their lives. As such, addressing the digital divide is not only a matter of improving access to technology but also a matter of ensuring equal representation in democratic processes (OECD, 2024).

Addressing these demographic disparities requires a multifaceted approach that not only addresses financial barriers but also the systemic inequities that contribute to the digital divide. One key solution is to increase investment in broadband infrastructure in underserved areas, especially rural and low-income urban neighborhoods, where internet access is limited. Expanding access to high-speed broadband is essential to ensure that all communities have equal opportunities to participate in the digital economy (Federal Communications Commission, 2024; U.S. Department of Agriculture, Rural Utilities Service, 2024).

Another critical solution is to make digital devices more affordable and accessible to minority communities. Subsidized programs that provide computers, tablets, or smartphones to low-income families can help ensure that technology is within reach for those who need it most (National Digital Inclusion Alliance, 2024). In addition, schools and community organizations can play a role by providing devices and internet access to students in need, ensuring that all children have the tools they need to succeed in school (U.S. Department of Education, Office of Educational Technology, 2024).

Efforts to close the digital divide must also include initiatives to promote digital literacy, particularly in minority communities. By providing targeted digital literacy training programs, community centers, libraries, and schools can help individuals build the skills necessary to navigate the digital world. Digital literacy is not only important for education but also for employment, healthcare, and civic participation. Empowering individuals with these skills is crucial for ensuring that they can take full advantage of the opportunities provided by technology (National Digital Inclusion Alliance, 2024; OECD, 2024).

Workforce development programs focused on digital skills training can help bridge the gap in access to digital jobs. By offering job training in digital fields such as coding, data analysis, and cybersecurity, these programs can provide individuals in minority communities with the skills they need to enter high-demand, high-paying sectors of the economy. Additionally, offering mentorship and internship opportunities can help individuals gain real-world experience and build professional networks, further enhancing their employment prospects (World Economic Forum, 2025).

Finally, policies that address the unique challenges faced by minority communities are essential for closing the digital divide. Policymakers should focus on creating laws and initiatives that provide equitable access to technology, including support for affordable internet services, digital devices, and training programs. It is also important to ensure that these policies are inclusive of all minority groups, including African American, Hispanic, and Indigenous populations, who are disproportionately affected by the digital divide. Only through targeted, systemic efforts can we begin to address the disparities that prevent these communities from fully participating in the digital economy and society (OECD, 2024; National Telecommunications and Information Administration, 2024a).

THE ROLE OF DIGITAL LITERACY IN INCLUSION

Another dimension of the digital divide lies in digital literacy, a critical factor for bridging the gap between access to technology and effective use. Simply having access to a device or internet connection is not enough if individuals lack the skills to use technology effectively. Digital literacy is a broad concept that includes basic tasks such as turning on a computer, navigating a smartphone, or sending an email. It also extends to more advanced skills such as using software programs, understanding how to search the internet efficiently, and knowing how to protect personal information online. Without these skills, individuals are unable to fully participate in the digital world, which has become central to modern life in areas such as education, work, healthcare, and social engagement (U.S. Department of Education, Office of Educational Technology, 2024; OECD, 2024).

For many people, particularly older adults, digital literacy remains an elusive skill set. Many older individuals may not have grown up with the internet or modern digital tools, and as such, they may feel intimidated or overwhelmed by the technology that has become integral to everyday life. For this group, the learning curve can be steep, as they may need to navigate unfamiliar devices and interfaces that change rapidly. Without the opportunity to acquire these skills, older adults can become excluded from digital spaces, making it harder for them to access critical services such as telehealth, online banking, or social connections through social media platforms. Moreover, they are at an increased risk of falling victim to online scams, misinformation, and digital exploitation, as they may not have the knowledge to recognize potential threats (National Digital Inclusion Alliance, 2024; OECD, 2024).

The digital literacy gap is not just limited to older adults but also affects individuals with lower levels of education and those living in low-income households. For individuals in these groups, the lack of digital literacy can have far-reaching consequences. Many low-income individuals do not have the opportunity to acquire digital skills at home, as they may not have access to the necessary technology or resources to engage in digital learning. Without formal education or access to training programs, these individuals can struggle to develop the skills necessary to navigate a world that increasingly requires digital proficiency. This lack of digital skills can further entrench poverty by limiting employment opportunities, access to educational resources, and the ability to manage personal affairs, such as paying bills or securing government benefits (Pew Research Center, 2024; National Digital Inclusion Alliance, 2024).

Education, in particular, is deeply impacted by gaps in digital literacy. As schools increasingly rely on technology for learning, students who lack the digital skills to engage with online resources are at a significant disadvantage. This challenge has been magnified by the COVID-19 pandemic, which forced many educational institutions to shift to remote learning. Students from lower-income households and rural areas were disproportionately affected by these changes, as they often lacked the technology and digital literacy to participate effectively in online classes. These gaps in digital literacy not only hinder academic success but also set students back in terms of developing skills needed in the modern workforce (U.S. Department of Education, Office of Educational Technology, 2024).

In the workforce, digital literacy is increasingly considered a prerequisite for many jobs. As more industries and sectors incorporate digital technologies into their operations, workers who lack digital skills are at risk of being left behind. Jobs that require familiarity with word processing software, spreadsheets, or customer relationship management tools are now common across various fields. Without the digital competence to perform such tasks, individuals may be excluded from job opportunities or face limited career advancement. This issue is particularly pronounced in low-income communities, where access to digital training may be scarce, and the resources for upskilling are limited. For individuals in these communities, the lack of digital literacy can perpetuate cycles of unemployment and economic hardship (World Economic Forum, 2025).

The gap in digital literacy also has significant implications for personal safety and security in the digital realm. As more personal and financial transactions move online, individuals who lack digital literacy are more vulnerable to cyber threats, including identity theft, online fraud, and phishing attacks. Without the knowledge to recognize suspicious emails or websites, individuals can easily fall victim to these dangers. In addition, many online platforms, such as social media and e-commerce websites, require users to understand complex privacy settings and terms of use. Individuals who lack digital literacy may unknowingly share personal information or become victims of online harassment, which can further erode their confidence in using digital tools (OECD, 2024).

While the digital literacy gap is evident across many demographic groups, it is particularly pronounced among racial and ethnic minorities. African American, Hispanic, and Indigenous populations often face a dual challenge of limited access to technology and a lack of digital skills. Historical and systemic inequalities have created barriers to education and economic advancement for these groups, which are compounded by the lack of access to digital literacy resources. For example, many minority communities live in areas with limited access to high-speed internet, and schools serving predominantly minority populations often lack the funding to provide adequate digital training for students. As a result, individuals in these communities may be more likely to lack the skills needed to participate in the digital economy, exacerbating existing inequalities (Pew Research Center, 2024; OECD, 2024).

The digital divide in digital literacy is also closely tied to the concept of digital exclusion, which refers to the marginalization of individuals or groups who are unable to fully participate in the digital world due to their lack of skills. Digital exclusion can result in the erosion of social capital, as individuals without digital skills may find it harder to stay connected with friends, family, and social networks. In an increasingly digital society, this exclusion can have profound psychological and emotional effects, as individuals feel left out or disconnected from the broader society. This isolation is particularly concerning for vulnerable groups, such as older adults, individuals with disabilities, and those living in poverty, who may already face social marginalization (OECD, 2024; National Digital Inclusion Alliance, 2024).

Bridging the digital literacy gap requires targeted, multi-pronged solutions. One important solution is to provide accessible and affordable digital literacy training programs, particularly in underserved communities. Libraries, community centers, and schools can play a vital role in offering digital literacy courses for people of all ages. These programs should focus on both basic skills, such as using a computer and navigating the internet, as well as more advanced topics such as online safety, digital communication, and using digital tools for work and education (National Digital Inclusion Alliance, 2024; U.S. Department of Education, Office of Educational Technology, 2024). By equipping individuals with these skills, society can help ensure that everyone has the opportunity to participate fully in the digital world.

Partnerships between public and private sectors are also essential in addressing the digital literacy gap. Internet service providers, technology companies, and nonprofits can collaborate to create initiatives that promote digital literacy in underserved communities. These initiatives can include offering free or discounted devices, providing internet access in low-income areas, and supporting digital literacy training programs. Through such partnerships, the resources and expertise of different sectors can be leveraged to ensure that digital literacy is promoted in a way that benefits everyone, particularly marginalized groups (National Digital Inclusion Alliance, 2024; OECD, 2024).

Another important strategy is to integrate digital literacy into the education system, starting from an early age. By embedding digital literacy into the K-12 curriculum, students can develop the skills they need to succeed in a digital world. Digital literacy should be taught not only as a technical skill but also as a tool for critical thinking, creativity, and responsible citizenship. Providing students with the opportunity to develop digital literacy skills from a young age can help reduce disparities later in life and create a more digitally inclusive society (U.S. Department of Education, Office of Educational Technology, 2024).

Incorporating digital literacy into adult education programs is also crucial, as many adults need to develop these skills to improve their job prospects and personal lives. Community-based adult education programs can offer digital literacy classes that are tailored to the specific needs of adult learners. These classes should be designed to accommodate different learning styles and should provide practical, hands-on training in using digital tools and resources. By offering digital literacy training to adults, we can help ensure that the workforce is equipped with the skills needed for success in a rapidly changing economy (OECD, 2024; World Economic Forum, 2025).

Finally, it is important to create a culture of lifelong learning around digital literacy. As technology continues to evolve, digital literacy must be treated as an ongoing skill that requires regular updating. Providing opportunities for individuals to continually refresh their digital skills through workshops, online courses, and other educational resources is essential for keeping up with new technologies and staying safe online. A culture of lifelong learning ensures that individuals of all ages can adapt to new digital tools and technologies, promoting a more inclusive digital society for everyone (U.S. Department of Education, Office of Educational Technology, 2024; OECD, 2024).

By focusing on digital literacy, we can help ensure that access to technology translates into meaningful opportunities for all individuals, regardless of their background. Building a digitally literate society is essential for closing the digital divide, empowering individuals to engage with technology in a responsible, effective, and informed way. It is a crucial step toward achieving digital equity and inclusion, where everyone, regardless of their socioeconomic status, age, or demographic background, can fully participate in the digital world and benefit from the opportunities it offers (OECD, 2024; International Telecommunication Union, 2024).

STRATEGIES FOR BRIDGING THE DIGITAL DIVIDE

Addressing the digital divide requires a multifaceted approach that focuses on both increasing access to technology and improving digital literacy. This divide, which has become even more apparent in recent years, poses significant barriers to social and economic participation, particularly for marginalized groups. The digital divide is not just about the availability of technology, but also about the skills required to use it effectively. Bridging this gap demands coordinated efforts from governments, businesses, educational institutions, and communities to ensure that all individuals, regardless of their background, can access and use technology to enhance their lives (OECD, 2024; ITU, 2024).

One of the most pressing issues in bridging the digital divide is expanding broadband infrastructure. Many rural and underserved urban areas continue to face significant challenges in accessing reliable internet services. This lack of access prevents individuals from fully participating in key areas such as education, healthcare, and employment, which increasingly rely on digital platforms. Governments must prioritize the expansion of broadband networks to these areas, ensuring that every citizen has access to affordable and reliable internet. This could include creating subsidies or grants for low-income families to support broadband access, as well as investing in publicly funded Wi-Fi hotspots in underserved communities to increase internet availability (NTIA, 2024).

The lack of affordable internet access is a significant barrier for many low-income households. Without the financial means to pay for broadband, these families are often excluded from the educational and economic opportunities that depend on digital connectivity. Public policy could help address this by introducing initiatives like low-cost internet service plans for qualifying households or offering tax incentives to private companies that expand their broadband networks into underserved areas. In addition, municipal governments can explore creative solutions, such as establishing free or low-cost internet programs in public spaces like libraries, community centers, and local parks, where people can access digital resources (Pew Research Center, 2024; OECD, 2024).

Furthermore, increasing access to technology means providing affordable devices such as laptops, tablets, and smartphones. While many technological devices have become more affordable over the years, the cost still remains prohibitive for some families, especially those living in poverty. Educational institutions, for instance, could work with tech companies to offer discounted or loaner devices to students who need them. This would ensure that all students, regardless of their economic status, have the tools they need to participate in remote learning, complete assignments, and access digital resources. Providing access to these devices should not be seen as a luxury, but as a necessity in the modern educational landscape (Pew Research Center, 2024).

In parallel with expanding access to technology, there is a pressing need to focus on improving digital literacy. Access to the internet and devices is only meaningful if individuals possess the skills to use them effectively. Digital literacy encompasses a range of skills, from basic functions such as using a computer or smartphone, to more advanced skills like navigating the internet, understanding online safety and privacy, and using productivity software. Without these skills, individuals are unable to take full advantage of the opportunities that technology offers. Therefore, efforts to close the digital divide must also focus on providing individuals with the tools they need to succeed in an increasingly digital world (NDIA, 2024; OECD, 2024).

Digital literacy programs should be introduced early in education to ensure that children are equipped with the necessary skills to thrive in a digital world. Schools have an essential role in embedding digital literacy into their curricula. By introducing technology-based learning at an early age, students will have the foundation they need to navigate a world where digital tools are integral to nearly every aspect of life. Additionally, schools can provide after-school programs or partnerships with local organizations to offer additional opportunities for students to develop their digital literacy skills, particularly in underserved communities where resources are scarce (Office of Educational Technology, 2024).

Educators also need to be equipped with the skills and knowledge to teach digital literacy effectively. Teacher training programs should include digital literacy as a core component, ensuring that educators can model and teach technology skills to their students. This includes not only the technical aspects of using technology but also the critical thinking skills necessary to evaluate digital information and navigate the complexities of online communication. By preparing teachers to incorporate digital literacy into their classrooms, we can help students build the competencies needed to succeed in the digital world, both academically and professionally (Office of Educational Technology, 2024).

Adult education is another crucial area in addressing the digital divide. Many adults, particularly those who have not had access to technology or digital training in their earlier years, may lack the skills necessary to engage with the modern workforce and society. Providing adults with opportunities to learn digital skills is essential, whether through community centers, libraries, or online courses. These programs can help adults develop basic skills such as using email, searching the internet, and understanding online security, as well as more advanced skills like using job search websites or managing personal finances online (NDIA, 2024).

Additionally, businesses and employers can play a significant role in bridging the digital divide by investing in the digital literacy of their employees. As digital tools become more prevalent in the workplace, workers must be able to use technology effectively to perform their jobs. Companies can offer digital training programs for employees to help them improve their skills and stay competitive in the job market. This is particularly important for low-income workers and those from marginalized communities, who may not have had access to the same technological resources or training opportunities as others. Employers who invest in their employees' digital literacy not only benefit from a more skilled workforce but also contribute to narrowing the digital divide (World Economic Forum, 2025).

Collaboration between governments, businesses, and non-profit organizations is essential to ensure that digital literacy programs are accessible and effective. Non-profit organizations can partner with schools and community centers to provide digital literacy training in underserved areas. These partnerships can help leverage resources, increase the reach of training programs, and ensure that they are tailored to the specific needs of the communities they serve. Moreover, public-private partnerships can help ensure that digital literacy programs are sustainable and that they reach the individuals who need them the most (NDIA, 2024).

The private sector also has a responsibility to support efforts to bridge the digital divide by making technology more affordable and accessible. Technology companies can contribute by providing discounted or subsidized devices for low-income households or offering free internet access to underserved communities. Additionally, tech companies can partner with educational institutions to develop digital literacy curricula and offer resources that support the development of digital skills. By making digital resources more accessible and providing the necessary training to use them, companies can help ensure that the benefits of technology are available to everyone (World Economic Forum, 2025).

At a broader level, digital literacy initiatives must address issues of equity and inclusion. While digital literacy is important for everyone, certain populations are disproportionately affected by the digital divide, including low-income individuals, people of color, and rural residents. To ensure that all individuals have equal access to digital opportunities, digital literacy programs must be designed to be culturally relevant and inclusive. This includes addressing language barriers, providing training for individuals with disabilities, and ensuring that digital content is accessible to all learners, regardless of their background or ability (OECD, 2024; ITU, 2024).

Finally, bridging the digital divide is not just about increasing access to technology or improving digital literacy; it also involves creating an environment where technology is used to enhance social and economic participation. Governments, businesses, and educational institutions must work together to foster an inclusive digital ecosystem where all individuals have the opportunity to use technology to improve their lives. This includes ensuring that online services, such as healthcare, education, and employment, are accessible to all and that individuals have the skills and resources to navigate them effectively (OECD, 2024).

By taking a comprehensive, coordinated approach to addressing the digital divide, we can create a more equitable and inclusive digital future. This requires collaboration across sectors, targeted investments in infrastructure and education, and a commitment to ensuring that everyone, regardless of their background, has the opportunity to thrive in a digital world. Only by working together can we ensure that technology becomes a tool for empowerment rather than exclusion, helping to close the gap between those who have access to its benefits and those who do not (OECD, 2024; ITU, 2024).

COMMUNITY-CENTERED DIGITAL INCLUSION INITIATIVES

Furthermore, digital inclusion efforts should be tailored to the specific needs of different communities to effectively address the challenges posed by the digital divide. A one-size-fits-all approach will not be sufficient to bridge the gap, as each community has its own unique set of barriers and needs. For instance, rural areas often face challenges related to infrastructure, including a lack of reliable internet access and limited availability of affordable digital devices. In these communities, the primary focus should be on providing affordable and accessible broadband services, as well as distributing devices that are suited to the needs of rural populations. This may require targeted investments in broadband expansion, such as building out infrastructure in remote areas and subsidizing internet access for families with limited resources (NTIA, 2024).

In addition to the challenge of physical infrastructure, rural areas may also have lower levels of digital literacy, as these communities often lack the resources to develop comprehensive technology training programs. Therefore, efforts to improve digital inclusion in these areas should also prioritize the development of digital literacy programs tailored to the unique needs of rural residents. These programs should teach individuals how to use the technology they now have access to, as well as build awareness of how digital skills can enhance their lives, from accessing healthcare services to pursuing educational opportunities and finding employment (NDIA, 2024).

For minority and low-income populations, digital inclusion initiatives must go beyond just access to technology. These communities often face a combination of barriers, including cultural and language challenges, that can hinder their ability to effectively engage with technology. Programs aimed at improving technology access and digital literacy in these communities should recognize and address these factors. This may involve offering digital literacy classes in multiple languages or incorporating cultural sensitivity into the design of technology programs. Involving community leaders and local organizations in the planning and delivery of digital inclusion programs can ensure that these initiatives are culturally relevant and resonate with the target audience (OECD, 2024).

Local partnerships are vital for ensuring the success of digital inclusion efforts in underserved communities. Nonprofit organizations, local government agencies, and educational institutions can work together to deliver targeted solutions that address both the technological and social barriers to digital participation. By engaging with these community partners, technology programs can be designed in ways that are more likely to be embraced by the local population. For example, community organizations can help recruit participants, build trust, and provide ongoing support to ensure that individuals remain engaged in digital literacy and technology access programs (NDIA, 2024).

Community engagement also plays a critical role in designing programs that are sustainable over the long term. Local leaders and residents should be actively involved in shaping digital inclusion initiatives to ensure that the programs are aligned with the specific needs and priorities of the community. This might involve conducting needs assessments or focus groups with community members to identify barriers and challenges to digital inclusion. It is also important that these programs be flexible, as the needs of different communities can change over time. By ensuring that the voices of local residents are heard and incorporated into program design, these initiatives are more likely to be successful and have lasting impact (OECD, 2024).

Another key consideration for digital inclusion in minority and low-income communities is ensuring that the technology provided is not only affordable but also practical and useful. It is important to understand the daily realities of community members and provide technology solutions that directly address their most pressing needs. For example, in low-income urban areas, access to mobile phones may be more feasible than providing computers or laptops, so mobile-friendly solutions should be prioritized. Programs should also focus on providing access to tools and platforms that are aligned with the needs of the community, whether for education, job searching, healthcare, or connecting with family and friends (Pew Research Center, 2024).

A comprehensive digital inclusion strategy should also integrate the development of digital skills in tandem with the provision of technology. Providing devices and internet access without supporting the development of digital literacy will not lead to lasting change. In many underserved communities, individuals may be hesitant to engage with new technology due to fear or lack of confidence in their abilities. Programs should address these concerns by offering structured, beginner-friendly digital literacy training that builds confidence and competency over time. These training sessions should include practical, hands-on learning that helps individuals build skills they can immediately apply in their daily lives (NDIA, 2024).

Additionally, a focus on long-term engagement is essential for ensuring that digital inclusion programs are not one-time solutions. Ongoing support and resources must be available to participants as they continue to develop their digital skills and adapt to new technologies. This might include establishing digital mentorship programs, creating community tech hubs where individuals can access support and resources, or offering refresher courses to help people stay up-to-date with emerging technologies. Sustaining engagement and providing continued learning opportunities ensures that individuals can maintain and build on the skills they have acquired, ultimately leading to greater social and economic inclusion (OECD, 2024).

Finally, it is crucial that digital inclusion efforts are designed with a focus on equity, ensuring that all members of society have equal access to the benefits of technology. This means actively addressing disparities in access to digital resources and ensuring that programs do not inadvertently perpetuate existing inequalities. For instance, outreach efforts should be mindful of the needs of vulnerable populations, such as elderly individuals, people with disabilities, and those who are geographically isolated. By intentionally focusing on equity and inclusivity, digital inclusion initiatives can ensure that everyone, regardless of their circumstances, has the opportunity to participate in the digital world (OECD, 2024; ITU, 2024).

WORKFORCE DEVELOPMENT AND ECONOMIC EMPOWERMENT

Workforce development programs are essential in addressing the digital divide, particularly as technology continues to shape the job market. The rise of automation, artificial intelligence, and digital platforms has created a demand for specialized skills, particularly in areas such as data analysis, software development, and digital marketing. However, individuals from disadvantaged communities, including low-income neighborhoods, rural areas, and minority groups, often face significant barriers in accessing the education and training required for these high-demand jobs. Without targeted workforce development programs, these populations may be excluded from the opportunities created by the digital economy, perpetuating cycles of poverty and limiting economic mobility (World Economic Forum, 2025).

To bridge this gap, workforce development initiatives must be expanded and designed with a focus on providing individuals from underrepresented communities with the technical skills and certifications needed for careers in technology. These programs should offer training in practical, marketable skills such as coding, cybersecurity, data management, and digital project management. Offering pathways to these types of certifications ensures that individuals gain not only the technical expertise required but also the credentials that employers recognize and value. By equipping individuals with these high-demand skills, workforce development programs can help level the playing field and provide new opportunities for people who might otherwise be left behind (World Economic Forum, 2025).

Additionally, workforce development programs should incorporate elements of career counseling, job placement assistance, and mentorship. This can ensure that participants not only acquire technical skills but also gain the confidence and network necessary to secure stable, well-paying jobs in the tech sector. For instance, partnering with local tech companies and organizations to create internship and apprenticeship opportunities can provide participants with hands-on experience and increase their chances of finding permanent employment. Furthermore, mentorship programs can connect trainees with industry professionals who can provide guidance, support, and career advice, helping to build long-term relationships that extend beyond the training period (World Economic Forum, 2025).

A key component of successful workforce development programs is making them accessible to individuals who face barriers to traditional education and training. This includes offering flexible training schedules, online courses, and programs that are designed to accommodate people with varying levels of prior education. Many individuals from disadvantaged backgrounds may already be working or managing family responsibilities, and therefore require training opportunities that allow them to learn at their own pace. Additionally, financial barriers should be addressed by providing scholarships, low-cost training options, or partnerships with employers who are willing to sponsor training programs for their future employees. These strategies can help make workforce development programs more inclusive and ensure that they reach the individuals who need them most (OECD, 2024).

Moreover, workforce development programs must not only focus on tech-specific skills but also consider the broader range of soft skills needed to thrive in the workplace. These include communication, problem-solving, and teamwork skills, which are essential for success in any profession. By integrating soft skills training into digital literacy and tech training, workforce development programs can provide a more holistic approach to preparing individuals for the workforce. This combination of technical and soft skills will ensure that participants are not only equipped to handle the technical aspects of their jobs but also prepared to succeed in the collaborative, fast-paced work environments common in the tech industry (OECD, 2024).

Ultimately, expanding and improving workforce development programs aimed at addressing the digital divide has the potential to transform the lives of individuals and communities. By providing individuals from disadvantaged backgrounds with the skills, support, and opportunities needed to succeed in the digital economy, these programs can open doors to upward mobility, economic empowerment, and greater social inclusion. In this way, workforce development plays a critical role in bridging the digital divide, ensuring that all individuals, regardless of their background, have an equal opportunity to participate in the modern workforce and share in the benefits of technological progress (World Economic Forum, 2025).

THE DIGITAL DIVIDE IN EDUCATION: ADDRESSING THE HOMEWORK GAP

The impact of the digital divide in education became particularly evident during the COVID-19 pandemic, as millions of students were forced to transition to remote learning. For many, this shift highlighted the stark disparities in access to technology and the internet. Students in low-income households, rural areas, or communities with limited infrastructure were disproportionately affected, struggling to participate in online learning due to the lack of reliable internet access and devices. The pandemic exacerbated the existing "homework gap," where students in underprivileged areas had little to no access to the technology needed to complete assignments, attend virtual classes, or engage in educational activities. As a result, these students faced significant learning setbacks, widening the achievement gap between them and their more affluent peers (Office of Educational Technology, 2024; Pew Research Center, 2024).

Addressing the digital divide in education requires a multi-pronged approach that prioritizes increasing access to the necessary digital tools for all students. This includes providing affordable or free devices, such as laptops and tablets, and expanding internet access to underserved areas, whether through broadband expansion, affordable connectivity plans, or community Wi-Fi initiatives. Schools, governments, and local communities must work together to ensure that no student is left behind due to a lack of technology. This could involve creating partnerships with internet service providers, offering subsidies for families, or leveraging federal and state funding to support initiatives that provide equitable access to digital resources. By ensuring that every student has the tools they need, we can mitigate the educational disparities amplified by the digital divide (Office of Educational Technology, 2024).

In addition to access to devices and internet, digital literacy must also be a key focus in addressing the educational divide. For students to benefit from the technology available to them, they must possess the skills to use it effectively. This includes basic competencies such as navigating the internet, utilizing software tools, and understanding online safety, as well as more advanced skills required for learning in a digital environment. Schools should integrate digital literacy into their curricula, starting from early education and continuing throughout students' academic careers. Providing students with the skills to navigate and leverage technology will not only empower them to succeed in online learning but also prepare them for future careers in an increasingly digital world. Therefore, a comprehensive approach that combines access to devices, reliable internet, and digital literacy education is essential to bridging the educational digital divide (Office of Educational Technology, 2024).

THE DIGITAL DIVIDE IN HEALTHCARE

The digital divide in healthcare is a growing concern, particularly as telemedicine has become an essential means of delivering healthcare services, especially in remote or underserved areas. Telemedicine allows patients to consult with healthcare providers via video calls, messaging, or other online platforms, reducing the need for in-person visits. This has proven to be a lifeline during times of crisis, such as the COVID-19 pandemic, when traditional healthcare settings were overwhelmed or inaccessible. However, individuals without access to reliable internet, digital devices, or the necessary technical skills are at risk of being excluded from these vital services. For those in rural areas or lower-income communities, the absence of broadband infrastructure or affordable devices can create barriers to accessing telehealth consultations, prescription renewals, and follow-up care. As a result, patients in these areas may face health disparities, with fewer opportunities to manage chronic conditions or receive timely medical advice (CDC/NCHS, 2024).

Bridging the digital divide in healthcare requires a comprehensive strategy that focuses on both access and usability. Expanding broadband infrastructure in rural and underserved urban areas is essential to ensure that more people can connect to telemedicine services. Additionally, healthcare providers need to ensure that their digital health platforms are accessible to all patients, regardless of their technological expertise. This means creating user-friendly interfaces, providing multilingual options, and developing tools that accommodate people with disabilities. Healthcare organizations should prioritize accessibility and ensure that their digital systems are designed to serve a wide range of users, including the elderly, people with limited digital literacy, and those with physical impairments. Furthermore, there should be ongoing efforts to make telehealth services affordable, with options for those who cannot pay for internet access or devices, to ensure that the digital divide does not prevent individuals from receiving necessary care (OECD, 2024; CDC/NCHS, 2024).

In addition to expanding access to technology, training is essential for empowering patients to use telemedicine effectively. Healthcare providers should offer guidance on how to navigate digital health tools, including tutorials, help desks, and patient outreach programs that teach individuals how to schedule virtual appointments, use telemedicine platforms, and understand the privacy and security features of online healthcare services. For older adults or those unfamiliar with digital technology, offering in-person or over-the-phone support can make a significant difference in their ability to participate in telemedicine. By making telehealth services accessible, user-friendly, and supported by educational initiatives, healthcare systems can ensure that all individuals, regardless of their socioeconomic status or digital skills, can access the healthcare services they need. Bridging the digital divide in healthcare is critical not only to improve patient outcomes but also to reduce health disparities that are exacerbated by unequal access to technology (CDC/NCHS, 2024).

THE ROLE OF BUSINESS IN PROMOTING DIGITAL INCLUSION

Businesses, particularly those in the tech industry, play a pivotal role in fostering digital inclusion by ensuring that technology is accessible to all individuals, regardless of their socioeconomic status. Many tech companies have the resources and expertise to make a significant impact in bridging the digital divide. By partnering with local communities and educational institutions, businesses can provide essential resources such as devices, internet access, and digital training to underserved populations. For example, major tech corporations can work with schools in low-income neighborhoods to provide students with laptops, tablets, or other devices, ensuring that they have the tools necessary for learning in today's digital world. These partnerships can also extend to offering affordable or subsidized broadband access, which would eliminate one of the key barriers to digital participation (World Economic Forum, 2025; OECD, 2024).

In addition to providing devices and internet access, businesses can play a significant role in creating training programs that equip individuals from marginalized communities with the digital skills needed for the modern workforce. Many technology companies already have the infrastructure to develop and offer online courses, certifications, and boot camps, which can help bridge the skills gap in the tech industry. By investing in these training initiatives, businesses ensure that individuals from underrepresented groups have the opportunity to gain technical expertise, such as software development, data analytics, or cybersecurity. This not only opens up opportunities for career advancement in the tech sector but also supports a more inclusive and diverse digital economy. As companies offer training and upskilling opportunities, they help create pathways for upward mobility and economic empowerment, particularly for those in historically disadvantaged communities (World Economic Forum, 2025).

Furthermore, businesses that actively address the digital divide contribute to building a more equitable and diverse workforce, which can have a ripple effect on innovation and productivity across industries. A workforce that includes individuals from different backgrounds, experiences, and perspectives is better positioned to solve complex problems and meet the needs of a global, interconnected market. By ensuring that everyone has access to the digital tools and skills needed to succeed, businesses are not only helping close gaps in access but also creating opportunities for a more diverse talent pool to thrive in the digital economy. This not only benefits businesses by expanding the available talent pool but also contributes to the broader goal of building a more inclusive society where everyone has the chance to participate in the opportunities that the digital age provides (World Economic Forum, 2025).

A COMMITMENT TO EQUITY IN TECHNOLOGY ACCESS

Finally, bridging the digital divide requires a fundamental commitment to equity, which involves ensuring that technology is not viewed as a luxury or privilege reserved for those who can afford it, but as a universal right accessible to all. This commitment to equity begins with recognizing that disparities in access to technology are not merely a result of economic or social status but are deeply intertwined with systemic issues like race, income inequality, and geographic location. Governments, organizations, and individuals must come together to create a collective approach that prioritizes making technology not just accessible, but also affordable and useful for everyone. This means that policy makers need to design initiatives that address both the availability of resources (such as devices and internet access) and the means to use them effectively, while also considering factors like education, language, and cultural context to ensure that technology is genuinely inclusive (OECD, 2024; ITU, 2024).

Governments must play a central role in this process by developing public policies that focus on expanding access to broadband internet, particularly in underserved rural and urban areas. Public investment in digital infrastructure should aim to ensure that every individual, regardless of their income or geographic location, has access to reliable and affordable internet. Additionally, public policy should emphasize support for community-based initiatives that empower local organizations to provide tailored digital literacy training and resources. By embedding equity in the fabric of technological policy, governments can help level the playing field, offering everyone an opportunity to participate fully in the digital world, from education to employment and beyond. Policies should also aim to address systemic inequalities that hinder access to technology for marginalized communities, whether those inequalities are based on race, class, or other factors (NTIA, 2024).

In addition to government involvement, businesses and community organizations must take responsibility for promoting digital equity. Companies, particularly those in the tech industry, should be held accountable for ensuring that their products and services are accessible to a broad range of individuals. This can include designing affordable devices, offering low-cost internet plans, and collaborating with non-profit organizations to provide technology access in underserved communities. Community organizations, on the other hand, can offer grassroots solutions that focus on connecting people to technology and providing the training they need to use it. By working together, these different stakeholders can create a more equitable digital ecosystem, one where access to technology is seen not just as a tool for economic advancement but as a right that enables individuals to thrive in an increasingly digital world (OECD, 2024).

CONCLUSION: BUILDING AN INCLUSIVE DIGITAL FUTURE

In conclusion, bridging the digital divide is not merely about providing access to technology but about fostering a more equitable society where all individuals have the opportunity to fully engage with and benefit from the digital world. It is essential to recognize that technology is not just a tool but a gateway to education, healthcare, economic opportunities, and civic participation. Addressing the digital divide requires a collaborative and multifaceted approach that combines efforts from governments, businesses, educational institutions, and community organizations. Expanding broadband infrastructure, particularly in underserved areas, is a critical step toward ensuring that all individuals, regardless of their geographic location or socioeconomic status, can access the resources they need. However, access alone is insufficient. Improving digital literacy, integrating technology training programs, and providing targeted workforce development opportunities are equally important to ensure that people can effectively navigate the digital landscape and participate in the growing digital economy (NTIA, 2024; OECD, 2024).

Moreover, ensuring that technology is accessible, affordable, and useful for everyone is key to closing the digital divide. This requires intentional efforts to remove barriers to access, such as cost and lack of digital skills, and to address disparities that disproportionately affect marginalized communities. Through policies that prioritize digital inclusion, investments in training programs, and partnerships between businesses, governments, and communities, we can create an ecosystem where technology becomes a tool for empowerment rather than a source of division. Only by addressing these disparities holistically and inclusively can we build a more equitable digital future, one that not only supports individual success but also strengthens the collective well-being of society as a whole (OECD, 2024; ITU, 2024; World Economic Forum, 2025).

TEN ESSAY QUESTIONS FOR CHAPTER 8

Bridging the Digital Divide—Equity and Inclusion in Technology Access

1. What is the digital divide, and how does it manifest in terms of access to technology, broadband internet, and digital literacy? Discuss the key factors that contribute to this divide.
2. Explain the impact of the digital divide on education. How did the COVID-19 pandemic highlight existing disparities, and what steps should be taken to address the homework gap?
3. Discuss the relationship between digital literacy and the digital divide. Why is it important to not only provide access to technology but also ensure that individuals possess the necessary skills to use it effectively?
4. In what ways do demographic factors such as income, geography, and race exacerbate the digital divide? How can policies and programs specifically target these disparities to promote digital inclusion?
5. How can governments and public policy play a role in bridging the digital divide? Discuss potential strategies, such as expanding broadband infrastructure and offering subsidies, to increase access to technology for underserved communities.
6. What is the role of businesses in fostering digital inclusion, and how can tech companies contribute to bridging the digital divide through partnerships with schools, nonprofits, and communities?
7. Discuss how digital inclusion efforts can be tailored to the unique needs of different communities, such as rural areas, minority populations, and low-income groups. How can cultural and language barriers be addressed in these programs?
8. How do workforce development programs contribute to closing the digital divide? What skills and certifications should be prioritized in training programs to prepare individuals from disadvantaged communities for success in the tech industry?
9. What are the challenges of ensuring digital equity in healthcare, and how can telemedicine help reduce disparities? What role can digital literacy play in making healthcare services more accessible to underserved populations?
10. In your opinion, why is it crucial to view technology access as a right rather than a privilege? How can equity in technology access be prioritized in public policy, business strategies, and community development to ensure that technology benefits all members of society?

REFERENCES

Centers for Disease Control and Prevention, National Center for Health Statistics. (2024). *Use of telemedicine among adults: United States, 2021–2023* (NCHS Data Brief No. 518). U.S. Department of Health and Human Services.

Federal Communications Commission. (2024). *Second report and order and second further notice of proposed rulemaking* (FCC 24-76). Federal Communications Commission.

International Telecommunication Union. (2024). *Measuring digital development: Facts and figures 2024*. International Telecommunication Union.

National Digital Inclusion Alliance. (2024). *Digital navigator services*. National Digital Inclusion Alliance.

National Digital Inclusion Alliance. (2024). *The Digital Navigator model: A decision-maker's toolkit*. National Digital Inclusion Alliance.

National Telecommunications and Information Administration. (2024a, November 19). *Biden-Harris Administration approves Texas' "Internet for All" initial proposal: All 56 states and territories now approved*. U.S. Department of Commerce.

National Telecommunications and Information Administration. (2024b). *Broadband Equity, Access, and Deployment (BEAD) Program: Fact sheet*. U.S. Department of Commerce.

National Telecommunications and Information Administration. (2024c). *Digital Equity Capacity Grant Program: Notice of Funding Opportunity*. U.S. Department of Commerce.

Organisation for Economic Co-operation and Development. (2024). *OECD digital economy outlook 2024*. OECD Publishing.

Pew Research Center. (2024, November 13). *Americans' use of mobile technology and home broadband*. Pew Research Center.

Pew Research Center. (2024). *Teens, social media and technology 2024*. Pew Research Center.

U.S. Department of Agriculture, Rural Utilities Service. (2024). *ReConnect Program overview*. U.S. Department of Agriculture.

U.S. Department of Education, Office of Educational Technology. (2024). *National Educational Technology Plan: A call to action for digital equity*. U.S. Department of Education.

World Economic Forum. (2025). *The Future of Jobs Report 2025*. World Economic Forum.

Sustainability and Technology: Innovation for the Planet

9

ABSTRACT

Chapter 9, *Sustainability and Technology: Innovation for the Planet*, explores the critical role of technology in addressing environmental challenges and advancing sustainability. It examines technological solutions such as renewable energy, smart agriculture, and digital tools for environmental monitoring, highlighting their transformative potential in combating climate change, resource depletion, and pollution. The chapter also delves into industry applications, consumer-focused innovations, and the integration of green technologies into daily life. While emphasizing success stories and emerging trends, it addresses barriers such as cost, policy gaps, and ethical considerations, advocating for a collaborative, equitable approach to sustainable innovation. This chapter provides a comprehensive analysis of how technology can drive global efforts toward a greener, more sustainable future.

INTRODUCTION

In recent years, the intersection of technology and sustainability has become a pivotal area of innovation, aiming to address pressing environmental challenges while creating a sustainable future (IPCC, 2023; UNEP, 2025). Rapid technological advancements have brought about transformative changes in how societies interact with their environments, offering tools and systems that can mitigate the impacts of climate change, reduce waste, and optimize resource use (IEA, 2025a; IRENA,

2025). This convergence of technology and sustainability represents not only a practical approach to environmental stewardship but also a necessary evolution to ensure the long-term viability of natural ecosystems and human societies (UN-Habitat, 2024).

The urgency of addressing environmental challenges has never been more apparent. Rising global temperatures, shrinking biodiversity, and escalating resource depletion demand immediate and effective action (IPCC, 2023; UNEP, 2025). Technology, with its capacity for large-scale implementation and continual innovation, is uniquely positioned to respond to these issues. By applying advanced technologies in sectors such as energy, agriculture, transportation, and waste management, humanity has the opportunity to reimagine its relationship with the planet and create systems that prioritize sustainability over exploitation (IEA, 2025a; FAO, 2024).

At the heart of this movement lies a dual purpose: to combat the detrimental effects of environmental degradation while simultaneously building systems that are resilient and adaptive. For example, renewable energy technologies like solar and wind power are not only reducing carbon emissions but are also decentralizing energy production, empowering communities to be less dependent on traditional power grids (IEA, 2025a; IRENA, 2025). These shifts reflect a broader trend in technological innovation aimed at decentralization, democratization, and greater efficiency—all essential components of a sustainable future (OECD, 2024).

The concept of sustainability itself has expanded with the advent of new technologies. No longer confined to conservation efforts, sustainability now encompasses systemic transformations in production, consumption, and governance. Technologies such as artificial intelligence (AI), blockchain, and the Internet of Things (IoT) are enabling unprecedented levels of data collection, analysis, and transparency, which in turn are fostering accountability and efficiency across industries (OECD, 2024; UN-Habitat, 2024). For instance, AI-powered systems can optimize energy consumption in buildings, while blockchain technology ensures the traceability of goods, reducing the environmental footprint of supply chains (IEA, 2025a).

While technological advancements have immense potential, they are not without limitations. The production and disposal of electronic devices contribute to e-waste, and the mining of rare earth elements for advanced technologies often comes with significant environmental and ethical costs (ITU & UNITAR, 2024; Circle Economy, 2025). Furthermore, the rapid pace of technological innovation can outstrip regulatory frameworks, creating gaps in oversight that may lead to unintended consequences. These realities underscore the importance of approaching technological solutions with a critical eye, ensuring that innovations genuinely align with the principles of sustainability rather than inadvertently exacerbating the problems they aim to solve (UNEP, 2025).

Another critical consideration is equity. Technological solutions must be accessible and inclusive, addressing the needs of both developed and developing regions. While wealthy nations can readily adopt advanced technologies, many communities in low-income countries lack the infrastructure and resources to implement such systems. Bridging this gap requires not only technological innovation but also international collaboration and investment to ensure that the benefits of sustainable technologies are distributed equitably (UNESCO & UN-Water, 2024; UN-Habitat, 2024).

Despite these challenges, the examples of successful integration of technology and sustainability are numerous and inspiring. From smart cities that optimize resource use to precision agriculture that reduces water and pesticide use, innovations are already reshaping industries and redefining what is possible (FAO, 2024; UN-Habitat, 2024). These success stories provide a roadmap for scaling sustainable technologies and adapting them to diverse contexts, proving that with the right strategies, technology can be a powerful ally in addressing global environmental challenges (IEA, 2025a; IRENA, 2025).

The transformation of industries through technological advancements is particularly evident in the energy sector. Renewable energy sources, supported by smart grid systems and efficient energy storage solutions, are disrupting traditional energy models (IEA, 2025a; IRENA, 2025). This shift not only reduces dependence on fossil fuels but also creates opportunities for local energy production, fostering greater resilience against environmental and economic uncertainties. Similarly, advancements in battery technologies are extending the viability of electric vehicles and renewable energy systems, addressing one of the critical barriers to widespread adoption (IEA, 2025b).

In agriculture, technology is enabling more sustainable practices that enhance productivity while reducing environmental impact. Precision farming, powered by sensors, drones, and data analytics, allows farmers to optimize resource use, minimize waste, and improve yields (FAO, 2024). Vertical farming and aquaponics are redefining traditional agricultural methods, making it possible to grow food in urban environments with minimal water and land use (FAO, 2024). These innovations are not only addressing food security challenges but also reducing the carbon footprint of agriculture, one of the largest contributors to global emissions (IPCC, 2023).

The role of technology in waste management is equally transformative. Advanced recycling systems, waste-to-energy technologies, and AI-driven sorting mechanisms are making it easier to divert waste from landfills and recover valuable materials (Circle Economy, 2025). Circular economy principles, supported by technological innovation, are encouraging businesses and consumers to rethink consumption patterns and prioritize reuse and recycling (OECD, 2024). These efforts are critical in addressing the growing problem of waste and resource depletion, creating pathways toward a more sustainable economy (UNEP, 2025).

Transportation is another area where technology is driving significant progress. Electric and hydrogen-powered vehicles, alongside innovations in public transit systems, are reducing emissions and improving urban air quality (IEA, 2025b; IEA, 2024). Autonomous vehicles and shared mobility platforms promise to further revolutionize transportation, making it more efficient and accessible while minimizing its environmental impact (UN-Habitat, 2024). The integration of these technologies with smart city infrastructure is creating a new paradigm for urban mobility that prioritizes sustainability and efficiency (UN-Habitat, 2024).

Water conservation and management, a pressing issue in many parts of the world, is also benefiting from technological advancements. Smart water systems, which monitor and manage water use in real-time, are reducing wastage and improving distribution efficiency (UNESCO & UN-Water, 2024). Desalination technologies, powered by renewable energy, are making it possible to provide clean water in regions facing severe water scarcity (IRENA, 2025). These innovations are addressing one of humanity's most fundamental needs while promoting sustainable water management practices (UNESCO & UN-Water, 2024).

The potential of technology to transform ecosystems extends beyond practical applications. Technological tools are also advancing our understanding of environmental systems, enabling more effective conservation efforts. Satellite imagery, remote sensing, and AI-driven models are providing insights into deforestation, wildlife migration patterns, and climate change impacts (IPCC, 2023). These technologies are equipping researchers and policymakers with the information needed to make informed decisions and implement targeted interventions (UNEP, 2025).

The interplay between technology and policy is another critical aspect of sustainability. Technological innovations must be supported by robust policies and regulations that ensure their responsible use and maximize their benefits (OECD, 2024; UNEP, 2025). Policymakers play a crucial role in creating an environment where sustainable technologies can thrive, providing incentives for innovation and enforcing standards that align with sustainability goals (IEA, 2025a).

As this chapter will illustrate, the integration of technology and sustainability is not without its complexities. Balancing innovation with ethical and environmental considerations requires a nuanced approach that recognizes the interconnectedness of systems and the need for holistic solutions (UNEP, 2025; IPCC, 2023). By examining the successes and shortcomings of current technological efforts, this analysis aims to provide a comprehensive understanding of how technology can contribute to a sustainable future.

Ultimately, the intersection of technology and sustainability represents a powerful opportunity to address some of the most pressing challenges of our time. Through critical analysis and a focus on innovation, this chapter highlights the transformative potential of technology while emphasizing the importance of mindful implementation (IEA, 2025a; IRENA, 2025). By leveraging the best of human ingenuity, it is possible to create systems that not only mitigate environmental harm but also foster resilience and equity, ensuring a sustainable future for generations to come (UN-Habitat, 2024; UNESCO & UN-Water, 2024).

Innovation for the planet: renewable energy, smart systems, and circular design for a sustainable future

Read through a sustainability lens, the image frames "innovation for the planet" as a systems challenge: shifting to renewables, electrifying demand, and orchestrating smart grids can cut emissions, but only if we pair them with storage, demand response, resilient infrastructure, and circular design that reduces material footprints across each product's life cycle. It highlights equity, too—who bears extraction, waste, and climate risk versus who reaps clean-tech benefits—so the practical prompt beneath the image is to measure success not just in megawatts added or tons avoided, but in durable, just outcomes: cleaner air and water, reliable energy access, good jobs, and less waste.

THE NEXUS OF TECHNOLOGY AND SUSTAINABILITY

The chapter begins by introducing the intricate relationship between technology and sustainability, emphasizing the pressing need to address climate change, resource scarcity, and environmental degradation (IPCC, 2023; UNEP, 2025). This interplay highlights the dual role of technology: as both a contributor to these challenges and a critical part of the solution. Historically, technological advancements have often prioritized economic growth over environmental considerations, leading to unintended consequences such as pollution, deforestation, and biodiversity loss. However, the current era presents a shift in focus, where innovation increasingly aims to balance human progress with planetary well-being (UN-Habitat, 2024).

One of the central themes explored is how renewable energy technologies are transforming the global energy landscape. Solar panels, wind turbines, and hydropower systems have become emblematic of efforts to reduce dependency on fossil fuels, thereby mitigating greenhouse gas emissions (IEA, 2025a; IRENA, 2025). These technologies not only address climate change but also promote energy independence and resilience. Innovations like floating solar farms and high-capacity battery storage systems demonstrate the potential for renewable energy to meet the growing energy demands of modern societies sustainably (IEA, 2025a).

In addition to energy, waste management has emerged as a critical area where technology plays a transformative role. Advanced recycling systems and waste-to-energy plants are helping to minimize landfill use while recovering valuable materials (Circle Economy, 2025). AI-powered sorting mechanisms are improving the efficiency of recycling processes, ensuring that fewer materials are wasted. These innovations align with the principles of a circular economy, where the lifecycle of products is extended, and resource extraction is minimized, fostering a more sustainable economic model (OECD, 2024).

Environmental monitoring tools have revolutionized our ability to understand and respond to ecological challenges. Technologies such as remote sensing, satellite imagery, and IoT-enabled sensors provide real-time data on deforestation, water quality, and atmospheric conditions (IPCC, 2023). These tools empower policymakers, researchers, and conservationists to make informed decisions, enabling targeted interventions that mitigate environmental harm. For instance, drones equipped with environmental sensors are now being used to monitor endangered species and combat illegal logging in remote regions (UNEP, 2025).

While technology offers immense potential for sustainable development, its role as a contributor to environmental challenges cannot be overlooked. The production and disposal of electronic devices contribute to e-waste, one of the fastest-growing waste streams globally (ITU & UNITAR, 2024). Additionally, the extraction of rare earth minerals required for many advanced technologies often results in significant environmental degradation and social injustices in mining regions.

Recognizing and addressing these trade-offs is essential to ensuring that technology serves as a tool for sustainability rather than exacerbating existing problems (UNEP, 2025).

The chapter also examines the role of artificial intelligence and data analytics in optimizing resource use. AI-driven systems are being deployed to enhance energy efficiency in buildings, predict crop yields in agriculture, and manage water distribution networks (FAO, 2024; IEA, 2025a). These technologies not only reduce waste but also improve the resilience of systems to environmental changes. For example, AI-powered platforms can forecast energy demand and adjust supply in real-time, ensuring that renewable resources are utilized effectively (IEA, 2025a).

Water scarcity, a growing concern in many parts of the world, has also been addressed through technological innovation. Smart water management systems use IoT devices to monitor usage and detect leaks, helping to conserve this vital resource (UNESCO & UN-Water, 2024). Desalination technologies, powered by renewable energy, are making it possible to turn seawater into potable water in arid regions (IRENA, 2025). These advancements illustrate how technology can provide practical solutions to critical sustainability challenges (UNESCO & UN-Water, 2024).

The chapter underscores the importance of integrating sustainability into urban planning through the concept of smart cities. By leveraging technologies such as IoT, AI, and renewable energy systems, smart cities aim to optimize energy use, reduce waste, and enhance the quality of life for residents (UN-Habitat, 2024). Urban areas, as centers of population and economic activity, are both significant contributors to and victims of environmental challenges. Smart city initiatives demonstrate how technology can create more sustainable and livable urban environments (UN-Habitat, 2024).

Transportation is another sector undergoing significant transformation due to technological advancements. Electric vehicles, hydrogen-powered buses, and autonomous public transit systems are reducing emissions and improving urban air quality (IEA, 2025b; IEA, 2024). Shared mobility platforms, supported by AI and real-time data, are changing how people move within cities, promoting more sustainable transportation habits (UN-Habitat, 2024). The integration of these technologies with broader urban infrastructure amplifies their impact, creating systems that are both efficient and environmentally friendly (UN-Habitat, 2024).

In agriculture, precision farming technologies are enabling farmers to optimize resource use and reduce environmental impact. Sensors, drones, and data analytics provide insights into soil health, weather patterns, and crop conditions, allowing for more targeted and sustainable farming practices (FAO, 2024). Innovations such as vertical farming and aquaponics are also addressing food security challenges while minimizing water and land use. These approaches illustrate how technology can reconcile the need for increased agricultural productivity with the imperative of environmental conservation (FAO, 2024; IPCC, 2023).

The chapter highlights the need for equitable access to sustainable technologies, particularly in developing regions. While advanced economies can readily adopt cutting-edge solutions, many communities lack the infrastructure or resources to implement them. Addressing this disparity requires international cooperation and investment to ensure that the benefits of sustainable technologies are distributed fairly (UN-Habitat, 2024; UNESCO & UN-Water, 2024). Collaborative initiatives can help bridge this gap, fostering global progress toward sustainability.

Ethical considerations are integral to the discussion of technology and sustainability. The development and deployment of technologies must align with ethical principles that prioritize the well-being of both people and the planet. This includes addressing issues such as data privacy, the environmental impact of manufacturing, and the social implications of automation (OECD, 2024; UNEP, 2025). By embedding ethical considerations into the design and application of technologies, it is possible to create solutions that are both effective and responsible.

The interplay between policy and technology is also explored, emphasizing the need for robust regulatory frameworks to support sustainable innovation. Policies that incentivize renewable energy adoption, promote research and development, and enforce environmental standards are crucial to maximizing the benefits of technological advancements (IEA, 2025a; UNEP, 2025). Governments, businesses, and civil society must work together to create an enabling environment where sustainable technologies can thrive.

Ultimately, the chapter illustrates that while technology alone cannot solve all environmental challenges, it is a powerful enabler of sustainable development when applied thoughtfully and equitably. By leveraging innovation to address climate change, resource scarcity, and environmental degradation, humanity can transition toward systems that prioritize sustainability (IPCC, 2023; UNEP, 2025). However, this requires a balanced approach that considers the trade-offs, addresses ethical concerns, and ensures inclusivity in access to technology.

In conclusion, the relationship between technology and sustainability is both complex and dynamic, characterized by opportunities and challenges. By critically examining the role of technological innovation in fostering sustainability, this chapter aims to provide insights into how industries, communities, and ecosystems can benefit from thoughtful integration of technology (UN-Habitat, 2024). As the world continues to face environmental challenges, the intersection of technology and sustainability remains a crucial frontier for creating a resilient and equitable future (IEA, 2025a; UNESCO & UN-Water, 2024).

TECHNOLOGICAL SOLUTIONS TO ENVIRONMENTAL CHALLENGES

This section explores groundbreaking technological innovations developed to tackle critical environmental challenges. It provides a focused analysis of advancements in renewable energy, carbon capture and storage (CCS), smart agriculture, and water management. Each innovation is examined through the lenses of environmental impact, scalability, and long-term viability, revealing their potential to transform industries and communities while addressing pressing global sustainability needs (UNEP, 2025).

Renewable energy is at the forefront of efforts to reduce dependency on fossil fuels, which are a major source of greenhouse gas emissions. Advances in solar power, including the development of high-efficiency photovoltaic cells and concentrated solar power systems, are making it increasingly feasible to harness the sun's energy at a larger scale (IEA, 2025a; IRENA, 2025). Similarly, offshore wind farms and innovations in wind turbine design are unlocking the potential of wind energy, even in regions with low wind speeds. Geothermal energy, often overlooked, offers a reliable and continuous energy source, with new drilling technologies expanding its accessibility to previously untapped areas (IEA, 2025a).

These renewable energy systems collectively contribute to reducing carbon emissions and mitigating climate change. However, challenges remain in their scalability, such as the need for improved energy storage solutions and grid infrastructure to handle intermittent supply. Emerging technologies like solid-state batteries and AI-driven grid management systems are addressing these issues, paving the way for a more resilient and sustainable energy future (IEA, 2025a).

Carbon capture and storage (CCS) represents another critical area of innovation aimed at mitigating climate change. CCS technologies capture carbon dioxide emissions from industrial processes and power plants, storing them underground or converting them into useful products (Global CCS Institute, 2024). Enhanced methods, such as direct air capture (DAC), are being developed to remove CO_2 directly from the atmosphere, expanding the scope of CCS applications. These technologies are particularly vital for industries like cement and steel production, where emissions are difficult to eliminate entirely (IPCC, 2023).

Despite its promise, CCS faces challenges related to cost, energy requirements, and public acceptance. The scalability of CCS technologies depends on continued investment and policy support, as well as advances in storage solutions that ensure long-term safety and efficiency (Global CCS Institute, 2024; UNEP, 2025). Nonetheless, CCS remains a crucial tool in the global effort to achieve net-zero emissions.

Smart agriculture technologies are transforming the way food is produced, with significant implications for resource conservation and food security. Precision farming, for example, uses GPS technology, drones, and IoT-enabled sensors to monitor soil health, weather conditions, and crop performance in real time (FAO, 2024). This data-driven approach allows farmers to optimize water, fertilizer, and pesticide use, minimizing waste and environmental impact.

IoT-enabled systems also enable better management of agricultural supply chains, reducing food loss and improving efficiency. Vertical farming and aquaponics are emerging as innovative solutions to urban food production, requiring less water and land while producing higher yields (FAO, 2024). These technologies address the dual challenge of feeding a growing population while preserving natural ecosystems.

However, the adoption of smart agriculture technologies varies widely between regions, with barriers such as high initial costs and lack of technical expertise. Ensuring equitable access to these tools is essential for their global scalability and impact. Governments and organizations must work collaboratively to provide resources, training, and infrastructure support to smallholder farmers and developing regions (FAO, 2024; UN-Habitat, 2024).

Water management innovations are crucial for addressing global water scarcity, a challenge exacerbated by population growth and climate change. Desalination technologies, powered increasingly by renewable energy, are enabling the conversion of seawater into potable water, providing a lifeline for arid regions (IRENA, 2025). Advanced water recycling systems are allowing cities to reclaim and reuse wastewater, reducing the strain on freshwater resources (UNESCO & UN-Water, 2024).

Smart water grids, which use IoT sensors to monitor and manage water distribution, are helping to detect leaks, prevent water loss, and ensure efficient usage (UNESCO & UN-Water, 2024). These technologies not only conserve water but also enhance the resilience of water infrastructure to environmental changes and extreme weather events. For example, AI-powered systems can predict water demand and optimize distribution during droughts or floods.

As with other innovations, scalability and cost are significant considerations for water management technologies. Desalination plants, for instance, require substantial energy inputs and can generate brine waste that must be managed sustainably. Continued research and development are necessary to improve efficiency and minimize environmental trade-offs, ensuring that these solutions are viable in the long term (IRENA, 2025; UNEP, 2025).

Each of these technological advancements demonstrates the potential for innovation to address specific environmental challenges. However, their success depends on factors such as policy support, investment, and the integration of ethical considerations into their design and deployment (OECD, 2024; UNEP, 2025). Ensuring that these technologies are accessible and inclusive is vital for their global impact, particularly in vulnerable regions most affected by environmental degradation.

The section concludes by emphasizing the interconnectedness of these innovations. Renewable energy systems power desalination plants and smart grids, while AI and IoT technologies underpin advancements in agriculture, water management, and carbon capture (IEA, 2025a; FAO, 2024). This synergy highlights the need for a holistic approach to sustainability, where technologies are developed and implemented in a way that maximizes their collective benefits.

Through critical analysis, this section illustrates the transformative potential of technology to address environmental challenges. By leveraging innovation responsibly and collaboratively, humanity can advance toward a future where economic growth and environmental sustainability coexist in harmony. While challenges remain, the examples presented here offer a glimpse into what is possible when technological ingenuity is directed toward the well-being of our planet (UNEP, 2025; IPCC, 2023).

THE ROLE OF GREEN TECHNOLOGIES IN INDUSTRY

Industries around the world are increasingly integrating green technologies into their operations to minimize environmental impact and align with circular economy principles (OECD, 2024a, 2024b, 2024c, 2024d; Circle Economy, 2025). This shift reflects a growing awareness of the need for sustainable practices to combat climate change, reduce waste, and optimize resource use. By examining green manufacturing, sustainable construction, and transportation innovations, this section highlights the transformative potential of these technologies while addressing the challenges of their implementation.

Green manufacturing represents a cornerstone of sustainable industrial practices, focusing on reducing energy consumption and waste while increasing efficiency (IEA, 2025a). Many manufacturers are transitioning to renewable energy sources, such as solar or wind, to power their facilities. Energy-efficient machinery and automation technologies are also being deployed to streamline production processes, lowering emissions and reducing energy use. Companies are adopting closed-loop systems, where industrial waste is recycled into raw materials for future production, aligning with circular economy principles (Circle Economy, 2025).

An example of green manufacturing is the automotive industry's shift toward more sustainable practices. Automakers are incorporating recycled materials in vehicle production, using renewable energy in manufacturing plants, and optimizing supply chains to reduce their carbon footprint (IEA, 2025b, 2025c). These efforts not only improve environmental outcomes but also appeal to environmentally conscious consumers, driving market demand for sustainable products.

Despite these advancements, barriers remain. The transition to green manufacturing often requires substantial upfront investment in new technologies and infrastructure, which can be prohibitive for small and medium-sized enterprises (SMEs). Additionally, regulatory inconsistencies and a lack of standardized practices can hinder broader adoption. Governments and international bodies play a crucial role in addressing these challenges by offering incentives, subsidies, and clear guidelines to support the shift toward sustainable manufacturing (OECD, 2024; UNEP, 2025).

Sustainable construction is another vital area where industries are leveraging green technologies to reduce environmental impact. The use of smart materials, such as self-healing concrete and thermal insulation systems, is revolutionizing construction practices by increasing durability and energy efficiency. Green building designs incorporate renewable energy systems, water recycling technologies, and advanced ventilation systems to create eco-friendly urban spaces (UN-Habitat, 2024).

The concept of net-zero buildings, which produce as much energy as they consume, is gaining traction worldwide. These structures combine energy-efficient designs with renewable energy sources, such as rooftop solar panels or geothermal heating, to minimize their environmental footprint (IEA, 2025a). Urban planners and architects are also integrating green spaces and vertical gardens into designs to improve air quality and promote biodiversity in urban environments (UN-Habitat, 2024).

However, the widespread adoption of sustainable construction practices faces challenges related to cost and scalability. Green building materials and technologies are often more expensive than conventional alternatives, creating financial barriers for developers and homeowners. Moreover, the lack of skilled professionals trained in green construction methods can slow the industry's transition to sustainability. Addressing these issues requires targeted education programs, financial incentives, and policy support to encourage the adoption of sustainable construction technologies (OECD, 2024; UNEP, 2025).

In the transportation sector, innovations such as electric vehicles (EVs), hydrogen fuel cells, and public transit solutions are addressing one of the largest sources of global greenhouse gas emissions (IEA, 2025b; IEA, 2024). EVs, powered by lithium-ion batteries, have become a symbol of the green transition in transportation. Advances in battery technology are improving vehicle range and charging efficiency, making EVs increasingly accessible to consumers.

Hydrogen fuel cells, another promising technology, offer a clean energy alternative for both personal and commercial vehicles. Hydrogen-powered transportation produces only water as a byproduct, making it an attractive option for reducing emissions in sectors like shipping and aviation (IEA, 2024). Public transit systems are also embracing green technologies, with cities adopting electric buses, light rail systems, and bike-sharing programs to reduce reliance on fossil fuel-based transportation (UN-Habitat, 2024).

While these innovations represent significant progress, their adoption is not without challenges. The high cost of EVs and the limited availability of charging infrastructure remain barriers to widespread consumer adoption. Similarly, hydrogen fuel cells face challenges related to production costs, storage, and distribution infrastructure. Public transit solutions require substantial investment in infrastructure and maintenance, as well as public buy-in to shift behavior away from car-centric transportation systems (IEA, 2025b; UN-Habitat, 2024).

Policy interventions and industry collaboration are essential to overcoming these barriers. Governments can play a critical role by implementing subsidies, tax incentives, and infrastructure investments to support the green transition in transportation. Private sector partnerships can also accelerate innovation and scale deployment by pooling resources and expertise (OECD, 2024; UNEP, 2025).

Public acceptance is another critical factor in the success of green technologies across industries. Consumer demand drives market trends, and industries must effectively communicate the benefits of sustainable practices to gain public support. Educational campaigns, transparent reporting, and collaborative efforts between governments, industries, and communities can help build trust and encourage the adoption of green technologies (UN-Habitat, 2024).

Furthermore, integrating green technologies into existing systems requires addressing systemic issues, such as outdated infrastructure and resistance to change. Industries must navigate these complexities while ensuring that their solutions are scalable, equitable, and inclusive. This involves designing technologies that are accessible to all socioeconomic groups and tailoring solutions to the specific needs of different regions and communities (UN-Habitat, 2024; OECD, 2024).

By embracing green technologies, industries have the potential to lead the global transition toward sustainability. These innovations not only mitigate environmental harm but also create economic opportunities, such as new markets for sustainable products and green jobs. However, achieving these outcomes requires a concerted effort to address the financial, regulatory, and social challenges associated with green technology adoption (UNEP, 2025; IEA, 2025a).

This section underscores the transformative potential of green technologies in shaping a sustainable future. By examining specific examples of innovation and analyzing their implementation challenges, it highlights the critical role of industries in driving environmental progress. Through collaboration, investment, and policy support, these technologies can pave the way for a future where economic growth and environmental stewardship go hand in hand (IEA, 2025a; UNEP, 2025).

DIGITAL TOOLS FOR ENVIRONMENTAL MONITORING AND MANAGEMENT

Digital technologies are transforming the way environmental monitoring and management are approached, bringing unprecedented opportunities for data-driven decision-making. The integration of artificial intelligence (AI), big data analytics, and remote sensing into environmental monitoring systems enables more precise, efficient, and real-time responses to environmental challenges. These tools allow for the collection, analysis, and interpretation of vast amounts of environmental data, which is essential for combating climate change, protecting biodiversity, and reducing pollution. This section explores key applications of these technologies, including climate modeling, biodiversity monitoring, and pollution detection, while also discussing the importance of these innovations for sustainability efforts (IPCC, 2023; UNEP, 2024).

One of the most notable ways digital technologies are being employed in environmental management is through climate modeling. AI-powered climate models are capable of analyzing complex climate data and predicting long-term patterns of climate change, such as temperature fluctuations, sea-level rise, and extreme weather events. These models are trained on vast datasets from a variety of sources, including satellite imagery, historical climate records, and real-time environmental data. By utilizing AI algorithms, these models can generate highly accurate predictions, helping scientists and policymakers understand the potential impacts of climate change on ecosystems, infrastructure, and human societies (WMO, 2024; NASA, 2024).

Moreover, AI-driven climate models enable the identification of effective mitigation strategies. For example, they can simulate the potential outcomes of different carbon reduction measures, such as the impact of renewable energy adoption, forest conservation, or carbon capture technologies. This predictive capability is invaluable for governments and organizations in creating targeted, evidence-based policies to reduce carbon emissions and mitigate the effects of climate change. However, while AI can significantly improve climate forecasting, challenges remain in ensuring the models account for the full complexity of global climate systems, including human and environmental variables that can change over time (IPCC, 2023; UNFCCC, 2023).

Biodiversity monitoring is another area where digital technologies are making a significant impact. Drones, satellite imagery, and AI-powered analytics are increasingly being used to track wildlife populations and monitor changes in habitats. Remote sensing technologies, such as high-resolution satellite imagery, allow for the monitoring of vast and often inaccessible areas, such as rainforests, wetlands, and marine environments. These technologies provide real-time insights into the health of ecosystems, including the detection of habitat loss, deforestation, and changes in biodiversity due to climate change or human activities (WWF, 2024; NASA, 2024).

Drones are particularly effective in collecting data in hard-to-reach or sensitive areas, where traditional methods of monitoring, such as ground surveys, are either difficult or dangerous. Equipped with cameras, thermal sensors, and GPS systems, drones can capture detailed images of wildlife populations and environmental conditions. These images are then analyzed using AI algorithms to identify species, track movements, and assess habitat quality. Additionally, drones can be used to deliver supplies, such as food or medical assistance, to endangered wildlife populations in remote regions, enhancing conservation efforts (WWF, 2024).

The use of digital tools for biodiversity monitoring not only helps in tracking the status of species and habitats but also contributes to the development of more effective conservation strategies. For instance, AI-driven models can analyze large datasets to identify patterns and trends in biodiversity, allowing researchers to predict areas at risk of biodiversity loss and prioritize conservation efforts accordingly. Furthermore, this data can be shared in real time with conservation organizations, governments, and local communities, ensuring that timely actions can be taken to protect vulnerable species and ecosystems (WWF, 2024; NASA, 2024).

Pollution detection is another critical area where digital technologies are making an impact, particularly in the areas of air and water quality monitoring. Sensors and Internet of Things (IoT) devices are increasingly being deployed to track environmental pollutants in real-time, providing immediate data on air and water quality in urban and rural areas. These sensors are often connected to centralized data systems, where the information they gather is analyzed to identify pollution hotspots and track changes in pollution levels over time. By providing real-time data, these systems enable rapid responses to pollution incidents, allowing for quicker identification of sources of contamination and the implementation of mitigation measures (WHO, 2024; US EPA, 2024a; UNESCO & UN-Water, 2024).

Air quality sensors, for example, are being used in cities to monitor pollutants such as particulate matter (PM2.5), nitrogen dioxide (NO2), and sulfur dioxide (SO2). This data is invaluable for public health officials in assessing the impacts of air pollution on human health and making informed decisions about public safety. In addition, real-time air quality monitoring helps governments enforce environmental regulations and take corrective actions when pollution levels exceed safe thresholds. Similarly, IoT devices that monitor water quality are used to detect pollutants such as heavy metals, pesticides, and sewage, which can have harmful effects on aquatic ecosystems and human health (WHO, 2024; US EPA, 2024a, 2024b).

One of the most promising aspects of pollution detection technologies is their ability to provide real-time feedback to local communities and industries. In some cases, these systems can trigger automatic alerts when pollution levels reach dangerous thresholds, notifying relevant authorities and enabling a swift response. This capability is particularly crucial in cases of industrial accidents or environmental disasters, where prompt action is essential to minimizing the damage. Furthermore, by collecting long-term data on pollution levels, these technologies contribute to a better understanding of pollution trends and the effectiveness of environmental policies (US EPA, 2024a; OECD, 2024a).

The integration of AI and big data analytics into pollution detection systems enhances their efficiency and accuracy. AI algorithms can process vast amounts of environmental data from multiple sensors, identifying patterns and trends that might not be immediately apparent. For instance, AI models can predict when and where pollution levels are likely to spike, allowing for proactive measures to be taken before hazardous levels are reached. Additionally, these technologies can be integrated with other smart systems, such as transportation networks or waste management infrastructure, to create more holistic, data-driven approaches to environmental management (OECD, 2024a; IEA, 2024b).

These digital technologies are not only transforming environmental monitoring but are also empowering governments, organizations, and communities to make informed decisions that support sustainability. By harnessing the power of AI, big data analytics, and remote sensing, stakeholders can create more effective and efficient strategies for mitigating environmental risks and promoting conservation. However, challenges such as data privacy concerns, infrastructure costs, and the need for cross-sector collaboration must be addressed to fully realize the potential of these technologies (OECD, 2024b; UNEP, 2024).

In addition, the successful integration of digital technologies into environmental management systems requires strong collaboration between stakeholders. Governments, tech companies, and environmental organizations must work together to ensure that these tools are accessible, affordable, and scalable. Public-private partnerships can facilitate the development and deployment of these technologies, ensuring that the benefits of digital innovation are widely distributed and that vulnerable communities have access to the resources needed to address environmental challenges (UNESCO & UN-Water, 2024; OECD, 2024b).

This section emphasizes the importance of digital technologies in advancing sustainability efforts, from climate modeling to pollution detection. These innovations not only enable real-time environmental monitoring but also provide the data necessary for informed decision-making, empowering stakeholders to take timely and effective action. As these technologies continue to evolve, they will play a crucial role in shaping the future of environmental conservation and sustainability, helping to safeguard the planet for future generations (IPCC, 2023; UNEP, 2024).

CONSUMER-FOCUSED SUSTAINABLE TECHNOLOGIES

In recent years, the rapid advancement of technology has been harnessed to promote sustainable consumer behavior, empowering individuals to make more environmentally conscious decisions in their daily lives. This section explores how technologies like smart home devices, eco-friendly products, and carbon footprint apps are helping to reduce environmental impact while fostering a greater sense of responsibility among consumers. As these innovations continue to evolve, they provide new opportunities for individuals to actively contribute to sustainability efforts, all while improving convenience and comfort in everyday life (IEA, 2024a; IRENA, 2024).

One of the most widely adopted technologies aimed at promoting sustainability is the use of smart home devices. These energy-efficient appliances and systems are designed to reduce household energy consumption, minimize waste, and enhance overall energy efficiency. Smart thermostats, for instance, allow homeowners to control heating and cooling systems remotely, adjusting temperatures based on occupancy patterns and optimizing energy usage. These devices are equipped with sensors that detect when a home is empty or when specific rooms are in use, enabling energy-saving adjustments without sacrificing comfort (ENERGY STAR, 2024; IEA, 2024a).

In addition to thermostats, smart lighting systems and appliances, such as refrigerators, washing machines, and water heaters, offer significant potential for reducing household energy consumption. These devices can be programmed to operate only when needed, and some models even learn the user's habits, making automatic adjustments to reduce unnecessary energy use. Smart home systems can also be integrated with renewable energy sources, such as solar panels, to optimize the use of clean energy and minimize reliance on fossil fuels. The combined effect of these technologies is a significant reduction in household carbon footprints and lower utility bills (IRENA, 2024; IEA, 2024b, 2024c).

Moreover, smart home devices often provide users with real-time data on their energy consumption, allowing for better monitoring and decision-making. Energy dashboards can provide insights into which appliances or systems are using the most energy, giving consumers the information they need to make informed changes in their behavior. This transparency helps foster a deeper understanding of energy use and encourages households to adopt more sustainable practices, such as reducing heating or cooling in unused rooms, using energy-efficient appliances, and turning off devices when not in use (ENERGY STAR, 2024; IEA, 2024a).

Another key area in promoting sustainable consumer behavior is the growing availability of eco-friendly products. In response to increasing environmental awareness, many companies are now offering goods that are biodegradable, reusable, or sustainably sourced. For example, products made from biodegradable materials, such as plant-based plastics, help reduce the environmental burden of waste, as these materials break down more easily than traditional plastics. This shift away from petroleum-based plastics is crucial in reducing pollution, particularly in oceans, where plastic waste is a significant threat to marine life (OECD, 2024c; UNEP, 2024).

In addition to biodegradable goods, reusable packaging has become a prominent feature in the drive for sustainability. Products like refillable water bottles, reusable grocery bags, and sustainable packaging solutions reduce the need for single-use plastic, which has become one of the major contributors to global plastic pollution. Companies are also exploring innovative packaging options, such as edible packaging and packaging made from seaweed or other plant-based materials, which offer environmentally friendly alternatives to traditional plastic wraps and containers (OECD, 2024c; UNEP, 2024).

Eco-friendly products also include sustainably sourced goods, such as fair trade coffee, clothing made from organic cotton, or furniture crafted from responsibly harvested wood. These products not only promote environmental sustainability but also support ethical labor practices and ensure that natural resources are managed in a way that benefits both people and the planet. Consumers who choose these products help reduce demand for environmentally harmful practices, such as deforestation, excessive pesticide use, and exploitation of workers in developing countries (FAO, 2023).

In tandem with the rise of eco-friendly products, mobile applications that track personal carbon footprints have become powerful tools for promoting sustainable behavior. Carbon footprint apps allow users to track their daily activities, from transportation choices to energy consumption, and calculate the associated carbon emissions. These apps can then suggest actionable steps to reduce environmental impact, such as taking public transportation, carpooling, or reducing energy use at home. By providing users with a clear understanding of the carbon footprint of their lifestyle choices, these apps empower individuals to take proactive steps toward sustainability (IPCC, 2023; UNEP, 2024).

Many carbon footprint apps also integrate with other platforms, such as smart home devices or fitness trackers, to provide a more comprehensive view of a person's environmental impact. For instance, an app might calculate the carbon emissions of a user's daily commute and suggest more sustainable alternatives like biking, walking, or using electric vehicles. Additionally, these apps often offer suggestions for reducing carbon footprints in various aspects of life, such as food choices, travel habits, and waste management. By providing personalized recommendations, carbon footprint apps make sustainability more accessible and actionable for users (IEA, 2024a; IEA, 2024b).

The effectiveness of these technologies in promoting sustainable behavior depends, in large part, on consumer awareness and education. While the availability of smart home devices, eco-friendly products, and carbon footprint apps is expanding, it is essential to ensure that consumers are aware of these options and understand how to use them effectively. Education campaigns, both online and offline, play a crucial role in raising awareness about the environmental benefits of these technologies. Through targeted messaging, consumers can learn how their everyday choices impact the environment and how they can make more sustainable decisions (UNEP, 2024).

Governments, non-profit organizations, and companies can also contribute to promoting sustainable consumer behavior by providing incentives and support for those who adopt these technologies. Financial incentives, such as tax rebates for purchasing energy-efficient appliances or subsidies for renewable energy installations, can encourage consumers to make environmentally friendly choices. In addition, businesses can play a role by offering discounts on eco-friendly products or providing educational resources about sustainability (IRENA, 2024; OECD, 2024b).

In this context, the role of social media and online platforms cannot be underestimated. Influencers, environmental activists, and sustainability advocates use digital platforms to spread awareness about eco-friendly products and the importance of reducing carbon footprints. As a result, consumers are increasingly exposed to ideas about sustainable living and motivated to adopt greener lifestyles. Social media also provides a platform for individuals to share their sustainability practices, creating a community-driven approach to environmental stewardship (UNEP, 2024).

While these technologies have the potential to promote sustainable consumer behavior, challenges remain in terms of accessibility, affordability, and consumer engagement. Smart home devices, for example, may be out of reach for low-income households due to upfront costs, even though they offer long-term savings. Similarly, while eco-friendly products are growing in popularity, they may still be more expensive than conventional alternatives, creating a barrier for some consumers. To overcome these challenges, it is essential to ensure that sustainable technologies are affordable and accessible to all, regardless of income level (OECD, 2024b; IRENA, 2024).

Overall, the integration of smart home devices, eco-friendly products, and carbon footprint apps into everyday life represents a significant step toward fostering sustainable consumer behavior. As these technologies continue to evolve, they offer promising solutions to help reduce individual carbon footprints and promote greater environmental responsibility. However, it is equally important to recognize that technology alone cannot solve the complex challenges of sustainability; consumer awareness, education, and collective action are essential to ensuring that these innovations are used effectively to create a more sustainable future (IPCC, 2023; UNEP, 2024).

BARRIERS TO TECHNOLOGICAL SUSTAINABILITY

Despite the promise of technology to drive sustainability, significant barriers remain that hinder the widespread adoption of green innovations and limit their long-term impact. One of the most substantial challenges is the economic barrier, particularly for small businesses and developing nations, where the high costs of developing and implementing sustainable technologies can be prohibitive. For instance, renewable energy infrastructure such as solar panels, wind turbines, and energy storage systems requires a substantial upfront investment. While these technologies can offer long-term savings and environmental benefits, the initial capital required can be a barrier, especially for small-scale enterprises or countries with limited financial resources. Furthermore, the costs associated with research and development for new green technologies can be a significant hurdle for smaller companies, which may lack the necessary funding or expertise to compete with larger corporations (IRENA, 2024; World Bank, 2024).

In developing nations, the situation is even more complex, as these countries often face competing priorities such as poverty reduction, healthcare, and education. Investing in green technologies may not be seen as a top priority when immediate developmental needs take precedence. For instance, nations with limited access to financing may struggle to afford the adoption of renewable energy systems or waste management technologies, which are essential for sustainable development. Without international aid or affordable financing mechanisms, it becomes even more challenging for these countries to transition to greener economies. As a result, there is a risk of widening the technology gap between developed and developing nations, leaving the latter at a disadvantage in the global shift toward sustainability (UNCTAD, 2024).

In addition to economic challenges, policy and regulation gaps also play a significant role in hindering technological innovation for sustainability. While some governments have introduced incentives for green technologies, many others have failed to establish consistent or comprehensive regulatory frameworks that can drive widespread adoption. Inconsistent policies across different regions create uncertainty, which can deter private investment and slow down the progress of green innovation. For example, subsidies for renewable energy might be available in some countries, but not in others, or they might be subject to changes in government leadership, leading to a lack of continuity in policy support. Without clear, long-term regulations that encourage innovation, businesses and innovators are less likely to invest in green technologies, especially when they are unsure of the political and financial stability of the market (OECD, 2024b).

Moreover, the lack of uniformity in regulations can also lead to market fragmentation, where different regions develop their own standards and rules. This creates additional barriers for companies trying to scale their technologies globally, as they must navigate a complex web of regulations that vary from one country to the next. The absence of international cooperation and standardization makes it more difficult for green technologies to achieve the scale necessary to make a significant impact on global sustainability goals. Inconsistent regulations can also discourage the creation of cross-border partnerships, limiting the ability to share knowledge, resources, and best practices in technology development and implementation (OECD, 2024b).

Another important factor to consider when discussing barriers to sustainable technology adoption is the technological limitations of some green solutions. While advancements in renewable energy, waste management, and water conservation are promising, some technologies still face challenges related to scalability and efficiency. For instance, solar and wind energy, though rapidly advancing, still face limitations in terms of storage capacity and energy output consistency. The intermittent nature of these energy sources can make it difficult to rely solely on them without backup systems or supplementary technologies. Additionally, while green technologies like electric vehicles (EVs) offer significant environmental benefits, they still face challenges in terms of range, charging infrastructure, and battery disposal (IEA, 2024b, 2024c).

Furthermore, some green technologies themselves can have unintended environmental consequences. For example, the production of solar panels and batteries often relies on mining for rare earth materials, which can be environmentally damaging if not carefully managed. The energy consumption of some green technologies is also a concern; for example, the production and recycling of electric vehicle batteries require large amounts of energy, potentially offsetting the benefits of reduced carbon emissions if the energy used comes from non-renewable sources. As these technologies continue to evolve, the challenge will be to ensure that their environmental impact is minimized across their entire lifecycle, from production to disposal (World Bank, 2023).

Technological obsolescence is another challenge that can limit the long-term viability of green innovations. The rapid pace of technological advancements means that solutions that are cutting-edge today may be outdated in just a few years. This issue can be particularly problematic for businesses and governments that invest heavily in certain technologies, only to find that newer, more efficient alternatives become available shortly thereafter. As a result, there is a risk that early investments in sustainability may not deliver the anticipated returns, leading to hesitation or reluctance to adopt new technologies. To counteract this, there needs to be a focus on creating flexible, adaptable technologies that can evolve alongside the rapid pace of innovation (OECD, 2024a).

To overcome these barriers, policy innovation is crucial. Governments must develop policies that provide long-term incentives for the development and adoption of sustainable technologies. These incentives should not only focus on financial support but also on creating a supportive regulatory environment that reduces uncertainty and encourages private sector investment. Furthermore, international collaboration is essential to ensure that green technologies can scale globally. By fostering cross-border partnerships and sharing knowledge and resources, countries can accelerate the transition to a more sustainable future (UNFCCC, 2023; OECD, 2024d).

Public-private partnerships are another key strategy for overcoming these barriers. By collaborating with private companies, governments can leverage the expertise, innovation, and financial resources of the private sector to accelerate the development and deployment of sustainable technologies. These partnerships can also help reduce the financial burden on governments, allowing them to focus on creating the right regulatory frameworks and incentives while the private sector drives technological innovation. Additionally, such partnerships can help ensure that the benefits of green technologies are accessible to all, including low-income communities and developing nations (IRENA, 2024; OECD, 2024b).

Finally, addressing the barriers to widespread sustainable impact requires a global effort. Climate change, resource depletion, and environmental degradation are global challenges that no single country can tackle alone. Therefore, international cooperation and collaboration are essential for driving innovation and creating solutions that can be scaled to meet the needs of the entire planet. By working together across borders, governments, businesses, and communities can share best practices, resources, and technologies to create a sustainable future for all. While the barriers to sustainability are significant, they are not insurmountable. Through policy innovation, collaboration, and a commitment to technological advancement, the world can move closer to achieving its environmental goals (UNFCCC, 2023; IPCC, 2023).

CASE STUDIES OF SUCCESSFUL TECHNOLOGICAL INNOVATION

Real-world case studies serve as valuable examples of how sustainable technologies can transform entire sectors and lead to positive environmental and economic outcomes. One such case is Denmark's remarkable success in wind energy production. Over the past few decades, Denmark has made significant strides in harnessing wind power, becoming a global leader in wind energy production. The country's commitment to renewable energy has been driven by a combination of political will, innovation, and a robust support system for green technologies. Denmark's extensive investment in wind energy infrastructure, along with favorable government policies such as subsidies and tax incentives for clean energy companies, has enabled the country to scale its wind power capabilities dramatically. Today, Denmark produces more than 40% of its electricity from wind, and it has set ambitious goals to achieve 50% wind energy by 2030. This transition has not only helped Denmark reduce its carbon emissions but also created thousands of green jobs, showcasing the economic potential of renewable energy technologies (Danish Energy Agency, 2024; Energinet, 2024).

The success of Denmark's wind energy industry offers valuable lessons for other countries looking to invest in renewable energy. Key to Denmark's achievements has been its long-term policy vision, where sustainability is prioritized across political administrations, ensuring that the country remains committed to its green energy goals. Additionally, Denmark has demonstrated the importance of creating a supportive ecosystem for innovation, with close collaboration between the government, research institutions, and private sector companies. These factors have enabled Denmark to build a strong wind industry that has driven technological advancements, reduced costs, and attracted international investment. For other countries, especially those with favorable wind conditions, adopting a similar approach could lead to similar successes in renewable energy production (Danish Energy Agency, 2024; Energinet, 2024).

Another compelling case study is Singapore's model for water sustainability. Given its limited natural freshwater resources, Singapore has turned to innovative technologies to ensure its water security. The country has implemented a range of advanced desalination technologies, which convert seawater into potable water, and water recycling systems, including NEWater, a high-grade reclaimed water. By investing in cutting-edge research and development, Singapore has been able to create a reliable and sustainable water supply despite its lack of natural freshwater sources. The success of Singapore's water sustainability model is not only due to technological innovation but also effective governance and public engagement. The government has worked to ensure that citizens understand the importance of water conservation and the role of technology in meeting the country's water needs. As a result, Singapore now has a resilient water supply that is largely independent of external sources, a model that is being studied and replicated by other nations facing similar water scarcity challenges (PUB, 2024).

The key takeaway from Singapore's approach to water sustainability is the importance of integrating technological innovation with sound policy and public cooperation. The country's ability to combine desalination, water recycling, and conservation efforts has enabled it to overcome natural resource constraints. For other countries facing water scarcity, Singapore's model demonstrates that technological solutions, if paired with supportive policies and public engagement, can ensure sustainable water management even in the most resource-constrained environments. Moreover, the success of Singapore's water sustainability strategies emphasizes the role of long-term planning and the need for investment in research and development to achieve lasting impact (PUB, 2024).

The transformative potential of sustainable technologies is also evident in the case of electric vehicles (EVs), with companies like Tesla leading the charge in reshaping the transportation sector. Tesla's innovation in electric vehicles has not only changed how we view cars but has also played a significant role in changing public and corporate attitudes toward sustainable transportation. Tesla's success can be attributed to several factors, including its cutting-edge battery technology, commitment to producing high-performance vehicles, and effective branding that has made EVs more attractive to consumers. The company's pioneering efforts in creating a viable alternative to traditional gasoline-powered cars have pushed other automakers to accelerate their own EV development, making electric vehicles increasingly mainstream (IEA, 2024c).

Tesla's impact extends beyond just the production of electric vehicles. The company has played a key role in changing the narrative around sustainable transportation, demonstrating that EVs can be both environmentally friendly and desirable. Tesla's marketing strategy has focused not only on the environmental benefits of EVs but also on their technological superiority, appealing to consumers who value innovation and performance. This shift in consumer attitudes has contributed to the growing demand for electric vehicles, and as more people adopt EVs, the automotive industry continues to evolve toward greener alternatives. Tesla has also contributed to the development of infrastructure necessary for EV adoption, including charging stations, further supporting the widespread use of electric vehicles (IEA, 2024c).

One of the most significant lessons from Tesla's success is the importance of innovation and public-private collaboration. Tesla's ability to revolutionize the automotive industry was made possible by both technological breakthroughs and strategic partnerships, including government incentives for clean energy and electric vehicles. This case illustrates how corporate leadership, technological innovation, and supportive policy frameworks can drive the widespread adoption of sustainable technologies. The rise of electric vehicles serves as a reminder that transforming entire industries is possible when there is both the vision to innovate and the political and financial support to bring those innovations to market (IEA, 2024b; IRENA, 2024).

In analyzing these case studies, several common factors emerge that contribute to the successful implementation of sustainable technologies. First and foremost is the role of government policy and regulation in creating an environment conducive to innovation. Whether it is through subsidies for renewable energy or incentives for electric vehicles, government support plays a crucial role in encouraging businesses to invest in green technologies. Additionally, public-private collaboration is key, as evidenced by the partnerships between governments and companies like Tesla, which have helped bring sustainable innovations to market more rapidly. Finally, public engagement and education are essential in ensuring that citizens not only understand the benefits of sustainable technologies but are also motivated to adopt them in their own lives (IEA, 2024b; IRENA, 2024).

While these case studies demonstrate the transformative potential of sustainable technologies, they also highlight the challenges that must be addressed to achieve global adoption. For instance, the scalability of Denmark's wind energy model, Singapore's water sustainability technologies, and Tesla's electric vehicles depends on overcoming economic, regulatory, and logistical barriers. Countries with different resources, infrastructures, and economies may face challenges in replicating these models without the proper financial, political, or technological support. Nonetheless, these case studies provide a roadmap for other nations to follow, offering both inspiration and practical insights into how to achieve sustainability through innovation (Danish Energy Agency, 2024; PUB, 2024; IEA, 2024c).

These real-world examples underscore the fact that sustainable technologies are not only necessary but also increasingly viable solutions to the world's environmental challenges. While the transition to sustainability is complex and multifaceted, the successes of Denmark, Singapore, and Tesla show that with the right combination of innovation, policy, and collaboration, it is possible to create a sustainable future for generations to come. These case studies not only inspire confidence in the transformative power of technology but also serve as important lessons for global policymakers, businesses, and communities seeking to advance sustainability (IEA, 2024b; IRENA, 2024).

THE ETHICAL DIMENSIONS OF SUSTAINABLE TECHNOLOGY

The ethical considerations of leveraging technology for sustainability are crucial in ensuring that the benefits of innovation are distributed fairly and responsibly. One of the primary concerns in this regard is equity in access to sustainable technologies. While the advancement of technologies such as solar energy, electric vehicles, and energy-efficient appliances holds significant potential to address global environmental challenges, these technologies are often inaccessible to marginalized communities, particularly those in developing countries or low-income areas. For instance, the high initial cost of solar panels or electric vehicles can prevent these communities from benefiting from green technologies. To ensure that sustainability efforts do not exacerbate existing inequalities, it is imperative to consider how these technologies can be made affordable and accessible to all. Governments, international organizations, and private companies must work together to develop policies and programs that address financial barriers, such as subsidies or financing options, and promote the dissemination of technologies in underserved regions. This approach would help ensure that marginalized populations are not left behind in the global transition toward sustainability (UN DESA, 2023; OECD, 2024d).

Equity in access is closely tied to the principle of responsible innovation, which calls for a thoughtful and proactive approach to the development of new technologies. While the promise of sustainable technologies is significant, it is also important to anticipate and mitigate any potential unintended environmental or social consequences. One such consequence is the issue of electronic waste (e-waste), which has become a growing problem as the adoption of renewable technologies, such as solar panels and electric vehicles, increases. The production, use, and disposal of these technologies can generate significant amounts of e-waste if not managed properly, contributing to pollution and resource depletion. For example, the disposal of batteries from electric vehicles or solar panels can lead to hazardous chemical leaks and other environmental issues if not recycled or disposed of in an environmentally responsible manner. It is essential that innovation in sustainability is accompanied by robust systems for recycling, waste management, and extended product life cycles. Responsible innovation involves not only the development of sustainable technologies but also the creation of systems that ensure their end-of-life impacts are minimized (Forti et al., 2024).

In addition to addressing the environmental impacts of technology, responsible innovation requires careful consideration of social and cultural factors. Technologies developed in one part of the world may not be universally applicable or may inadvertently harm certain communities. For example, the widespread use of biofuels, which are often seen as a sustainable alternative to fossil fuels, can have negative consequences for food security in developing countries if agricultural land is diverted away from food production to biofuel crops. This scenario highlights the need for ethical considerations to extend beyond just environmental impacts to include the social and economic implications of new technologies. As technologies continue to evolve, it is essential to engage diverse stakeholders in the innovation process, ensuring that the needs, values, and concerns of affected communities are taken into account (FAO, 2023).

Another significant ethical consideration in the adoption of sustainable technologies is global collaboration. While many developed countries have the financial resources and technological infrastructure to implement green technologies, developing nations often lack the necessary capital and expertise. This disparity creates an ethical imperative for wealthier nations to support the efforts of developing countries in adopting sustainable technologies. This support could take the form of financial aid, knowledge transfer, and capacity-building initiatives. By fostering international partnerships, developed nations can help bridge the technological divide and ensure that developing countries are not left behind in the global push for sustainability. Furthermore, collaboration between nations is critical in addressing global challenges such as climate change, biodiversity loss, and resource depletion, which require coordinated efforts at the international level. As the world becomes increasingly interconnected, global collaboration is not only an ethical obligation but also a practical necessity for achieving sustainability goals (UNFCCC, 2023; OECD, 2024d).

The principle of global collaboration also brings to light the need for shared responsibility in addressing environmental issues. Climate change, for instance, is a global problem that disproportionately affects vulnerable populations, particularly in developing countries that are less equipped to cope with its impacts. Therefore, developed nations, which have historically been the largest contributors to greenhouse gas emissions, have a moral responsibility to take the lead in mitigating the effects of climate change. This includes not only reducing emissions domestically but also supporting the efforts of developing nations through technology transfer and financial assistance. By fostering an inclusive and cooperative approach to sustainability, the global community can ensure that all countries have an equal opportunity to participate in and benefit from the transition to a greener economy (UNFCCC, 2023).

Finally, integrating ethics into the design and deployment of technological solutions is essential for creating a sustainable future that is not only environmentally sound but also just and equitable. Ethics must be woven into every stage of technological development, from initial design to end-of-life disposal. This includes considering the potential social, environmental, and economic impacts of new technologies and ensuring that all stakeholders are consulted and included in the decision-making process. Additionally, the ethical framework guiding the development and deployment of sustainable technologies should be transparent, accountable, and adaptable to changing circumstances. By prioritizing ethical considerations in the technological revolution for sustainability, we can ensure that innovations not only solve pressing environmental challenges but also contribute to the creation of a fair and just society for all (OECD, 2024a; UN DESA, 2023).

FUTURE DIRECTIONS IN TECHNOLOGY AND SUSTAINABILITY

The final section delves into emerging trends that have the potential to significantly shape the future of sustainability. Biomimicry, for instance, represents an innovative approach to design by drawing inspiration from natural processes and systems. This concept, which involves mimicking nature's strategies for solving environmental challenges, is gaining traction in the development of sustainable technologies. From energy-efficient building designs inspired by termite mounds to water purification systems that replicate the way plants filter water, biomimicry offers novel solutions that not only address environmental challenges but also create technologies that are inherently sustainable. By leveraging the wisdom embedded in natural systems, biomimicry encourages the development of products and processes that are resource-efficient, resilient, and aligned with the principles of sustainability. This approach has the potential to transform industries by creating products that minimize waste and energy consumption while maximizing efficiency (Biomimicry Institute, 2024).

Decentralized energy systems also represent a key emerging trend in the pursuit of sustainability. Traditional energy grids, often centralized and reliant on fossil fuels, are being replaced by community-based renewable energy systems that promote energy independence. These decentralized grids harness local resources such as solar, wind, or biomass, enabling communities to generate and manage their own energy. Not only do these systems reduce dependence on nonrenewable energy sources, but they also empower local communities, reduce energy costs, and enhance resilience against disruptions. With advancements in energy storage and smart grid technologies, decentralized energy systems are becoming increasingly viable, particularly in remote or underserved areas. This shift towards community-based energy solutions is a critical step in building a sustainable energy future that is both equitable and resilient (IRENA, 2024).

Another emerging area of focus is sustainable AI, which seeks to reduce the environmental impact of artificial intelligence systems while also integrating sustainability into AI applications. AI has the potential to contribute significantly to sustainability efforts by optimizing resource use, improving efficiency in various sectors, and enabling data-driven decision-making. However, the energy consumption of AI systems, particularly those related to training large models, has raised concerns about their carbon footprint. To address these concerns, researchers are exploring ways to develop more energy-efficient algorithms and hardware, as well as integrating sustainability metrics into AI applications. Sustainable AI aims not only to reduce the environmental impact of AI itself but also to use AI to solve complex environmental problems, such as climate modeling, waste management, and conservation efforts. By aligning AI development with sustainability principles, this emerging trend can accelerate the global transition to a more sustainable future. Continued innovation, research, and cross-sector collaboration will be essential to unlocking the full potential of these technologies and ensuring that they contribute to the achievement of global sustainability goals (OECD, 2024a; UNEP, 2024).

CONCLUSION

The chapter concludes by reaffirming the vital role that technology plays in shaping a sustainable future. Technological innovations have already begun to drive meaningful change across industries, communities, and ecosystems, proving that they hold the potential to mitigate environmental crises such as climate change, resource depletion, and biodiversity loss. However, achieving global sustainability is not without its challenges. From economic barriers and technological limitations to regulatory gaps and public resistance, these obstacles must be addressed to ensure that technological solutions are deployed effectively. Nevertheless, the advances already underway in renewable energy, smart agriculture, green manufacturing, and other sectors show that, with the right investments and strategies, technological innovation can pave the way for a more sustainable and resilient world (IPCC, 2023; IEA, 2024b; IRENA, 2024).

The transformative potential of green technologies offers hope for a brighter future, where environmental concerns are tackled through sustainable practices and efficient resource management. Yet, to realize this potential, it is imperative to foster ethical practices that guide the development and deployment of these technologies. This includes ensuring that innovations are developed with equity in mind, so that marginalized communities are not left behind, and that technologies do not cause unintended harm, such as contributing to e-waste or exacerbating inequality. By promoting responsible innovation, stakeholders can ensure that technology not only addresses environmental challenges but also respects social and ethical principles, aligning sustainability goals with human well-being and environmental stewardship (Forti et al., 2024; UN DESA, 2023).

Equitable access to these technologies is another crucial factor in achieving global sustainability. It is not enough to simply develop sustainable technologies; they must be made accessible to all people, especially in developing nations where resources and infrastructure may be limited. Collaboration between governments, industries, and communities will be essential to overcoming these barriers. By supporting policies that incentivize sustainable innovation and fostering partnerships that promote the widespread adoption of green technologies, we can harness the power of innovation to benefit both people and the planet. In this way, technology can be a catalyst for a global movement toward a more sustainable, equitable, and prosperous future for all (UNFCCC, 2023; OECD, 2024d).

CHAPTER 9 ESSAY PROMPTS

Sustainability and Technology—Innovation for the Planet

1. How can technological innovation address the key environmental challenges of climate change, resource depletion, and biodiversity loss? Provide specific examples of technologies that contribute to sustainability.
2. What role do renewable energy technologies, such as solar, wind, and geothermal energy, play in reducing reliance on fossil fuels? Discuss the advantages and limitations of these technologies in achieving global sustainability.
3. Explain the concept of the circular economy. How can green manufacturing practices, such as recycling industrial waste and adopting energy-efficient machinery, contribute to this model?
4. Discuss the ethical considerations in the development and deployment of sustainable technologies. How can equity in access and responsible innovation help ensure that technological solutions benefit all communities?
5. How do digital technologies, such as artificial intelligence (AI) and big data analytics, enhance environmental monitoring and management? Provide examples of how these technologies are applied in climate modeling, pollution detection, or biodiversity conservation.
6. What are the key barriers to the widespread adoption of sustainable technologies? Discuss the economic, policy, and technological challenges that need to be addressed to accelerate global sustainability efforts.
7. Examine the role of decentralized energy systems in promoting energy independence and sustainability. How can community-based renewable energy grids contribute to reducing carbon emissions and empowering local communities?
8. What are the potential environmental consequences of technological innovations, such as e-waste from renewable technologies? How can the issue of e-waste be mitigated while still advancing sustainability through technological innovation?
9. How has Denmark's success in wind energy production demonstrated the transformative potential of sustainable technologies? What lessons can be learned from Denmark's experience and applied globally to advance renewable energy adoption?
10. Discuss the emerging trend of biomimicry in sustainable technology design. How can this approach to innovation, which draws inspiration from natural processes, help solve environmental challenges in industries like construction, manufacturing, or agriculture?

REFERENCES

Biomimicry Institute. (2024). *Annual impact report*. Biomimicry Institute.

Circle Economy. (2025). *Circularity gap report 2025*. Circle Economy Foundation.

Danish Energy Agency. (2024). *Energy statistics 2023*. Danish Ministry of Climate, Energy and Utilities.

ENERGY STAR. (2024). *Connected and smart thermostats: Savings and performance*. U.S. Environmental Protection Agency.

Energinet. (2024). *Electricity statistics 2023: Production and consumption*. Energinet.

Food and Agriculture Organization of the United Nations. (2024). *The state of food and agriculture 2024*. FAO.

FAO. (2023). *The state of food and agriculture 2023*. Food and Agriculture Organization of the United Nations.

Forti, V., Baldé, C. P., Kuehr, R., & Bel, G. (2024). *The global e-waste monitor 2024*. International Telecommunication Union & UNITAR.

Global CCS Institute. (2024). *Global status of CCS 2024*. Global CCS Institute.

IEA. (2024a). *World energy outlook 2024*. International Energy Agency.

IEA. (2024b). *Tracking clean energy progress 2024*. International Energy Agency.

IEA. (2024c). *Global EV outlook 2024*. International Energy Agency.

International Energy Agency. (2024). *Global hydrogen review 2024*. IEA.

International Energy Agency. (2025a). *World energy outlook 2025*. IEA.

International Energy Agency. (2025b). *Global EV outlook 2025*. IEA.

International Renewable Energy Agency. (2025). *Renewable energy statistics 2025*. IRENA.

IRENA. (2024). *World energy transitions outlook 2024*. International Renewable Energy Agency.

Intergovernmental Panel on Climate Change. (2023). *Climate change 2023: Synthesis report*. IPCC.

NASA. (2024). *Earth observation for biodiversity and ecosystems*. National Aeronautics and Space Administration.

OECD. (2024a). *AI, the environment and the climate: Practical tools*. Organisation for Economic Co-operation and Development.

OECD. (2024b). *Innovation, investment and policy for net-zero transitions*. Organisation for Economic Co-operation and Development.

OECD. (2024c). *Global plastics outlook: Policy scenarios to 2060—2024 update*. Organisation for Economic Co-operation and Development.

OECD. (2024d). *Climate finance provided and mobilised by developed countries: 2013–2022*. Organisation for Economic Co-operation and Development.

Organisation for Economic Co-operation and Development. (2024). *OECD digital economy outlook 2024*. OECD Publishing.

PUB, Singapore's National Water Agency. (2024). *Our water, our future* (including NEWater and desalination). PUB.

UN-Habitat. (2024). *World cities report 2024: The value of sustainable urbanization*. UN-Habitat.

UN DESA. (2023). *Sustainable development goals report 2023*. United Nations, Department of Economic and Social Affairs.

UNESCO & UN-Water. (2024). *United Nations world water development report 2024: Water for prosperity and peace*. UNESCO.

UNEP. (2025). *Emissions gap report 2024*. United Nations Environment Programme.

United Nations Environment Programme. (2025). *Emissions gap report 2025*. UNEP.

UNFCCC. (2023). *Technical synthesis report of the first global stocktake under the Paris Agreement*. United Nations Framework Convention on Climate Change.

UNCTAD. (2024). *Technology and innovation report 2024*. United Nations Conference on Trade and Development.

US EPA. (2024a). *Air sensor toolbox for communities and citizens*. U.S. Environmental Protection Agency.

US EPA. (2024b). *Water quality monitoring: Sensors and IoT applications*. U.S. Environmental Protection Agency.

WHO. (2024). *Ambient (outdoor) air pollution: Health impacts—Fact sheet*. World Health Organization.

WMO. (2024). *State of the global climate 2023*. World Meteorological Organization.

World Bank. (2023). *Minerals for climate action: The mineral intensity of the clean energy transition* (updated brief). World Bank Group.

World Bank. (2024). *Financing sustainable infrastructure in developing countries*. World Bank Group.

WWF. (2024). *Living planet report 2024*. World Wide Fund for Nature.

Globalization, Technology, and Cultural Exchange

10

ABSTRACT

Chapter 10 explores the profound impact of globalization and technology on cultural exchange, examining how technological advancements facilitate cross-cultural interactions and global connectivity. The chapter highlights how digital platforms, social media, and emerging technologies such as virtual reality and artificial intelligence enable the rapid dissemination and hybridization of cultures worldwide. It delves into the dual role of technology in both promoting cultural diversity and raising concerns about cultural homogenization. The chapter also addresses the challenges and ethical considerations associated with global cultural exchange, including equity in access, preservation of local cultures, and the digital divide. Ultimately, it emphasizes the need for a balanced approach to ensure that globalization fosters mutual respect, understanding, and inclusivity across cultures.

INTRODUCTION

In the modern era, globalization has become a defining force that shapes economies, societies, and cultures (Giddens, 1990; Held & McGrew, 2007; Steger, 2017). The world has become more interconnected than ever before, leading to unprecedented levels of interaction between people, businesses, and governments across borders. This global connectivity has created both opportunities and challenges, as individuals and organizations increasingly rely on each other for resources, knowledge, and ideas. However, while the term "globalization" often carries economic connotations, its effects extend far beyond trade and finance—it also has profound implications for culture, identity, and social interactions (Appadurai, 1996; Tomlinson, 1999).

One of the driving forces behind the acceleration of globalization is technology, which has significantly transformed how people communicate, collaborate, and access information (Castells, 2010; van Dijck, 2013). Advances in communication technologies, such as the internet, mobile devices, and social media platforms, have shattered geographical boundaries, making it easier for people from different cultures to connect and exchange ideas (boyd, 2014; Jenkins, 2006). These technologies have democratized access to knowledge and cultural expressions, allowing individuals in even the most remote regions of the world to participate in global conversations (Cunningham & Craig, 2019).

At the heart of this transformation lies the interplay between technology and cross-cultural interactions. Through technological innovations, people are not only able to share their own cultural practices and traditions with a global audience but also engage with and learn from other cultures (Kraidy, 2005; Pieterse, 2009). Digital platforms, such as social media, video-sharing websites, and online forums, have provided a virtual space for cultural exchange that transcends physical distances. This has resulted in a more interconnected world where cultural influences flow freely across borders, influencing art, music, fashion, food, and language (Robertson, 1995; Tomlinson, 1999).

Moreover, technology plays a significant role in the way global identities are being constructed and negotiated in the modern world. As people from different cultural backgrounds increasingly interact, new hybrid identities emerge that reflect both local and global influences (Hall, 1996; Pieterse, 2009). These identities are not static but are constantly evolving as individuals navigate the complexities of cultural exchange in a globalized world. Technology, in this sense, acts as both a facilitator and a catalyst for the creation of these new global identities (Appiah, 2006; Delanty, 2009).

The emergence of global communities and virtual networks has also led to the rise of "digital cosmopolitanism," a phenomenon where individuals identify more with a global community than with their local or national identity (Appiah, 2006; Beck, 2006). This shift is particularly evident among younger generations, who have grown up with the internet and digital technologies. These individuals often view themselves as global citizens and engage with cultures from around the world in ways that were unimaginable just a few decades ago (boyd, 2014).

However, the rapid spread of technology and cultural exchange also raises important questions about cultural homogenization and the preservation of cultural diversity. While globalization has brought the world closer together, it has also raised concerns about the erosion of local cultures and traditions in the face of dominant global forces (Tomlinson, 1999; Steger, 2017). The pervasive influence of Western culture, particularly through media and technology, has led to the spread of global consumerism and the standardization of lifestyles and values (Ritzer, 2004; Schiller, 1991). This phenomenon has sparked debates about cultural imperialism and the need to safeguard cultural heritage in a rapidly changing world (Smith, 2006).

One of the most notable examples of technology-driven cultural exchange is the spread of popular culture through platforms like YouTube, Netflix, and Instagram. These platforms have allowed for the global dissemination of music, films, and television shows, which often transcend cultural boundaries and gain massive international followings (Jenner, 2018; Lobato, 2019). At the same time, these platforms have also enabled the emergence of new forms of cultural expression, such as viral trends, meme culture, and user-generated content, which reflect the blending of global and local influences (Jenkins et al., 2013; Nieborg & Poell, 2018).

The rise of e-commerce and online marketplaces has further facilitated cultural exchange by providing a platform for artisans, small businesses, and creators to share their products and services with a global audience (Cunningham & Craig, 2019). This has not only increased the visibility of traditional crafts and indigenous goods but has also allowed consumers to access a wider array of culturally diverse products than ever before. At the same time, the growth of global supply chains and international trade has contributed to the spread of consumer goods and services, creating a more uniform global marketplace (Steger, 2017).

The impact of technology on cross-cultural interactions is also evident in the realm of education, where online learning platforms and virtual classrooms have opened up new opportunities for people from different cultures to interact and learn from one another (Altbach & Knight, 2007; Laurillard, 2014). This globalized form of education encourages the exchange of ideas and perspectives, fostering greater understanding and empathy across cultural divides. It also provides an opportunity for students to explore subjects related to globalization, technology, and cultural exchange, further expanding their global awareness (DeBoer et al., 2014).

However, as with any technological advancement, the increasing reliance on digital tools and platforms has its drawbacks. The digital divide, for example, remains a significant barrier to full participation in global cultural exchange (van Dijk, 2020). While people in developed countries may have access to the latest technologies and platforms, those in rural or economically disadvantaged areas often lack the infrastructure and resources necessary to engage with the digital world. This inequality in access to technology exacerbates existing social and cultural disparities and limits the potential for inclusive cultural exchange (Noble, 2018).

Furthermore, the ethical implications of cross-cultural interactions in the digital age must be considered. The internet and social media platforms can facilitate the spread of both positive and harmful content (Gillespie, 2018). While these platforms have democratized access to information, they have also become spaces where misinformation, hate speech, and cultural appropriation can thrive (Rogers, 2006; Young, 2010). This raises important questions about the responsibility of technology companies, governments, and individuals in ensuring that digital platforms are used for constructive and respectful cultural exchange (Couldry & Hepp, 2017).

Despite these challenges, the potential for technology to foster global understanding and cultural exchange remains vast. As technological advancements continue to reshape the global landscape, the ways in which we interact with different cultures will evolve, creating new opportunities for collaboration, innovation, and mutual respect (Castells, 2010). The challenge moving forward is to ensure that these opportunities are harnessed in a way that promotes inclusivity, cultural preservation, and ethical engagement across borders (Appiah, 2006; Smith, 2006).

Ultimately, the relationship between globalization, technology, and cultural exchange is complex and multifaceted. While technology has undoubtedly facilitated greater interaction between cultures, it has also brought to the forefront important ethical and social issues that require careful consideration (Tomlinson, 1999; Steger, 2017). The ongoing evolution of this relationship will likely continue to shape the global landscape in profound ways, influencing everything from individual identities to global governance. As such, understanding the interplay between these forces is crucial for navigating the challenges and opportunities that lie ahead in the digital age (Held & McGrew, 2007).

This chapter will examine the various facets of globalization and technology, exploring how these forces intersect and interact to create new cultural dynamics. By analyzing the opportunities and challenges presented by this intersection, it will provide insights into how technology can both support and challenge cultural exchange in the 21st century. Through critical reflection on these issues, the chapter will seek to offer a balanced perspective on the role of technology in shaping the future of global cultural interactions (Delanty, 2009).

THE ROLE OF TECHNOLOGY IN GLOBAL CONNECTIVITY

Technology has been a primary driver of globalization, reducing geographical barriers and enabling real-time communication and exchange of information (Castells, 2010). The advent of the internet, mobile devices, and social media platforms has revolutionized how people interact, share knowledge, and access culture. No longer constrained by physical borders, individuals and organizations can instantly connect with others on the other side of the globe, exchange ideas, and collaborate on projects that would have been unimaginable a few decades ago (Jenkins, 2006; van Dijck, 2013). These advances have not only accelerated the pace of globalization but have also transformed its very nature, making it more interconnected, instantaneous, and accessible to a broader audience (boyd, 2014).

Global connections, local voices: how platforms, AI, and networks accelerate cultural exchange while testing equity, identity, and cultural diversity

Read through a culture-and-identity lens, the image suggests that platforms and AI broaden who can share and remix stories across borders, even as algorithmic curation can flatten diversity; the task is to pair translation and accessibility with provenance, community governance, and fair creator economics so global exchange amplifies local voices instead of erasing them.

Technologies like cloud computing, high-speed internet, and artificial intelligence (AI) have opened new avenues for cross-cultural dialogue, allowing for the seamless sharing of information, music, art, literature, and even traditions (Couldry & Hepp, 2017; Steger, 2017). Cloud computing has made it possible for individuals and organizations to store and share vast amounts of data without the limitations of physical storage, fostering a greater flow of information across borders. AI has further enhanced this process by enabling more efficient data analysis, real-time translation, and content personalization, creating opportunities for deeper cultural engagement (Cunningham & Craig, 2019). These technologies have created a more fluid environment for global conversations, enabling cultural products to reach new audiences and for people to share their perspectives with the world (Jenkins et al., 2013).

Through platforms like Facebook, Twitter, Instagram, YouTube, and others, people can engage in virtual exchanges, participate in global conversations, and contribute to cultural production in ways that transcend national boundaries (van Dijck, 2013). Social media, in particular, has become a cornerstone of global communication, facilitating the exchange of ideas and fostering the growth of online communities that bridge cultural divides (boyd, 2014). These platforms not only provide individuals with the ability to share personal experiences and opinions but also enable collective participation in the shaping of global culture. For instance, viral trends, memes, and digital art movements spread quickly across borders, allowing users to collectively contribute to cultural expressions that span the globe (Jenkins et al., 2013; Nieborg & Poell, 2018).

The digitalization of culture, including music, films, literature, and fashion, has not only made cultural products more accessible but has also helped create hybrid forms of culture that blend influences from multiple regions and traditions (Pieterse, 2009; Tomlinson, 1999). The availability of global media content at the touch of a button has democratized access to culture, enabling people in far-flung locations to explore and appreciate traditions that were once distant and inaccessible. This has led to the emergence of new cultural forms that fuse diverse influences, creating innovative blends of music, art, and fashion that resonate with audiences worldwide (Robertson, 1995).

The global popularity of K-pop is a testament to the power of technology in facilitating cultural exchange, with millions of fans around the world consuming and even contributing to the cultural phenomenon, despite being geographically distant from its origin (Jin, 2016). K-pop artists have used digital platforms like YouTube, Twitter, and Instagram to connect with their global fan base, creating a community of fans who actively participate in the promotion of K-pop through fan-driven initiatives, streaming parties, and social media campaigns (Cunningham & Craig, 2019; Jenkins et al., 2013). This global fanbase not only consumes K-pop music but also generates content related to the genre, from fan art and dance covers to translations and cultural analyses, further amplifying the global reach of the genre (Jenner, 2018).

In addition to music, the rise of digital streaming platforms has had a profound impact on the global film and television industry. Platforms like Netflix, Hulu, and Amazon Prime have revolutionized how content is distributed, enabling the worldwide release of films and shows that were once limited by regional broadcasting schedules and formats (Lobato, 2019). As a result, content from diverse countries and cultures can now be enjoyed by global audiences, creating a more inclusive and interconnected global entertainment landscape. For example, popular shows like *Money Heist* from Spain and *Sacred Games* from India have gained international followings, illustrating the power of technology to break down cultural barriers in entertainment (Jenner, 2018).

The expansion of online educational platforms has also played a crucial role in promoting cultural exchange. MOOCs (Massive Open Online Courses) and other online learning tools have made it possible for individuals from diverse backgrounds to access educational content on a variety of subjects, including language, history, art, and culture (Altbach & Knight, 2007; Laurillard, 2014). These platforms foster cross-cultural dialogue by bringing together students from different countries and providing them with opportunities to learn about each other's cultures in an academic setting. For instance, platforms taught by universities around the world allow students from different regions to engage with global perspectives and gain a deeper understanding of cultural contexts (DeBoer et al., 2014).

Furthermore, the proliferation of digital technologies has allowed for the rapid dissemination of news and information, making it easier for people to stay informed about global events. Social media platforms and news websites provide real-time updates on events and issues happening around the world, creating a global awareness of cultural, political, and social phenomena (Gillespie, 2018; Tufekci, 2017). This has led to increased solidarity and support for various causes, as people from different cultures and regions can connect over shared interests and concerns. For example, movements like Black Lives Matter and the global climate strike have garnered widespread international support, largely facilitated by social media and digital communication (Tufekci, 2017).

The digitalization of cultural products has also led to the growth of online marketplaces and e-commerce platforms, providing a space for individuals and small businesses to promote and sell culturally significant products on a global scale (Cunningham & Craig, 2019). This has led to a rise in demand for handmade, artisanal, and locally produced goods that reflect unique cultural traditions. Online platforms allow artisans from various countries to showcase their work and sell it to consumers worldwide. This has not only increased the visibility of cultural products but has also contributed to the preservation and promotion of traditional crafts and artistic practices that may have otherwise been at risk of fading away (Smith, 2006).

While technology has undoubtedly facilitated cultural exchange, it has also raised concerns about cultural homogenization. As global media content becomes increasingly accessible, there is a fear that local cultures may be overshadowed by dominant global influences, particularly those from Western countries (Schiller, 1991; Ritzer, 2004). The proliferation of Hollywood films, American pop music, and Western fashion trends has led some to worry that indigenous cultures and local traditions may be eroded or marginalized in the face of global consumer culture. This concern is particularly relevant in countries with rich cultural heritages that are at risk of being overshadowed by more commercially dominant cultures (Tomlinson, 1999).

In response to these concerns, many cultures have sought to preserve their unique identities while engaging in global exchange. This has led to a phenomenon known as "cultural preservation in the digital age," where communities use digital platforms to document, share, and celebrate their cultural practices (Cameron & Kenderdine, 2007; Parry, 2007). Social media platforms, for example, are often used to promote indigenous languages, traditional crafts, and local customs, allowing younger generations to connect with their cultural heritage and share it with the world. This process of digital cultural preservation ensures that even as cultures engage in global exchange, they are also able to maintain and celebrate their distinctiveness (Champion, 2015; Smith, 2006).

One of the key aspects of cross-cultural exchange facilitated by technology is the ability for individuals to participate in global conversations around social, political, and environmental issues. Platforms like Twitter and Facebook allow people from different cultures to engage in discussions, share their views, and collaborate on initiatives that address global challenges (Tufekci, 2017). This form of digital activism has become an important tool for raising awareness about issues like climate change, human rights, and social justice, enabling people from diverse backgrounds to come together in support of a common cause (Couldry & Hepp, 2017).

However, the widespread use of technology in cultural exchange also presents challenges related to ethics and responsibility. The ease with which content can be shared and disseminated raises questions about the ownership of cultural products and the potential for cultural appropriation (Rogers, 2006; Young, 2010). In some cases, cultural elements from one society may be adopted or commercialized by another without proper understanding or respect for the context in which they originated. This has led to debates around cultural appropriation, intellectual property rights, and the ethical responsibility of individuals and companies involved in global cultural exchange (Scafidi, 2005).

Despite these challenges, technology remains a powerful tool for fostering cross-cultural understanding and collaboration. By providing platforms for people to share their stories, express their creativity, and engage with others, technology enables the creation of a global dialogue that transcends national and cultural boundaries (Jenkins, 2006; Castells, 2010). As the world becomes increasingly interconnected, the role of technology in shaping cultural exchange will only continue to grow, offering new opportunities for collaboration, innovation, and mutual understanding (Cunningham & Craig, 2019).

In conclusion, technology has played a pivotal role in facilitating cross-cultural interactions and global connectivity. The digital age has made it easier than ever for people to connect, share ideas, and engage with cultures from all corners of the globe. While challenges related to cultural homogenization and ethical considerations persist, the opportunities for cultural exchange and the creation of hybrid identities continue to expand (Pieterse, 2009; Tomlinson, 1999). As technology continues to evolve, it will undoubtedly shape the future of cultural exchange, creating new possibilities for global collaboration and understanding (Held & McGrew, 2007).

CROSS-CULTURAL INTERACTIONS AND THE SHAPING OF GLOBAL IDENTITY

Globalization and technology have given rise to an increasingly interconnected world where cultural exchange is no longer a one-way process (Steger, 2017). As cultures interact and influence one another, they shape a new, more fluid sense of global identity. This transformation has been propelled by technological advancements, particularly in communication, which have dramatically reshaped how people engage with the world around them (Castells, 2010; boyd, 2014). In the past, cultural exchange typically occurred through traditional forms like trade, migration, or travel. These exchanges were often slow and limited by geography and accessibility. Today, however, the speed and scope of technological communication have made cultural exchange instantaneous and pervasive (van Dijck, 2013). People can now access cultural products from virtually anywhere and, in turn, share their own cultural experiences with others. This accessibility has fostered a broader understanding of global diversity, allowing people to consume and interact with cultures beyond their immediate surroundings (Appadurai, 1996; Jenkins, 2006).

The dynamic of cultural exchange facilitated by technology has led to the emergence of a globalized culture that transcends national boundaries (Robertson, 1995; Tomlinson, 1999). This new form of interconnectedness has created opportunities for individuals to form identities that are both local and global, drawing on multiple influences from different parts of the world (Hall, 1996; Pieterse, 2009). The digital age has transformed how cultural practices and ideas spread, making them accessible to a global audience. The accessibility of cultural products—whether they are food, music, literature, fashion, or technology—has led to an unprecedented flow of ideas and practices between cultures. The result is a more diverse and interconnected world where cultural lines are becoming increasingly blurred, and people are able to engage with a range of different influences from different parts of the world (Kraidy, 2005; Delanty, 2009).

However, this newfound interconnectedness has also raised important questions about the preservation of cultural uniqueness and identity. While cultural exchange has led to an increase in diversity in many societies, enriching local cultures with foreign influences, it has also led to concerns about the homogenization of culture (Ritzer, 2004; Schiller, 1991). The spread of global brands like McDonald's, Coca-Cola, and Apple, as well as Hollywood films, has been criticized for contributing to cultural imperialism and undermining local cultural practices. The dominance of certain global cultures—especially those from Western nations—has raised concerns that the traditions, languages, and customs of indigenous cultures may be eroded or marginalized (Smith, 2006; Tomlinson, 1999). As global media and consumer culture continue to dominate, the fear is that local cultures may lose their distinctiveness in the face of an overwhelming global narrative.

This dual role of technology in cultural exchange is significant. While technology facilitates the global spread of dominant cultures, it also provides platforms for the preservation and promotion of local cultures (Cameron & Kenderdine, 2007; Parry, 2007). Social media and digital platforms allow for the sharing of regional music, art, language, and traditions on a global scale. The internet, and platforms like YouTube, Instagram, and TikTok, have become powerful tools for people to showcase their unique cultural identities. These platforms enable individuals to share indigenous languages, traditional dances, and local crafts with a worldwide audience, allowing cultures that might have been at risk of fading to maintain relevance in the global conversation (Champion, 2015; Smith, 2006). As such, these technologies help people reclaim and assert their cultural identities, fostering a sense of pride and connection to their heritage.

Through these digital platforms, previously marginalized groups can also assert their cultural identities and ensure that their traditions and languages are not lost in the face of globalization (Couldry & Hepp, 2017; Noble, 2018). For instance, indigenous communities around the world have used online spaces to promote their languages, art, and history, fostering greater awareness of their cultures and encouraging younger generations to engage with their roots. Similarly, social media has become an important tool for activism, as marginalized communities use these platforms to raise awareness about issues like cultural preservation, environmental justice, and social rights (Tufekci, 2017). In this way, technology not only aids in the global dissemination of cultural practices but also serves as a tool for resistance against cultural erasure and the erosion of indigenous cultures.

While technology can contribute to the preservation of cultural identities, it also presents challenges in terms of cultural appropriation. The ease with which cultural elements can be borrowed, altered, and commercialized raises questions about the ethics of cultural exchange (Rogers, 2006; Young, 2010). In many instances, cultural elements from marginalized or indigenous groups are adopted by larger, often more powerful, cultures without the consent or understanding of the original creators. This can lead to the commodification of sacred traditions or symbols and can perpetuate a form of cultural exploitation (Scafidi, 2005). Issues of cultural appropriation have become especially prominent in the fashion, music, and entertainment industries, where cultural symbols are often stripped of their original meaning and sold to a global audience.

The digital age, while providing a platform for marginalized voices, has also brought about new forms of digital colonialism. Large tech companies dominate the global marketplace, shaping the way information is shared and consumed (Couldry & Mejias, 2019). These companies, often headquartered in Western countries, set the agenda for global conversations and create content that tends to reflect a Western-centric perspective. This dominance can sometimes result in the suppression or underrepresentation of non-Western cultures, leading to a cultural imbalance in the digital space. Additionally, the algorithms that govern social media platforms can reinforce dominant cultural norms, amplifying content that aligns with mainstream values while sidelining minority voices and perspectives (Pariser, 2011; Striphas, 2015).

On the other hand, digital technology has also allowed for the diversification of content and the creation of alternative media spaces where people can express their cultural identities outside of the mainstream. For example, the rise of online streaming platforms has allowed for the proliferation of non-Western film and television content, from Korean dramas to Bollywood films, which has had a significant impact on global cultural exchange. These platforms have provided a space for cultures that were once marginalized in global media to reach international audiences. In doing so, they have contributed to a more inclusive global culture that better represents the diversity of the world's people (Lobato, 2019; Jenner, 2018).

Moreover, technology has opened up new forms of collaboration that further facilitate cultural exchange. Online communities and platforms have enabled people from different parts of the world to collaborate on projects that reflect their shared cultural interests (Jenkins et al., 2013; Shirky, 2008). Crowdsourcing platforms, for instance, allow individuals to fund and support cultural projects that may not have access to traditional funding channels. This has led to an explosion of creative works that reflect diverse cultural perspectives and have the potential to influence global trends. Technology, in this sense, has democratized the cultural production process, enabling individuals from all over the world to create and share content that speaks to their unique experiences (Cunningham & Craig, 2019).

In addition, the globalization of technology itself has created a new form of cultural exchange. As technology spreads across borders, it carries with it the values, ideas, and practices of the countries and cultures that develop it. Silicon Valley, for example, has become a global cultural hub, influencing the development of technologies and shaping how people around the world interact with the digital environment. At the same time, technology is also being shaped by local cultural contexts, as different regions of the world adapt and modify technologies to suit their own needs and values. This two-way exchange is leading to the creation of technology that reflects a more diverse set of cultural norms and priorities (Winner, 1980; Jasanoff, 2004).

The growing interconnectedness of the world through technology has also resulted in the blending of cultural traditions. Hybrid cultural forms that blend global and local influences have emerged in a variety of fields, including food, fashion, and music (Pieterse, 2009; Robertson, 1995). The fusion of different culinary traditions, for instance, has led to the creation of new food trends that combine ingredients and cooking techniques from multiple cultures. Similarly, fashion designers often blend traditional clothing styles from different cultures to create new, innovative designs that appeal to a global audience. This fusion of cultures has also extended to music, where genres like reggaeton and Afrobeat have gained international popularity by blending elements of Caribbean, African, and Western music styles (Kraidy, 2005).

While hybrid cultural forms can enrich global culture, they also raise concerns about the commercialization of cultural elements. The commercialization of hybrid identities can sometimes lead to the dilution of cultural significance, as traditional practices are adapted for profit without regard for their original context (Young, 2010; Scafidi, 2005). The growing demand for hybrid cultural products can sometimes result in the commodification of cultural elements, with little regard for the authenticity or significance of those traditions. This has led to debates about the ethical implications of cultural hybridization, especially when it comes to preserving the integrity of indigenous cultures (Smith, 2006).

One area where globalization and technology have had a particularly significant impact is in the realm of education. The proliferation of online learning platforms has provided new opportunities for individuals to learn about different cultures and languages. Through platforms like Duolingo, Coursera, and edX, people can access educational content from institutions and instructors around the world, enabling them to broaden their understanding of global cultures (Altbach & Knight, 2007; Laurillard, 2014). These platforms have made it easier for people to study abroad virtually, participate in cultural exchange programs, and engage with different cultural perspectives in ways that were not possible before (DeBoer et al., 2014).

Technology has also transformed the way cultural heritage is preserved and shared. Digital archives and databases have made it possible to store and share cultural artifacts and historical records that might otherwise be lost. Museums and cultural institutions are increasingly turning to digital technologies to preserve and showcase their collections, making them accessible to a global audience. Virtual reality (VR) and augmented reality (AR) technologies are being used to create immersive cultural experiences that allow people to explore cultural sites and artifacts remotely. These digital tools have made it possible to preserve cultural heritage for future generations, ensuring that the richness of diverse cultural traditions is not lost in the digital age (Cameron & Kenderdine, 2007; Champion, 2015; Parry, 2007).

Despite the challenges that globalization and technology pose to cultural preservation, they also offer new opportunities for cultural exchange and collaboration. As cultures continue to interact, influence, and blend with one another, technology will continue to play a central role in shaping the future of global cultural dynamics. The digital age has ushered in an era of unprecedented cultural connectivity, where individuals from all walks of life can share their stories, experiences, and traditions with the world. As technology continues to evolve, it will undoubtedly deepen the interconnectedness of the world's cultures, fostering a more inclusive and diverse global identity (Castells, 2010; Couldry & Hepp, 2017).

In conclusion, globalization and technology have fundamentally altered the dynamics of cultural exchange. The rise of digital platforms and online communication has enabled the rapid dissemination of cultural products, giving rise to a new, hybrid global culture that blends influences from different parts of the world. While concerns about cultural homogenization and appropriation persist, the opportunities for cultural preservation, collaboration, and hybridization are vast. Technology plays a crucial role in both facilitating and mitigating the effects of globalization, providing a platform for marginalized voices while also allowing for the creation of new cultural forms that reflect the complexities of our interconnected world. Ultimately, the challenge lies in ensuring that cultural exchange remains respectful, inclusive, and mindful of the diversity and richness of the world's many cultures (Appiah, 2006; Smith, 2006; Tomlinson, 1999).

THE IMPACT OF TECHNOLOGY ON CULTURAL INDUSTRIES

The rise of digital platforms has transformed cultural industries such as music, film, literature, and fashion by making them more globally accessible and interconnected (Hesmondhalgh, 2013; Cunningham & Craig, 2019). These platforms have reshaped how cultural products are created, distributed, and consumed, giving artists and creators the ability to reach a worldwide audience without the need for traditional gatekeepers such as record labels, film studios, or publishers (Nieborg & Poell, 2018). The democratization of cultural production has led to a surge in diverse content creation, as people from different backgrounds and regions now have the means to showcase their work on a global stage (Jenkins et al., 2013).

For instance, musicians can upload their tracks directly to streaming platforms like Spotify, Apple Music, or SoundCloud, bypassing the traditional music industry infrastructure and reaching listeners from all over the world. In the past, an artist would have to sign with a record label to gain exposure, but now anyone with a digital connection and the right tools can share their music with a vast global audience. While this expands access, creators face visibility challenges amid content abundance and platform logics that shape discovery and compensation (Eriksson et al., 2019; Prey, 2020).

Similarly, independent filmmakers have gained unprecedented access to international audiences through platforms like YouTube, Vimeo, or Netflix. These digital platforms have allowed filmmakers to bypass traditional film distribution channels, such as movie theaters and television networks, which were often dominated by a few major production companies (Lobato, 2012; Lobato, 2019). Now, anyone with a camera and editing software can create and share a film that has the potential to be seen by millions. This shift has made it possible for independent filmmakers from marginalized or underrepresented communities to share their stories and perspectives on a global scale, fostering a more diverse cinematic landscape (Jenner, 2018). The growth of digital streaming services has also allowed global audiences to access films from different countries and cultures, further promoting cross-cultural exchange through visual storytelling (Lobato, 2019).

In literature, digital platforms like Kindle Direct Publishing and Wattpad have revolutionized the publishing industry. Writers no longer have to rely on traditional publishers to distribute their work, giving rise to a new era of self-publishing. Independent authors now have the ability to publish e-books and audiobooks, reaching readers from around the world (Ramdarshan Bold, 2018; Rehberg Sedo, 2011). This has led to the rise of genres and voices that might not have fit into the traditional literary canon, increasing the diversity of literature available to readers. Moreover, online platforms have facilitated a closer connection between writers and their audiences, allowing readers to provide feedback, interact with authors, and even shape the direction of a story—fostering a more participatory and inclusive literary world (Jenkins, 2006).

While the democratization of cultural production has led to greater diversity and creativity, it has also posed significant challenges for established industries. The rapid expansion of digital content has resulted in the fragmentation of cultural consumption. In the past, consumers relied on a limited number of platforms or media outlets to access content, but now audiences have access to an overwhelming amount of content from across the globe. With this increased variety, it has become harder for individual creators to stand out amidst the noise. Algorithms and recommendation systems can help surface content, but they often prioritize material that aligns with platform commercial interests, leaving independent creators at a disadvantage (Nieborg & Poell, 2018; Striphas, 2015).

Moreover, the proliferation of online platforms has raised concerns about intellectual property rights and the fair compensation of creators. As more and more cultural products are shared and distributed online, the risk of unauthorized copying and redistribution has increased. Digital content can be easily reproduced, shared, and pirated, making it challenging for creators to protect their work and ensure they receive proper compensation for their efforts (Lessig, 2004; Lobato, 2012). While some platforms have introduced monetization options for creators, payout structures and curation practices can privilege larger creators and commercially aligned content (Eriksson et al., 2019; Prey, 2020).

In the fashion industry, technology has played a significant role in driving global trends. The advent of online shopping, social media influencers, and global fashion weeks has enabled fashion trends to spread rapidly across borders (Khamis et al., 2017). A style or design that originates in one part of the world can be instantly disseminated to consumers in other regions, creating a truly global fashion marketplace. Social media platforms have empowered influencers and fashion bloggers to set trends and reach millions of followers, often dictating what is fashionable and desirable. Online shopping platforms have further facilitated the global spread of fashion by offering consumers access to a wide range of products from international designers and brands.

However, this fast-paced global exchange of fashion has also led to ethical concerns about sustainability. The rise of fast fashion—the rapid production of cheap, trendy clothing—has contributed to significant environmental degradation, including overproduction, increased textile waste, and the depletion of natural resources (Niinimäki et al., 2020; Fletcher, 2014). Many fast-fashion brands rely on inexpensive labor in developing countries, where workers are often subjected to poor working conditions and low wages (Joy et al., 2012). The environmental and social costs of fast fashion have become a major issue, with critics arguing that the fashion industry must undergo a radical transformation to prioritize sustainability and ethical labor practices.

As digital platforms continue to play an integral role in the fashion industry, they also raise questions about the commodification of culture. Cultural appropriation in fashion has become an increasingly prominent issue, as global brands borrow elements from traditional or indigenous cultures without acknowledgment or proper compensation (Scafidi, 2005; Rogers, 2006; Young, 2010). This practice has led to calls for greater respect and recognition of cultural heritage in fashion, with designers and brands urged to acknowledge the origins of their inspirations and collaborate with the communities whose cultures they are borrowing from (Smith, 2006).

At the same time, technology has enabled fashion brands to embrace more inclusive and diverse representations of beauty and identity. Digital platforms have allowed a wider range of body types, skin tones, and cultural backgrounds to be represented in fashion campaigns, contributing to a more inclusive and reflective global fashion landscape. The rise of virtual fashion and digital avatars has also opened up new possibilities for self-expression, enabling individuals to explore and create their own unique fashion identities in virtual (Rodríguez-García et al., 2024) spaces. Digital technology has enabled designers to experiment with new materials and techniques that may not have been possible in the physical world, leading to innovations in sustainable fashion. For example, digital fashion shows and virtual fitting rooms allow consumers to try on clothes virtually, reducing the need for physical samples and thereby decreasing waste (Fletcher, 2014).

Despite these advancements, the global reach of digital platforms also raises concerns about the homogenization of fashion. The rapid spread of global trends has made it more difficult for local fashion cultures to retain their distinctiveness, as consumers in different parts of the world adopt similar styles and aesthetics. This has led to debates about the loss of cultural identity in the face of global fashion trends. While some view the blending of fashion influences as a celebration of cultural exchange, others worry that it threatens the diversity of local fashion traditions. The global dominance of Western fashion trends, in particular, has been seen as a form of cultural imperialism, where global brands and designers shape the fashion tastes of consumers worldwide, overshadowing indigenous and regional styles (Schiller, 1991; Ritzer, 2004).

Fashion is not the only cultural industry affected by the rise of digital platforms. In the realm of literature, digital self-publishing has allowed writers to bypass traditional publishers and directly reach readers. While this has democratized access to literature, it has also raised questions about the quality and credibility of self-published works. Traditional publishers typically have editorial processes that ensure the quality of content, while self-publishing platforms lack these gatekeeping mechanisms. This has led to concerns about the rise of substandard content. However, self-publishing has also created opportunities for new voices and niche genres that might not fit into mainstream publishing (Ramdarshan Bold, 2018; Rehberg Sedo, 2011).

In the music industry, streaming services have revolutionized how people consume music. While these platforms offer vast catalogs of songs from across the globe, they have also led to the devaluation of music, with artists earning minimal royalties for their work (Eriksson et al., 2019; Prey, 2020). The ease with which music can be shared and streamed has made it difficult for musicians to generate revenue, leading some to argue that the traditional music industry business model is unsustainable. As a result, many musicians are turning to alternative revenue streams, such as live performances, merchandise, and brand partnerships, to sustain their careers. Streaming platforms have also introduced new challenges regarding data privacy and the use of algorithms to determine what content is promoted to users, which can create homogenized listening experiences (Striphas, 2015).

While these challenges exist, digital platforms also provide unique opportunities for cultural industries to collaborate and innovate in ways that were previously not possible. In the realm of music, for example, artists can collaborate remotely, producing songs or albums with collaborators from different countries and cultures. Similarly, filmmakers and writers can use digital platforms to collaborate with international talent, expanding the scope of their projects and enriching their content with diverse perspectives. Technology has also enabled cultural industries to explore new forms of storytelling, such as interactive media and virtual reality experiences. These innovations have the potential to reshape how cultural content is consumed and experienced in the future (Jenkins, 2006; Champion, 2015).

In conclusion, the rise of digital platforms has had a profound impact on cultural industries across the globe. The ability to create and distribute content without the need for traditional gatekeepers has democratized cultural production, leading to greater diversity and innovation. However, this democratization has also brought about new challenges, including the fragmentation of cultural consumption, concerns about intellectual property, and the ethical implications of fast fashion and cultural appropriation. As the global cultural landscape continues to evolve, it will be crucial to address these challenges while embracing the opportunities that digital platforms offer. Technology has the potential to foster greater cross-cultural exchange, support marginalized voices, and drive innovation in the cultural industries, but it also requires a mindful approach that respects cultural heritage and promotes sustainability (Hesmondhalgh, 2013; Couldry & Hepp, 2017; Smith, 2006).

THE ROLE OF TECHNOLOGY IN EDUCATION AND CULTURAL AWARENESS

Technology plays a crucial role in fostering cultural awareness and understanding through education. Over the past few decades, digital platforms and online resources have revolutionized how individuals access information about different cultures, languages, and histories. Traditional barriers to cultural exchange, such as geographic distance and limited access to information, are being dismantled by the growing reach of technology. Today, an individual in any corner of the world can learn about a different culture, communicate with someone from another country, or study a foreign language—all from the comfort of their own home (UNESCO, 2023; Helm, 2024).

Language learning apps like Duolingo, Babbel, and Memrise are making it easier for people to acquire new languages, opening doors to communication and cross-cultural interactions. These apps offer interactive, gamified lessons that allow users to progress at their own pace, making learning more accessible and enjoyable. As individuals learn new languages, they are not just gaining a new skill; they are also developing a deeper understanding of the cultural nuances embedded in language. Language is often a window into a society's values, customs, and worldviews, and by mastering a new language, users gain a better appreciation for the cultures from which they originate (Essafi, Belfakir, & Moubtassime, 2024).

Moreover, the rise of online education has made studying abroad virtually a reality. Universities and educational institutions now offer a wide array of online courses and degree programs that allow students to engage with global perspectives without leaving their home countries. Through these virtual programs, students can study topics ranging from international relations to global history, all while interacting with peers and professors from diverse cultural backgrounds. This form of education promotes cross-cultural learning and broadens students' understanding of the world by exposing them to ideas and practices that they may not encounter in their local environments (Hackett et al., 2023).

The ability to engage with different cultures through technology has become particularly important in a world that is increasingly interconnected. As globalization accelerates, understanding and appreciating cultural differences are crucial for fostering peaceful coexistence and collaboration. Technology serves as a bridge between cultures, allowing people to connect and communicate across borders. For instance, video conferencing tools like Zoom and Skype enable individuals from different countries to collaborate on projects, exchange ideas, and participate in dialogues about issues of mutual concern. This type of interaction fosters not only professional relationships but also personal connections that build empathy and understanding between individuals of diverse backgrounds (Deng, Shen, & Chan, 2021; Chen, 2022).

In the context of cross-cultural communication, technology facilitates the exchange of ideas, values, and experiences, ultimately fostering tolerance, empathy, and mutual respect. Online platforms such as social media, blogs, and podcasts provide individuals with the opportunity to share their stories, perspectives, and traditions. This exchange of ideas helps to break down stereotypes, challenge misconceptions, and promote a more nuanced understanding of different cultures. For example, social media influencers and activists from marginalized communities are using platforms like Instagram and Twitter to amplify their voices and raise awareness about issues such as racial injustice, gender equality, and indigenous rights. These digital spaces provide a powerful platform for promoting cultural awareness and advocating for positive social change (UNESCO, 2023; OECD, 2024).

However, despite the potential of technology to foster cultural understanding, there are significant challenges that need to be addressed. One of the most pressing issues is the digital divide—the gap between those who have access to technology and the internet and those who do not. While technology has enabled many people to connect across borders, this access is not universally available. In many parts of the world, especially in developing countries, access to reliable internet services and digital devices is limited or nonexistent. According to the International Telecommunication Union (ITU), nearly half of the world's population still lacks access to the internet. This digital divide creates a barrier to cultural exchange and understanding, as individuals in underserved regions are unable to fully participate in the global digital conversation (International Telecommunication Union, 2024).

The digital divide is not only a matter of access to technology but also one of digital literacy. In some areas, even where technology is available, people may not possess the skills necessary to navigate the digital landscape effectively. Without the knowledge of how to use digital tools and platforms, individuals are unable to take full advantage of the opportunities for cultural exchange and education that technology provides. This lack of digital literacy can limit their ability to engage in global conversations, access online courses, or communicate with people from other cultures (UNESCO, 2021; OECD, 2024).

Another challenge in using technology for cultural awareness is the potential for misinterpretation or distortion of cultural practices and traditions. While technology facilitates the sharing of cultural content, it can also lead to oversimplifications or misrepresentations. For example, videos or images shared online may not always capture the full context of a cultural practice, leading to misunderstandings or even cultural appropriation. The risk of misinterpretation is particularly high when individuals engage with cultures they are not familiar with, without fully understanding the cultural context or significance of certain practices. Therefore, it is essential for digital platforms to encourage cultural sensitivity and provide educational resources to help users better understand the cultures they are engaging with (UNESCO, 2023; Helm, 2024).

Moreover, the rapid spread of information on social media and other digital platforms can sometimes contribute to the amplification of harmful stereotypes or biased narratives. While these platforms allow for greater diversity of voices, they also make it easier for misinformation, fake news, and hate speech to spread. In the context of cultural exchange, this can undermine efforts to promote understanding and tolerance. For example, viral videos or posts that perpetuate negative stereotypes about a particular culture or group can contribute to the stigmatization of that community. Therefore, it is important for digital platforms to actively combat misinformation and create environments that foster positive, respectful dialogue (OECD, 2024; UNESCO, 2023).

Despite these challenges, the potential of technology to foster cultural understanding remains immense. Many organizations and initiatives are working to bridge the digital divide and increase access to technology in underserved communities. For example, nonprofit organizations like One Laptop per Child and the Digital Literacy Project are providing digital devices and training to children and adults in developing countries. These initiatives are helping to empower individuals and communities by providing them with the tools and skills needed to engage with the digital world, thus opening up new opportunities for cross-cultural exchange and education (International Telecommunication Union, 2024).

In addition to increasing access to technology, it is essential to ensure that digital platforms are designed with inclusivity and diversity in mind. Platforms that prioritize diverse voices, represent various cultural perspectives, and promote cultural sensitivity are more likely to foster meaningful cross-cultural communication. For instance, many social media platforms now offer features that allow users to translate content into different languages, making it easier for people from different linguistic backgrounds to communicate. These platforms also offer opportunities for users to share their personal stories and experiences, helping to create a more inclusive digital space (Meta AI, 2023).

Technology also plays a role in preserving and promoting cultural heritage. Digital tools are increasingly being used to document and share traditional knowledge, languages, and practices that may otherwise be at risk of disappearing. Projects such as the Endangered Languages Project and the UNESCO Memory of the World initiative are using digital technology to preserve and promote cultural heritage. These projects provide a platform for communities to share their cultural stories, traditions, and languages with the world, ensuring that they are not lost to history. In this way, technology serves as a tool for both cultural preservation and cultural exchange (Endangered Languages Project, 2025; UNESCO, 2025).

The role of technology in fostering cultural awareness is not limited to the digital realm. For instance, virtual reality (VR) and augmented reality (AR) technologies are opening up new possibilities for immersive cultural experiences. VR, in particular, allows users to virtually visit historical sites, museums, and cultural landmarks from around the world, providing them with a first-hand experience of different cultures. Similarly, AR technology is being used in educational settings to bring cultural artifacts and historical events to life, offering students a more interactive and engaging way to learn about other cultures. These technologies have the potential to enhance cross-cultural education by allowing individuals to experience cultures in a more immersive and meaningful way (A Comprehensive Review of VR in Cultural Heritage, 2024).

One of the most significant impacts of technology on cultural understanding is its ability to connect individuals in real-time across vast distances. Social media platforms like Twitter, Facebook, and Instagram have allowed people from different cultures to communicate instantaneously, exchange ideas, and share their experiences. This real-time interaction has created a global conversation in which individuals can discuss shared interests, challenges, and solutions. The ability to connect in this way has not only promoted cultural exchange but has also created global networks of support, particularly in times of crisis. For example, during natural disasters or humanitarian emergencies, technology has enabled individuals from around the world to offer assistance, share resources, and provide emotional support to affected communities (UNESCO, 2023; OECD, 2024).

While the benefits of technology for fostering cultural understanding are clear, it is important to approach digital engagement with caution and responsibility. Users must be mindful of the complexities of cross-cultural communication, and digital platforms should strive to promote respectful, informed interactions. Additionally, governments, organizations, and educators must work together to address the digital divide and ensure that access to technology is equitable. Only by doing so can technology truly fulfill its potential as a tool for cultural awareness and global understanding (International Telecommunication Union, 2024; UNESCO, 2023).

In conclusion, technology has the power to bridge cultural divides and promote understanding in ways that were once unimaginable. Through online platforms, language learning apps, virtual education, and digital cultural exchange, individuals can connect with others across the globe, learning about their histories, languages, and traditions. While challenges such as the digital divide, misinformation, and cultural misrepresentation remain, the overall impact of technology on cultural awareness is overwhelmingly positive. As access to technology continues to grow, and as platforms evolve to be more inclusive and sensitive to cultural nuances, technology will play an increasingly important role in fostering empathy, tolerance, and global citizenship (OECD, 2024; UNESCO, 2023).

THE FUTURE OF GLOBALIZATION AND CULTURAL EXCHANGE IN A TECHNOLOGICAL AGE

As technology continues to evolve, so too does the nature of globalization and cultural exchange. With innovations like virtual reality (VR) and augmented reality (AR), individuals now have opportunities to engage with cultures and experiences in ways that were previously unimaginable.

These technologies allow people to "visit" different parts of the world and immerse themselves in cultures, history, and environments—all from the comfort of their own homes. For example, VR can transport users to ancient historical sites such as the Pyramids of Egypt or the Great Wall of China, providing them with an immersive experience that is both educational and awe-inspiring. Similarly, AR can bring cultural artifacts and historical events to life, creating interactive learning experiences that were not possible in traditional educational settings (A Comprehensive Review of VR in Cultural Heritage, 2024).

The potential of VR and AR in cultural exchange is particularly significant in the fields of education and tourism. In education, VR can provide students with virtual field trips to museums, cultural landmarks, and natural wonders, allowing them to gain a deeper understanding of the world's diverse cultures and histories. These virtual experiences can supplement traditional classroom learning, offering students a richer, more engaging way to connect with the content. For instance, students studying ancient civilizations can virtually walk through ruins or interact with digital reconstructions of historical sites, giving them a more tangible sense of the past. In tourism, VR can offer prospective travelers a preview of destinations, helping them make informed decisions about where to travel. It also allows individuals who may not have the financial means or physical ability to travel to explore the world in a way that was once beyond their reach (Chen, 2022; Helm, 2024).

Moreover, the technology has great potential in cultural preservation. As communities face the challenges of modernity, cultural practices, languages, and traditions are at risk of being lost. VR and AR provide innovative solutions for documenting and sharing cultural heritage. For example, indigenous tribes could use VR to preserve and share their traditions, rituals, and languages with the world. By creating digital archives of these cultural practices, VR and AR offer an opportunity to safeguard traditions and ensure they are passed down to future generations. In this way, technology not only enhances our ability to experience other cultures but also helps preserve the cultures themselves for the benefit of humanity (UNESCO, 2025; Endangered Languages Project, 2025).

Artificial intelligence (AI) is another technology poised to transform cultural exchange. AI can analyze vast amounts of cultural data, providing insights into global trends, preferences, and behaviors. This analysis can enhance cultural exchange by making it easier for individuals to discover new music, films, literature, and art that they may otherwise not encounter. For example, AI-powered recommendation systems on streaming platforms like Spotify, Netflix, and YouTube are already shaping how users discover content. These systems analyze a user's viewing or listening habits and suggest similar content, offering a personalized and curated cultural experience. In the future, AI systems may become even more advanced, able to facilitate culturally relevant exchanges by considering regional preferences, languages, and historical context when recommending content (OECD, 2024).

In the realm of cross-cultural communication, AI holds the potential to break down language barriers, enabling individuals to communicate seamlessly regardless of their native tongue. Real-time translation (Barrault et al., 2023) tools, powered by AI, can help bridge the gap between speakers of different languages, facilitating more direct and meaningful exchanges. As AI-driven language translation tools become more accurate and sophisticated, they will allow people from different cultures to interact with one another in ways that were previously hindered by language differences. This could lead to more inclusive conversations and collaborative efforts across the globe, helping to promote mutual understanding and empathy (Meta AI, 2023).

While the potential benefits of technology in promoting cultural exchange are vast, the future of globalization and cultural exchange is not without its challenges. One of the most significant challenges that must be addressed is the ethical and political issues arising from technological advancements. As digital platforms increasingly shape global culture, questions about privacy, surveillance, data ownership, and the power of tech giants must be carefully considered. In many cases, the personal data of individuals is collected and analyzed without their full understanding or consent, raising concerns about privacy rights and the potential for exploitation. Additionally, the centralization of power in the hands of a few major tech companies could lead to a concentration of influence over the global digital landscape, limiting the diversity of voices and perspectives represented online (UNESCO, 2023).

One of the most pressing concerns about the rise of digital platforms is the lack of equitable representation of all cultures. While technology enables global cultural exchange, it also creates an imbalance in which dominant cultures, particularly those from wealthier, more developed nations, have greater access to platforms and resources for cultural dissemination. This can result in the marginalization of less economically developed cultures, whose voices may struggle to be heard on the global stage. Efforts to address this issue must include creating more inclusive platforms, promoting diverse cultural content, and ensuring that all cultures are given an equal opportunity to participate in the global conversation (OECD, 2024; UNESCO, 2023).

Additionally, the proliferation of digital platforms has raised concerns about the commodification of culture. As more cultural products are digitized and distributed online, there is a risk that cultural experiences may be reduced to mere entertainment or consumer products. This could lead to the erosion of cultural authenticity, as cultural practices are adapted or watered down to appeal to a global market. It is essential to ensure that cultural exchanges facilitated by technology remain respectful of the nuances and complexities of different cultures, avoiding the tendency to homogenize or commercialize these practices for profit (UNESCO, 2023; Helm, 2024).

The digital divide remains another significant barrier to equitable cultural exchange. While technology has made cross-cultural communication more accessible, access to the internet and digital platforms is not universal. In many parts of the world, individuals still face significant barriers to accessing the internet, whether due to economic factors, lack of infrastructure, or government censorship. This digital divide creates a disparity in cultural exchange, as individuals in underserved regions may have limited opportunities to participate in the global conversation. Closing this divide will be crucial in ensuring that the benefits of technology are shared by all people, regardless of their location or socioeconomic status (International Telecommunication Union, 2024).

Another issue to consider is the environmental impact of digital technologies. As more cultural content is created, shared, and consumed digitally, the energy consumption associated with data storage and transmission increases. The rapid expansion of digital platforms and content creation has led to a rise in energy consumption and electronic waste, which has significant environmental consequences. As the digital world grows, it will be essential to address the sustainability of these technologies, ensuring that they are developed and used in ways that minimize their environmental impact while continuing to support cultural exchange and communication (International Energy Agency, 2024).

Despite these challenges, the role of technology in promoting cultural exchange and globalization is undeniably transformative. As digital platforms continue to evolve and as new technologies like AI, VR, and AR are integrated into cultural experiences, the potential for greater cross-cultural understanding and connection increases. However, it is crucial to approach the development and use of these technologies with a sense of responsibility and awareness of the ethical, political, and environmental issues at play. By doing so, we can create a global digital ecosystem that fosters inclusive, respectful, and meaningful cultural exchange (UNESCO, 2023; OECD, 2024).

In conclusion, the interplay between technology, globalization, and cultural exchange is a dynamic and multifaceted process. Technological advancements such as AI, VR, and AR have the potential to break down barriers, foster cultural understanding, and make global cultural experiences more accessible. At the same time, these advancements must be carefully managed to ensure that they are used ethically and equitably. As we continue to explore the possibilities of technology in fostering cultural exchange, it is essential to prioritize inclusivity, respect, and sustainability. By doing so, we can harness the power of technology to create a more connected, empathetic world that celebrates cultural diversity and promotes global understanding (OECD, 2024; UNESCO, 2023).

CONCLUSION

In conclusion, technology has become a driving force in the evolution of globalization and cultural exchange, creating new opportunities for individuals to engage with diverse cultures and experiences. Innovations such as virtual reality, augmented reality, and artificial intelligence are reshaping how we learn about, interact with, and appreciate the world around us. These technologies not only offer immersive, interactive experiences but also hold the potential to break down barriers, allowing individuals from different cultural backgrounds to communicate and collaborate in unprecedented ways.

However, as we embrace the transformative power of technology, it is crucial to acknowledge and address the challenges that accompany these advancements. Ethical issues, such as privacy concerns, data ownership, and the representation of marginalized cultures, must be carefully considered to ensure that technological progress is inclusive and equitable. The digital divide, environmental impact, and the risk of commodifying culture all highlight the need for thoughtful, responsible engagement with technology.

As we move forward, it will be essential to strike a balance between innovation and responsibility, ensuring that technology serves as a tool for promoting cultural understanding, empathy, and cooperation. By fostering an environment that respects cultural diversity and encourages meaningful exchanges, we can create a global digital landscape that celebrates the richness of human experience and fosters a more connected, understanding, and sustainable world. In this way, technology can continue to enhance the process of globalization and cultural exchange, paving the way for a future that values inclusivity, respect, and mutual benefit (UNESCO, 2023; International Energy Agency, 2024).

CHAPTER 10 ESSAY PROMPTS

Globalization, Technology, and Cultural Exchange

1. How has technology contributed to the acceleration of globalization, and in what ways has it transformed cultural exchange?
2. Discuss the role of digital platforms in democratizing cultural production and consumption. What are the positive and negative impacts of this shift on global cultures?
3. Examine the potential risks and benefits of cultural homogenization due to the spread of global brands and Western media. How can technology help preserve local cultures in the face of globalization?
4. In what ways has technology facilitated cross-cultural communication, and how has this influenced the development of new global identities?
5. Analyze how virtual reality (VR) and augmented reality (AR) technologies are reshaping the way people experience and engage with different cultures. What are the implications for education, tourism, and cultural exchange?
6. What ethical challenges arise from the widespread use of technology in cultural exchange, particularly in relation to privacy, surveillance, and data ownership?
7. How can artificial intelligence (AI) contribute to cultural exchange by analyzing global trends and preferences? What are the potential risks and rewards of using AI to personalize cultural experiences?
8. Consider the impact of social media platforms on cultural exchange. How have platforms like Facebook, Twitter, and Instagram both facilitated and complicated cross-cultural interactions?
9. Evaluate the role of technology in supporting or hindering the preservation of indigenous cultures. How can technology be used to ensure the survival of endangered cultural practices?
10. How does globalization and technology challenge the traditional notions of national identity and cultural boundaries? Discuss the implications of these changes for the future of global culture.

REFERENCES

Barrault, L., Chung, Y.-A., Cora Meglioli, M., Dale, D., Dong, N., Duquenne, P.-A., Elsahar, H., Gong, H., Heffernan, K., Hoffman, J., Klaiber, C., Li, P., Licht, D., Maillard, J., Rakotoarison, A., Sadagopan, K. R., Wenzek, G., Ye, E., … Wang, S. (2023). *SeamlessM4T: Massively multilingual & multimodal machine translation* (arXiv Preprint No. 2308.11596). arXiv. https://arxiv.org/abs/2308.11596 arXiv

Chen, S. (2022). Facilitating the development of intercultural competence via virtual internationalization. *Frontiers in Communication, 7*, 794107. https://doi.org/10.3389/fcomm.2022.794107

Deng, L., Shen, Y. W., & Chan, J. W. W. (2021). Supporting cross-cultural pedagogy with online tools: Pedagogical design and student perceptions. *TechTrends, 65*(5), 760–770. https://doi.org/10.1007/s11528-021-00631-9

Endangered Languages Project. (2025). *Endangered Languages Project: Overview and resources.* https://endangeredlanguages.com/ Endangered Languages Project

Essafi, H., Belfakir, L., & Moubtassime, M. (2024). Investigating mobile-assisted language learning apps: Babbel, Memrise, and Duolingo as a case study. *Journal of Curriculum and Teaching, 13*(2), 197–[end pagination varies]. https://doi.org/10.5430/jct.v13n2p197

Hackett, S. E., Janssen, J., Beach, P., Perreault, M., Beelen, J., & van Tartwijk, J. (2023). The effectiveness of Collaborative Online International Learning (COIL) on intercultural competence development in higher education. *International Journal of Educational Technology in Higher Education, 20*, 5. https://doi.org/10.1186/s41239-022-00373-3 SpringerOpen

Helm, F. (2024). *Intercultural learning in virtual exchange.* Cambridge University Press. https://doi.org/10.1017/9781009385589

International Energy Agency. (2024). *Electricity 2024.* IEA. https://www.iea.org/reports/electricity-2024 IEA

International Telecommunication Union. (2024). *Measuring digital development: Facts and Figures 2024.* ITU. https://www.itu.int/dms_pub/itu-d/opb/ind/D-IND-ICT_MDD-2024-4-PDF-E.pdf ITU

Rodríguez-García, E., Guillén-Sanz, A., Checa, D., & Bustillo, A. (2024). A systematic review of virtual 3D reconstructions of cultural heritage in immersive virtual reality. *Multimedia Tools and Applications, 83*, 89743–89793. https://doi.org/10.1007/s11042-024-18700-3

OECD. (2024). *Facts not Fakes: Tackling disinformation, strengthening information integrity.* OECD Publishing. https://doi.org/10.1787/d909ff7a-en

UNESCO. (2021). *Media and information literacy curriculum for educators and learners: Think critically, click wisely!* UNESCO Publishing. https://www.unesco.org/mil4teachers/en/curriculum

UNESCO. (2023). *Guidelines for the governance of digital platforms: Safeguarding freedom of expression and access to information through a multistakeholder approach.* UNESCO Publishing. https://unesdoc.unesco.org/ark:/48223/pf0000387339

UNESCO. (2025). *Memory of the World Programme: Overview.* https://www.unesco.org/en/memory-world

The Future of Work: Ethics and Automation

11

ABSTRACT

Chapter 11 explores the ethical implications of automation in the future of work, examining its impact on employment, human dignity, and societal well-being. As automation technologies like artificial intelligence and robotics continue to evolve, concerns about job displacement, the erosion of human purpose, and the widening of economic inequalities are becoming increasingly pressing. The chapter delves into how these technologies challenge traditional notions of work and human value, while also addressing the ethical issues surrounding privacy, bias, and accountability in automated systems. It emphasizes the need for thoughtful policy, regulation, and global cooperation to ensure that automation benefits society as a whole, fostering a future where technology enhances human potential and promotes equitable opportunities.

INTRODUCTION

- The rapid advancement of automation technologies, including artificial intelligence (AI), robotics, and machine learning, is reshaping the global workforce in profound ways. (World Economic Forum, 2023; International Monetary Fund, 2024)
- Automation has already begun to replace tasks traditionally performed by humans in industries such as manufacturing, agriculture, logistics, and even service sectors. (McKinsey Global Institute, 2023; U.S. Bureau of Labor Statistics, 2024)

- This transformation promises increased efficiency and productivity, but it also raises significant ethical questions about the future of work. (Organisation for Economic Co-operation and Development, 2023)
- As machines take on more roles within the economy, the fundamental nature of employment, the value of human labor, and societal well-being are all being called into question. (International Labour Organization, 2023)
- The displacement of jobs and the growing reliance on machines to perform tasks traditionally carried out by humans are reshaping economic and social structures globally. (World Economic Forum, 2023; Stanford Institute for Human-Centered Artificial Intelligence, 2024)
- The result is an urgent need for a thorough examination of the implications of automation, especially with regard to its impact on employment, dignity, and social equality. (OECD, 2023; IMF, 2024)
- The impact of automation on employment is perhaps the most immediate concern. (ILO, 2023; WEF, 2023)
- As robots, AI, and other automated systems become more sophisticated, they are able to perform tasks faster, more accurately, and at lower costs than human workers. (McKinsey Global Institute, 2023)
- In certain industries, this has led to significant job losses, with workers being displaced by machines that can complete tasks more efficiently. (U.S. Bureau of Labor Statistics, 2024)
- For example, the rise of self-checkout kiosks and automated warehouses has already reduced the need for cashiers and manual laborers. (WEF, 2023)
- While the automation of low-skill jobs has long been a reality, advances in AI and robotics are now encroaching on more skilled professions as well. (OECD, 2023)
- Autonomous vehicles, for instance, could displace truck drivers and delivery personnel, while AI-driven software may replace legal assistants, journalists, and even medical professionals. (Stanford HAI, 2024; WEF, 2023)
- The scale and speed of job displacement caused by automation raise concerns about the long-term sustainability of the job market, especially in regions where retraining and education programs may not be readily accessible. (ILO, 2023; OECD, 2023)
- The growing prevalence of automation also introduces critical ethical questions about the future of work and human dignity. (OECD, 2023)

- At the core of these concerns is the question of what it means to be human in an age where machines can perform an increasing number of tasks previously thought to require human intelligence and creativity. (Pew Research Center, 2023)

- If machines can replace workers in areas such as driving, diagnosing diseases, and composing music, what role will human beings have in the workforce? (Stanford HAI, 2024)

- Some argue that automation could lead to a devaluation of human work, where people are left to occupy only jobs that machines cannot perform, potentially diminishing their sense of purpose and self-worth. (Pew Research Center, 2023)

- The shift towards a technology-driven economy also raises concerns about inequality. (IMF, 2024)

- Automation's benefits, such as increased productivity and efficiency, tend to accrue to the owners of the technology, leading to a concentration of wealth and power in the hands of a few large corporations. (IMF, 2024; WEF, 2023)

- This economic divide could exacerbate existing social inequalities, leaving marginalized groups and lower-income workers more vulnerable to job displacement and economic insecurity. (ILO, 2023)

- A critical challenge in the future of work is finding ways to ensure that automation benefits society as a whole, not just a select few. (OECD, 2023)

- One proposed solution is the concept of universal basic income (UBI), a policy that provides all citizens with a guaranteed income regardless of employment status. (Stockton Economic Empowerment Demonstration, 2021)

- Proponents of UBI argue that as automation continues to displace workers, a basic income could provide economic security and help people navigate the challenges of a changing job market. (SEED, 2021)

- However, critics of UBI argue that it could disincentivize work, create dependency on government support, and be financially unsustainable. (IMF, 2024)

- Others argue that the focus should be on policies that promote job creation in new industries, such as renewable energy, healthcare, and technology, where automation could complement human work rather than replace it. (OECD, 2023)

- Education and retraining programs will play a crucial role in helping workers adapt to the demands of an automated economy. (ILO, 2023; OECD, 2023)

- As automation renders some skills obsolete, it is essential to equip workers with the knowledge and capabilities to thrive in new and emerging fields. (OECD, 2023)
- However, the accessibility and quality of education and retraining programs vary widely depending on geographic location, socioeconomic status, and individual circumstances. (ILO, 2023)
- This raises another ethical challenge: how can we ensure that all workers, regardless of background, have access to the skills needed to succeed in the future economy? (OECD, 2023)
- Failure to address this issue could result in a widening gap between those who are able to transition into high-skilled jobs and those who are left behind in low-wage, insecure positions. (IMF, 2024)
- Another key ethical concern is the impact of automation on worker autonomy and decision-making. (OECD, 2023)
- In many industries, automation is being introduced as a tool to increase efficiency, reduce costs, and improve safety. (WEF, 2023)
- However, the widespread adoption of AI and robotics could also lead to a loss of control over work processes and decision-making. (Pew Research Center, 2023)
- As machines take over more tasks, workers may find themselves relegated to monitoring systems or performing tasks that are less engaging and offer fewer opportunities for creative problem-solving. (OECD, 2023)
- This shift could result in the erosion of job satisfaction, as workers feel disconnected from the work they are doing and the decisions that shape their roles. (Pew Research Center, 2023)
- Furthermore, the increasing reliance on automation raises important questions about accountability and transparency. (European Union, 2024)
- If an automated system makes a mistake or causes harm, who is responsible? (EU, 2024)
- In the case of autonomous vehicles, for example, who should be held accountable if a self-driving car causes an accident? (EU, 2024)
- While automation may improve efficiency, it also introduces complexities in determining liability when things go wrong. (EU, 2024)

Automation and human dignity: productivity gains, job shifts, and accountability to share the benefits

Read through a dignity-and-work lens, the image frames automation as a productivity engine that also reorganizes power on the job: it can remove drudgery and raise safety, yet it risks de-skilling, surveillance, and loss of autonomy if humans become "operators of opaque systems" rather than decision-makers. The practical prompt beneath the image is to pair innovation with counterweights—skills pathways into newly created roles, worker voice in design and deployment, transparent and auditable algorithms with human override, and fair sharing of gains through wages, time, and benefits—so efficiency advances without eroding human dignity.

- These questions become even more pressing as AI systems become more autonomous and capable of making decisions without human intervention. (Stanford HAI, 2024)
- Ethical frameworks and regulations must be developed to ensure that companies and governments take responsibility for the actions of automated systems, and that transparency is maintained in the development and deployment of these technologies (EU, 2024; World Health Organization, 2021).

AUTOMATION ALSO CHALLENGES EXISTING LABOR LAWS AND PROTECTIONS. (ILO, 2023)

- In a world where many jobs are being replaced by machines, how should the rights of workers be safeguarded? (OECD, 2023)
- Traditional labor protections, such as minimum wage laws, overtime pay, and collective bargaining, may no longer apply to workers who are replaced by robots or who are employed in gig economy jobs facilitated by technology. (OECD, 2023)
- The rise of automation in various sectors calls for a rethinking of labor laws and the development of new frameworks that address the rights of workers in an increasingly automated world. (ILO, 2023)
- For instance, how can we ensure that workers who are displaced by automation are provided with adequate support, such as unemployment benefits, retraining programs, and job placement assistance? (ILO, 2023)
- As automation continues to reshape the workforce, global cooperation and policy coordination will be crucial. (IMF, 2024)
- Automation is not confined to any single country or region; its effects are felt worldwide. (WEF, 2023)
- However, different countries are adopting automation technologies at different rates, and some nations are more equipped to manage the challenges posed by automation than others. (OECD, 2023)
- Developing countries, in particular, face unique challenges, such as limited access to technology, inadequate infrastructure, and a lack of skilled workers to take on new roles. (IMF, 2024)

- Global efforts to address these disparities will be essential in ensuring that the benefits of automation are distributed equitably and that no country or community is left behind. (ILO, 2023)

- One area where automation can be particularly beneficial is in addressing global challenges, such as climate change, healthcare, and poverty. (International Energy Agency, 2024; WHO, 2021)

- By automating processes in industries such as renewable energy, agriculture, and healthcare, we can increase efficiency and reduce waste, contributing to sustainability goals. (IEA, 2024)

- For instance, automated systems in agriculture can help optimize irrigation and reduce the use of pesticides, while AI-driven healthcare technologies can improve diagnosis and treatment outcomes. (WHO, 2021)

- However, the implementation of these technologies must be done thoughtfully, with careful consideration of the ethical implications and the potential for unintended consequences. (WHO, 2021)

- Finally, the future of work in an automated world requires a deep philosophical reflection on the nature of work itself. (Pew Research Center, 2023)

- As automation takes over more tasks, the concept of work may need to be redefined. (OECD, 2023)

- Traditionally, work has been seen as a means of survival and self-fulfillment, but in a world where automation frees us from many routine and labor-intensive tasks, we may need to reconsider how we derive meaning and purpose from our work. (Pew Research Center, 2023)

- Instead of focusing solely on productivity and efficiency, society may need to embrace new models of work that prioritize creativity, personal growth, and human connection. (OECD, 2023)

- In conclusion, automation presents both opportunities and challenges for the future of work. (WEF, 2023)

- While the benefits of increased productivity and efficiency are clear, the ethical questions it raises about employment, human dignity, and societal well-being cannot be ignored. (ILO, 2023)

- Thoughtful policies, global cooperation, and a rethinking of our approach to work will be necessary to ensure that automation serves to enhance human potential and create a more equitable and sustainable future for all. (IMF, 2024; OECD, 2023)
- Addressing the ethical implications of automation will require a concerted effort from governments, businesses, and civil society to develop frameworks that protect workers, promote human dignity, and ensure that technology works for the benefit of everyone. (ILO, 2023)

AUTOMATION AND EMPLOYMENT: A CHANGING LANDSCAPE

- One of the most significant concerns regarding automation is its impact on employment. (WEF, 2023; McKinsey Global Institute, 2023)
- Historically, technological advancements have led to changes in the types of jobs available, often rendering certain tasks obsolete but creating new ones. (WEF, 2023)
- For example, the rise of the automobile displaced carriage makers, but it also created new opportunities in the automotive industry. (OECD, 2023)
- However, the current wave of automation, driven by artificial intelligence (AI), robotics, and smart technologies, presents a unique challenge. (Stanford HAI, 2024)
- Jobs that once required human skill, decision-making, and emotional intelligence are now at risk of being automated, particularly in industries such as manufacturing, transportation, healthcare, and even professional services like law and accounting. (OECD, 2023)
- The rapid pace of automation means that entire sectors of the workforce are facing obsolescence, leaving workers with fewer opportunities to remain employed in their current fields. (WEF, 2023)
- This is particularly evident in industries that have traditionally relied on human labor for routine tasks. (U.S. Bureau of Labor Statistics, 2024)
- For instance, in manufacturing, robots have already replaced assembly line workers in many factories, performing tasks such as welding, painting, and assembly with precision and efficiency. (McKinsey Global Institute, 2023)
- While this has improved production rates, it has also led to significant job losses, especially for workers without the skills to transition to other fields. (ILO, 2023)

- As AI and robotics continue to improve, more sophisticated tasks previously thought to be outside the realm of automation are now becoming vulnerable. (OECD, 2023)
- Similarly, the transportation industry is facing dramatic changes due to the rise of autonomous vehicles. (WEF, 2023)
- The introduction of self-driving trucks and delivery drones threatens to eliminate millions of jobs in sectors such as long-haul trucking, taxi services, and courier delivery. (IMF, 2024)
- These roles, which once offered stable employment opportunities for a wide range of workers, are now at risk of disappearing, leaving many workers uncertain about their future prospects. (ILO, 2023)
- The displacement of workers in these industries could have cascading effects on related fields such as vehicle maintenance, insurance, and logistics, further deepening the economic disruption. (OECD, 2023)
- Beyond blue-collar jobs, automation is also making its way into traditionally white-collar professions, including law, accounting, and finance. (Stanford HAI, 2024)
- AI-driven software is now capable of analyzing legal documents, conducting research, and even drafting basic legal contracts, tasks that were once performed by paralegals and junior lawyers. (Stanford HAI, 2024)
- Similarly, in accounting, AI tools can process financial statements, perform audits, and generate tax filings with little to no human input. (OECD, 2023)
- These advancements are likely to reduce the demand for human workers in these professions, particularly in lower-level positions. (OECD, 2023)
- While higher-skilled roles that require complex decision-making and judgment may still be safe for the time being, the trend suggests that even skilled professionals will face increasing pressure to adapt to an automated world. (IMF, 2024)
- While automation has the potential to increase productivity and efficiency across many industries, it also risks displacing millions of workers, especially those employed in low-skilled and repetitive roles. (ILO, 2023)
- These workers are often the most vulnerable to technological disruption, as they tend to have fewer educational qualifications and fewer opportunities to acquire new skills. (OECD, 2023)

- Many of these workers may find it difficult to transition into emerging industries, particularly if they lack the training or resources needed to upskill. (ILO, 2023)
- For example, a truck driver who loses their job due to self-driving trucks may not have the skills to move into a tech-related field, such as software development or cybersecurity, which may require years of retraining. (WEF, 2023)
- The displacement of workers due to automation is not just a temporary challenge—it is a structural shift in the economy that could lead to long-term unemployment if not addressed. (IMF, 2024)
- As automation technologies become more affordable and accessible, they are likely to be adopted by businesses across all sectors, further reducing the need for human labor. (McKinsey Global Institute, 2023)
- This could lead to a situation where large swaths of the population are unable to find work, creating widespread social and economic instability. (OECD, 2023)
- In addition to the economic consequences, the displacement of workers due to automation raises important questions about the social and emotional well-being of those affected. (Pew Research Center, 2023)
- Work has long been a key component of human identity, providing individuals with a sense of purpose and a means of contributing to society. (Pew Research Center, 2023)
- Losing one's job due to automation can lead to a loss of dignity and self-worth, particularly if there are limited opportunities for workers to transition into new roles. (ILO, 2023)
- For individuals who have spent their lives in industries that are being automated, such as manufacturing or transportation, the psychological toll of job displacement can be profound, leading to issues such as depression, anxiety, and social isolation. (Pew Research Center, 2023)
- To mitigate the negative impacts of automation on employment, it is essential to develop strategies that focus on reskilling and upskilling workers. (ILO, 2023; OECD, 2023)
- As industries evolve, new jobs will be created, particularly in sectors such as healthcare, renewable energy, and technology. (WEF, 2023)
- However, these jobs often require a higher level of education and technical expertise than those traditionally available in industries vulnerable to automation. (OECD, 2023)

- For workers who have been displaced by automation, access to affordable retraining programs will be crucial. (ILO, 2023)
- Governments also have a critical role to play in addressing the challenges of automation-induced displacement. (IMF, 2024)
- One potential solution is the implementation of a universal basic income (UBI), which would provide a guaranteed income to all citizens regardless of employment status. (SEED, 2021)
- UBI could help cushion the economic impact of job losses due to automation, providing workers with the financial stability needed to retrain and reskill. (IMF, 2024)
- However, UBI remains a controversial proposal, with critics arguing that it could discourage work and place a significant strain on government resources. (IMF, 2024)
- In addition to UBI, other social safety nets, such as unemployment insurance, paid retraining programs, and job placement services, could help workers transition to new opportunities in an automated economy. (ILO, 2023)
- Policymakers will need to carefully balance the need to support displaced workers with the goal of fostering innovation and economic growth. (OECD, 2023)
- Creating an environment in which workers are equipped to adapt to the changing landscape of work will be essential to minimizing the negative effects of automation on society. (ILO, 2023)
- Furthermore, businesses have a responsibility to consider the ethical implications of automation on their workers. (OECD, 2023)
- While automation can increase efficiency and profitability, companies should also be mindful of the impact their decisions have on their employees. (WEF, 2023)
- Ethical business practices should include investing in training and development programs to help workers transition into new roles, offering severance packages for displaced workers, and providing opportunities for workers to take on more complex and rewarding tasks as automation takes over repetitive ones. (ILO, 2023)
- As the workforce becomes increasingly automated, labor unions will play an important role in advocating for workers' rights and protections. (ILO, 2023)
- Unions can help ensure that workers are not left behind as automation transforms industries, pushing for fair wages, job security, and access to retraining programs. (OECD, 2023)

- They can also play a key role in advocating for policies that protect workers' dignity and well-being in the face of technological change. (ILO, 2023)
- In the longer term, it will be necessary to rethink the concept of work itself. (Pew Research Center, 2023)
- If automation significantly reduces the need for human labor, society may need to reconsider its approach to work and leisure. (OECD, 2023)
- Instead of focusing solely on productivity, there may be a shift toward valuing human creativity, innovation, and personal fulfillment. (Pew Research Center, 2023)
- This could lead to the development of new social and economic models, such as a reduced workweek or a focus on work that contributes to societal well-being, such as caregiving, education, and community service. (OECD, 2023)
- In conclusion, while automation holds the potential to revolutionize industries and increase efficiency, it also presents significant challenges for workers and society. (WEF, 2023)
- As automation continues to displace jobs across a wide range of sectors, it is essential to develop strategies that support workers in transitioning to new roles, ensure that the benefits of automation are widely shared, and protect the dignity of those affected by job displacement. (ILO, 2023; IMF, 2024)
- Addressing these challenges will require cooperation between businesses, governments, and workers to create a future of work that is equitable, sustainable, and human-centered. (OECD, 2023)

THE CHALLENGE TO HUMAN DIGNITY AND PURPOSE

- Beyond the economic concerns, automation poses ethical questions related to human dignity and purpose. (OECD, 2023)
- Work has long been a central part of human identity, providing individuals not only with financial stability but also a sense of contribution and purpose. (Pew Research Center, 2023)
- It is through work that many individuals find their role in society, establish a sense of community, and develop personal pride. (Pew Research Center, 2023)

- As automation increasingly takes over tasks traditionally performed by humans, the nature of work itself is being called into question. (Stanford HAI, 2024)
- If machines can perform tasks faster, more efficiently, and without the need for rest, what does that mean for the dignity of human labor? (OECD, 2023)
- This dilemma highlights the tension between technological progress and the fundamental need for human involvement in the workforce. (Pew Research Center, 2023)
- The loss of jobs to automation may not only erode financial security but also challenge individuals' sense of self-worth. (Pew Research Center, 2023)
- For many, work is not just a means to an economic end; it is a way of participating in society, contributing to a common goal, and feeling valued. (Pew Research Center, 2023)
- The idea that a machine could perform a task previously done by humans might lead to a loss of confidence and feelings of inadequacy, as individuals struggle to find their place in a world where their traditional roles are disappearing. (ILO, 2023)
- This is particularly true for workers who have spent their entire lives in fields that are now increasingly automated, such as manufacturing, driving, or clerical work. (ILO, 2023)
- The personal implications of job loss extend far beyond financial hardship and include psychological and social effects such as depression, anxiety, and a sense of purposelessness. (Pew Research Center, 2023)
- Moreover, as automation continues to replace human workers, it raises questions about what constitutes meaningful work in an increasingly automated world. (OECD, 2023)
- In many cases, the jobs created by automation may require specialized skills or higher levels of education, creating a divide between those who can adapt and those who cannot. (IMF, 2024)
- The demand for workers in technical or creative fields, such as software development, artificial intelligence, and design, is rising, but these roles often require years of education and training. (OECD, 2023)
- Meanwhile, workers displaced by automation may struggle to find meaningful employment, especially if they lack the necessary skills to transition to these new fields. (ILO, 2023)

- The situation becomes particularly challenging for individuals in low-income or marginalized communities, who may not have access to the resources needed to retrain or reskill. (ILO, 2023)
- This issue is especially pronounced for older workers, who may have spent decades in a particular trade or occupation and may find it difficult to start anew in a rapidly changing job market. (OECD, 2023)
- For example, a factory worker who loses their job due to automation may not have the time or financial resources to retrain for a job in a tech-related field, particularly if they have little to no experience with computers or coding. (ILO, 2023)
- The result is a growing gap between those who can adapt to the changing demands of the labor market and those who cannot, leading to increased inequality and social division. (IMF, 2024)
- The challenge, therefore, is not just about the number of jobs lost but also about the value placed on human labor and the dignity associated with meaningful, fulfilling work. (OECD, 2023)
- As more tasks are automated, society may begin to question the very notion of what it means to "work" and how the fruits of that labor are distributed. (Pew Research Center, 2023)
- If machines can do much of the work, what is left for humans to contribute? (Stanford HAI, 2024)
- This question brings to the forefront the issue of human dignity and the need to ensure that people continue to feel valued, regardless of the changing nature of work. (OECD, 2023)
- The erosion of meaningful work also has significant social implications. (Pew Research Center, 2023)
- Employment has long been a key source of identity and community for many individuals, providing not only financial stability but also a sense of belonging. (Pew Research Center, 2023)
- Workplaces serve as social hubs, offering individuals a chance to interact with others, form relationships, and contribute to a shared purpose. (OECD, 2023)

- As automation displaces human workers, there is a risk that communities will become fragmented, with people losing their connections to both their work and their fellow human beings. (ILO, 2023)
- The breakdown of these social networks could contribute to social isolation, alienation, and a general decline in well-being, which may ultimately lead to greater social unrest. (Pew Research Center, 2023)
- Addressing the ethical concerns of automation requires a broader societal dialogue about the role of work in human life and how to ensure that all individuals, regardless of their skill level or education, can find meaningful ways to contribute to society. (OECD, 2023)
- Governments, businesses, and educators must work together to create policies and initiatives that promote lifelong learning, skill development, and the redistribution of work. (ILO, 2023)
- For example, instead of focusing solely on creating jobs that mimic traditional employment structures, there could be a shift toward creating new forms of work that provide personal fulfillment and contribute to societal well-being, such as caregiving, environmental protection, and artistic expression. (OECD, 2023)
- At the same time, society must reconsider the relationship between work and economic security. (IMF, 2024)
- The traditional model of employment, where individuals work for a company and receive a salary or wage, may no longer be the only or best way to ensure that people are able to meet their needs. (OECD, 2023)
- Universal basic income (UBI) has been proposed as a potential solution to this problem, providing individuals with a guaranteed income regardless of their employment status. (SEED, 2021)
- UBI would allow individuals to pursue work that is meaningful to them without the pressure of financial instability, potentially mitigating some of the negative effects of job displacement due to automation. (IMF, 2024)
- However, the implementation of UBI raises difficult questions about funding, societal values, and the role of work in defining human dignity. (IMF, 2024)
- Another solution to the ethical concerns of automation lies in redefining the concept of labor itself. (OECD, 2023)

- As automation continues to take over routine, low-skilled tasks, human labor could be increasingly focused on areas that require creativity, emotional intelligence, and complex problem-solving. (Stanford HAI, 2024)
- These are areas where humans continue to excel and where the unique qualities of human beings—such as empathy, intuition, and critical thinking—cannot easily be replicated by machines. (Pew Research Center, 2023)
- By shifting the focus of work towards these inherently human capabilities, society could create a future where humans are not merely performing repetitive tasks but engaging in work that is intellectually and emotionally fulfilling. (OECD, 2023)
- As we confront the ethical implications of automation, it is essential to ensure that the future of work remains inclusive, equitable, and human-centered. (ILO, 2023)
- This means taking active steps to support workers in their transition to new roles and industries, promoting access to education and training, and encouraging the development of work that aligns with human dignity and purpose. (OECD, 2023)
- It also means recognizing the value of work that is often overlooked, such as caregiving, volunteerism, and community-building, which contribute immeasurably to societal well-being but are often undervalued in traditional economic models. (OECD, 2023)
- The impact of automation on human dignity also calls for a broader cultural shift in how we view work and success. (Pew Research Center, 2023)
- Society must move away from a narrow focus on productivity and profit as the ultimate measures of success and instead embrace a more holistic understanding of human well-being. (OECD, 2023)
- This would involve valuing work that contributes to the emotional, social, and environmental health of society, as well as recognizing the worth of individuals who may not participate in the workforce in the traditional sense. (Pew Research Center, 2023)
- By rethinking how we define work and success, we can ensure that automation enhances, rather than diminishes, human dignity. (ILO, 2023)
- Ultimately, the future of work in an age of automation will depend on our collective ability to adapt to these technological shifts while safeguarding the values that make work meaningful. (OECD, 2023)
- This will require careful thought, ethical decision-making, and a commitment to ensuring that the benefits of automation are shared equitably. (IMF, 2024)

- By embracing new ways of working and redefining the role of labor in society, we can create a future where human dignity remains intact and where all individuals have the opportunity to lead fulfilling lives, regardless of their employment status. (ILO, 2023)
- In conclusion, automation poses significant ethical challenges, particularly regarding the dignity of human labor. (OECD, 2023)
- As machines increasingly take over routine tasks, it is essential to ensure that humans continue to find purpose and fulfillment in their work. (Pew Research Center, 2023)
- This requires a thoughtful approach to reskilling, job creation, and the redistribution of labor, as well as a broader cultural shift in how we view work and its role in our lives. (ILO, 2023)
- By addressing these ethical concerns, we can create a future of work that is both technologically advanced and human-centered, where individuals are valued for their unique contributions to society. (WEF, 2023)

SOCIETAL WELL-BEING AND THE RISKS OF INEQUALITY

- The widespread implementation of automation presents significant risks to societal well-being, particularly in terms of inequality. (IMF, 2024; OECD, 2023)
- As technology rapidly advances, automation is reshaping industries and creating both opportunities and challenges. (WEF, 2023)
- While it can drive increased efficiency and productivity, it also raises the specter of growing inequality. (IMF, 2024)
- Automation disproportionately impacts low-wage, low-skilled workers, who are most vulnerable to job displacement, while individuals with the capital, skills, and education to develop and implement new technologies stand to benefit the most. (ILO, 2023)
- This growing divide risks leaving a significant portion of society behind, deepening existing social and economic disparities. (OECD, 2023)
- In many industries, automation is already beginning to replace tasks that were traditionally performed by humans, especially in fields such as manufacturing, transportation, and retail. (U.S. Bureau of Labor Statistics, 2024)

- These jobs are often low-skilled and offer limited opportunities for advancement. (ILO, 2023)
- As machines and artificial intelligence (AI) increasingly take over these roles, many workers are left without clear alternatives. (OECD, 2023)
- This results in a scenario where the benefits of automation accrue to a small group of wealthy tech entrepreneurs, investors, and highly educated workers, while the majority of the workforce faces job insecurity, reduced wages, or even permanent displacement. (IMF, 2024)
- This exacerbates income inequality, as those who own or have access to capital reap the rewards of increased productivity, while others are left to struggle in an economy where opportunities are scarce. (OECD, 2023)
- The impact of automation on employment is especially pronounced in low-wage sectors that rely on human labor for routine, repetitive tasks. (ILO, 2023)
- For example, in industries such as fast food, retail, and warehousing, automation technologies like robotic systems and AI-driven tools are already replacing jobs that once provided stable income for many workers. (WEF, 2023)
- At the same time, new job opportunities in the tech and service sectors tend to require advanced skills or specialized education that many displaced workers may not possess. (OECD, 2023)
- As a result, the unemployed may be unable to secure new jobs that offer comparable wages or job security, leading to long-term financial instability and social unrest. (IMF, 2024)
- The economic benefits of automation, if not distributed equitably, are likely to be concentrated in the hands of a few large corporations and their executives. (IMF, 2024)
- These companies, often the drivers of technological innovation, may see significant increases in profits due to the reduced cost of labor and greater efficiency provided by automation. (OECD, 2023)
- However, the wealth generated by these technologies may not trickle down to the broader population. (IMF, 2024)
- Instead, it could further entrench the power and wealth of corporations, leaving a growing proportion of the population excluded from the economic benefits of technological progress. (OECD, 2023)

- This concentration of wealth could lead to increased political and economic instability, as large portions of society struggle to make ends meet. (IMF, 2024)
- Furthermore, the displacement of workers due to automation could lead to a growing divide between the educated elite and those left behind in low-skilled occupations. (ILO, 2023)
- Workers without access to retraining or reskilling programs may find themselves trapped in a cycle of unemployment or underemployment, with few opportunities to improve their circumstances. (OECD, 2023)
- As the labor market becomes increasingly polarized, the gap between those who can thrive in the new economy and those who cannot could lead to a fractured society, where access to opportunities and quality of life is determined by one's ability to adapt to technological changes. (IMF, 2024)
- In addition to economic inequality, automation poses risks to the social fabric of society. (Pew Research Center, 2023)
- As large segments of the population experience job displacement and economic hardship, feelings of alienation and disillusionment may grow. (Pew Research Center, 2023)
- Work has long been a source of personal identity, dignity, and community. (Pew Research Center, 2023)
- As automation takes over more tasks, many individuals may find themselves without the meaningful work that once provided a sense of purpose. (OECD, 2023)
- This loss of purpose, combined with economic uncertainty, could contribute to increased social unrest, mental health issues, and a general decline in societal well-being. (Pew Research Center, 2023)
- To address the risks posed by automation, policymakers must consider a range of approaches to ensure that the benefits of automation are distributed equitably. (IMF, 2024; OECD, 2023)
- One such approach is universal basic income (UBI), a system that provides all citizens with a guaranteed income regardless of their employment status. (SEED, 2021)
- UBI could help mitigate the economic impact of job displacement by providing a financial safety net for those who lose their jobs to automation. (IMF, 2024)

- This approach could also reduce economic inequality by ensuring that all individuals, regardless of their employment situation, have a basic level of financial security. (OECD, 2023)
- However, the implementation of UBI would require significant adjustments to the current economic system, including new taxation models and government spending priorities. (IMF, 2024)
- In addition to UBI, investments in education, job retraining, and lifelong learning programs will be crucial in equipping workers with the skills needed to thrive in an increasingly automated world. (ILO, 2023; OECD, 2023)
- Workers who are displaced by automation must be given the tools and resources to transition into new roles that are less susceptible to automation. (ILO, 2023)
- This might involve retraining programs that focus on emerging fields such as technology, healthcare, or green energy, where there is expected to be growing demand for human labor. (OECD, 2023)
- By providing access to education and training, society can help workers stay relevant in a rapidly changing job market and prevent the deepening of inequality. (ILO, 2023)
- Governments and businesses must also collaborate to create new models of work that are less reliant on traditional full-time employment. (OECD, 2023)
- As automation reduces the demand for human labor in certain sectors, there may be opportunities to create new forms of work that focus on areas such as creativity, innovation, caregiving, and community building. (OECD, 2023)
- These jobs, while often undervalued in traditional economic systems, are essential for the well-being of society and could provide meaningful work for individuals who are displaced by automation. (ILO, 2023)
- By shifting the focus of the economy towards these forms of labor, it may be possible to create more inclusive and sustainable employment opportunities for all. (OECD, 2023)
- Another potential solution to mitigate the risks of automation is the concept of a "job guarantee." (ILO, 2023)
- This idea proposes that the government would act as an "employer of last resort," providing jobs for anyone who is willing and able to work. (OECD, 2023)

- These jobs could focus on public services, infrastructure development, or environmental sustainability, and would provide individuals with a steady income and a sense of purpose. (ILO, 2023)
- By guaranteeing employment, society could ensure that everyone has access to meaningful work, even in the face of automation-driven job displacement. (OECD, 2023)
- To foster a more inclusive future, it will also be necessary to rethink the nature of work itself. (OECD, 2023)
- Rather than simply focusing on economic output and productivity, we must begin to consider how work can contribute to broader societal goals, such as environmental sustainability, social equity, and community well-being. (IMF, 2024)
- By shifting the focus away from the narrow pursuit of profit and growth, it may be possible to create a more just and equitable society where work is valued not just for its economic contributions but also for its social and cultural impact. (OECD, 2023)
- Ultimately, to address the risks posed by automation, societies must adopt a comprehensive, long-term approach that includes a combination of economic, educational, and social reforms. (IMF, 2024; OECD, 2023)
- By prioritizing equity, access to opportunity, and the value of human dignity in the future of work, it is possible to ensure that automation enhances societal well-being rather than exacerbating inequality. (ILO, 2023)
- Only through collaboration between governments, businesses, and communities can we create a future where technology benefits all members of society, not just the privileged few. (WEF, 2023)
- The future of work in an era of automation will undoubtedly present significant challenges, but it also offers the opportunity for positive transformation. (WEF, 2023)
- With the right policies, investments, and cultural shifts, we can create a world where technology serves to uplift, rather than divide, humanity. (IMF, 2024)
- By embracing innovation while safeguarding human well-being, we can ensure that the benefits of automation are shared equitably, creating a more just and prosperous society for future generations. (OECD, 2023)
- In conclusion, the widespread implementation of automation poses substantial risks to societal well-being, especially in terms of economic inequality. (IMF, 2024)

- As technology continues to advance, policymakers must prioritize equitable solutions such as universal basic income, job retraining, and inclusive forms of work to ensure that the benefits of automation are widely shared. (ILO, 2023; OECD, 2023)

- By focusing on these solutions, societies can mitigate the risks of job displacement, economic instability, and social unrest, creating a future where automation enhances, rather than diminishes, overall societal well-being. (WEF, 2023)

ETHICAL CONSIDERATIONS: RESPONSIBILITY AND ACCOUNTABILITY

As automation technologies continue to evolve and integrate into various industries, ethical considerations related to responsibility and accountability have become increasingly important. One of the most pressing questions is who should be held accountable when an autonomous system makes a mistake or malfunctions. For example, when a self-driving car causes an accident or an AI system makes a critical error in diagnosing a medical condition, determining responsibility is not always straightforward. Should the manufacturer, the software developer, the operator, or the AI system itself be held liable? These situations highlight the need for clear and comprehensive legal frameworks that can address accountability in an era of advanced automation. Without such frameworks, there is a risk that victims of AI errors will be left without recourse or compensation. (European Union, 2024; Council of Europe, 2024; National Institute of Standards and Technology, 2023; World Health Organization, 2021)

The ethical implications of automation extend beyond accountability for accidents and mistakes. As AI systems are integrated into everyday life, concerns about privacy and consent are also at the forefront. For instance, AI systems often rely on vast amounts of personal data to function effectively. In healthcare, facial recognition, and even social media algorithms, these systems analyze sensitive information about individuals, raising questions about how that data is collected, stored, and used. Who owns the data, and how can individuals ensure their privacy is protected in an increasingly automated world? The use of personal data by AI systems must be carefully regulated to prevent abuse and ensure that individuals retain control over their own information. (Information Commissioner's Office, 2023; European Data Protection Board, 2022; UNESCO, 2023)

Equally concerning is the potential for bias in AI systems. Since many AI models are trained on historical data, they can inadvertently reinforce existing biases and inequalities. For example, facial recognition software has been shown to have higher error rates for people of color, leading to concerns about its application in law enforcement and security. If AI systems are trained on biased data or without diverse perspectives, they can perpetuate stereotypes and discriminatory practices, further marginalizing already vulnerable communities. It is therefore essential to ensure that AI systems are designed and tested in a way that actively seeks to identify and mitigate biases. This can be achieved through inclusive data sets, rigorous testing, and ongoing audits to assess the impact of AI on different demographic groups. (National Institute of Standards and Technology, 2022; European Data Protection Board, 2022; Stanford Institute for Human-Centered Artificial Intelligence, 2024)

The issue of fairness in AI is not only a question of preventing bias but also about ensuring that automation benefits all members of society equitably. As AI becomes more pervasive in decision-making processes, such as hiring, lending, and criminal justice, there is a growing concern about how these systems may disproportionately affect certain groups. For example, algorithms used in hiring may inadvertently favor candidates from certain demographic backgrounds, leading to discriminatory hiring practices. Similarly, AI-driven predictive policing systems could reinforce existing patterns of racial profiling, exacerbating inequalities in the criminal justice system. To address these concerns, AI systems must be designed with fairness in mind, ensuring that they do not inadvertently perpetuate social inequalities. (U.S. Equal Employment Opportunity Commission, 2023; Organisation for Economic Co-operation and Development, 2023; Group of Seven, 2023)

Automation's impact on decision-making extends beyond concerns about bias and fairness. As more and more processes are automated, the role of human judgment is being diminished in critical areas such as healthcare, education, and criminal justice. AI systems may be able to analyze data and make decisions more quickly than humans, but they lack the ability to understand the nuances of individual situations or the emotional intelligence needed to navigate complex social interactions. In healthcare, for example, AI systems may be able to diagnose diseases based on patterns in medical data, but they cannot provide the empathy and understanding that human doctors offer their patients. Similarly, in education, AI can assess students' learning progress, but it cannot replace the emotional support and guidance that teachers provide. In criminal justice, AI systems used for sentencing or parole decisions may overlook important human factors, such as the potential for rehabilitation or the socio-economic circumstances of the individual involved. (World Health Organization, 2021; Stanford Institute for Human-Centered Artificial Intelligence, 2024)

These limitations of automation raise the question of whether machines should be allowed to replace human workers in areas that require empathy, judgment, and understanding. While automation can improve efficiency and reduce costs, it may not always be the best choice for situations where human experience and compassion are necessary. In fields like caregiving, for example, AI and robotics may be used to assist or replace human workers in providing care to elderly or disabled individuals. While robots may be able to perform basic tasks, such as administering medication or monitoring vital signs, they cannot replicate the warmth, understanding, and emotional connection that human caregivers provide. Striking the right balance between technological efficiency and the need for human compassion will be crucial as automation continues to permeate sensitive areas of life. (Organisation for Economic Co-operation and Development, 2023; World Health Organization, 2021)

One of the key challenges in the ethical deployment of automation is determining where the line should be drawn between what is appropriate for automation and what should remain a human responsibility. In many cases, automation may be best suited for tasks that are repetitive, dangerous, or require high levels of precision, such as in manufacturing or logistics. However, tasks that require complex decision-making, emotional intelligence, or ethical reasoning may be better suited for human workers. The challenge, therefore, is to ensure that automation is deployed in a way that enhances human well-being, rather than replacing the valuable qualities that humans bring to the workplace. This will require careful consideration of the roles that humans and machines should play in society. (National Institute of Standards and Technology, 2023; ISO/IEC, 2023a)

Another ethical concern with automation is the potential for job displacement and the loss of livelihood for millions of workers. As machines and AI systems take on more tasks, many workers may find themselves out of a job, particularly those in industries where automation is most easily implemented, such as manufacturing, retail, and transportation. This displacement could result in economic hardship for individuals who lack the skills or education necessary to transition into new roles. In light of this, policymakers must consider ways to ensure that displaced workers are supported through retraining programs, social safety nets, and opportunities for new types of work that leverage human creativity, judgment, and interpersonal skills. (International Labour Organization, 2023; Organisation for Economic Co-operation and Development, 2023)

The growing reliance on automation also raises questions about the social implications of reducing human involvement in work. As work becomes increasingly automated, there may be a shift in societal values, with an increased emphasis on leisure, creativity, and personal fulfillment. While this could lead to a more balanced and fulfilling life for many, it also raises concerns about the potential for social isolation and a loss of purpose for those who have traditionally found meaning in their work. The challenge will be to ensure that automation is used in a way that promotes human flourishing, rather than undermining the sense of purpose that work has provided for generations. (International Labour Organization, 2023; Organisation for Economic Co-operation and Development, 2023)

Moreover, ethical frameworks for automation must account for the environmental impact of AI and robotics. While automation has the potential to improve efficiency and reduce waste in certain industries, it also poses environmental risks. The production, maintenance, and disposal of robots and AI systems can contribute to resource depletion and environmental pollution. Additionally, as automation leads to more efficient production processes, it may result in increased consumption and waste. It is essential that ethical guidelines for automation consider the broader environmental impact of these technologies, promoting sustainability and responsible resource management. (International Energy Agency, 2024)

Ultimately, the ethics of automation will require ongoing dialogue among developers, policymakers, workers, and society at large. As automation continues to reshape industries and society, it is essential to address the ethical implications of these technologies in a way that promotes fairness, accountability, and human well-being. This will involve creating clear guidelines for responsibility and accountability, ensuring that AI and automation systems are transparent and free from bias, and finding ways to balance technological efficiency with human empathy and judgment. By prioritizing these ethical considerations, we can help ensure that automation serves the greater good and contributes to a more just and equitable future. (European Union, 2024; National Institute of Standards and Technology, 2023; Group of Seven, 2023; ISO/IEC, 2023b)

In conclusion, automation presents both immense opportunities and significant ethical challenges. As AI and robotics become increasingly integrated into various aspects of life, it is crucial to develop frameworks that address responsibility, accountability, privacy, bias, and the loss of human judgment. By engaging in a thoughtful and inclusive conversation about these issues, we can shape a future where automation is used responsibly and ethically, enhancing human well-being while avoiding harmful consequences. (Council of Europe, 2024; UNESCO, 2023)

THE NEED FOR ETHICAL GUIDELINES AND REGULATION

In light of the numerous ethical challenges posed by the rise of automation, the development of comprehensive ethical guidelines and regulatory frameworks is more important than ever. Governments, businesses, and technology developers must collaborate to create standards that prioritize the well-being of individuals and society as a whole. These frameworks will need to address a wide range of issues, from job displacement and income inequality to privacy, transparency, and fairness in AI systems. Given the rapid pace of technological development, these ethical guidelines must not be static; they should be adaptable and able to evolve as new technologies emerge, ensuring that the societal impact of these innovations is properly assessed and addressed in real-time. (European Union, 2024; Department for Science, Innovation and Technology, 2023; ISO/IEC, 2023a)

A key area where regulatory bodies will need to take action is in sectors where the stakes are particularly high, such as hiring, criminal justice, and healthcare. In these fields, AI systems and automated decision-making tools can have life-altering consequences. For example, AI-driven hiring algorithms may inadvertently introduce biases that exclude qualified candidates from underrepresented groups, or predictive policing tools might disproportionately target certain communities. Similarly, in healthcare, AI systems used for diagnosing diseases or recommending treatments must be designed to reduce errors and ensure that human oversight remains central in life-critical decisions. Governments must implement regulations that ensure these technologies are used responsibly, mandating transparency and fairness to avoid harmful outcomes. (U.S. Equal Employment Opportunity Commission, 2023; World Health Organization, 2021; European Data Protection Board, 2022)

In addition to setting regulatory standards for these high-stakes applications, policies should be developed to incentivize businesses to use automation technologies in ways that benefit both workers and employers. Rather than solely focusing on the efficiency gains that automation can bring, policymakers should encourage companies to adopt technologies that have positive social outcomes, such as creating new job opportunities or enhancing worker capabilities. For instance, tax incentives could be offered to companies that invest in reskilling programs for employees displaced by automation, or that develop job roles which require human oversight of AI systems. These initiatives would help bridge the gap between technological progress and its socio-economic consequences, ensuring that the benefits of automation are more evenly distributed. (International Labour Organization, 2023; Organisation for Economic Co-operation and Development, 2023)

Moreover, businesses should be encouraged to consider the broader social implications of the technologies they develop. Companies that build AI and automation systems have a responsibility to ensure that these tools do not exacerbate existing inequalities or contribute to societal harm. This could involve implementing internal guidelines for ethical development, conducting regular impact assessments, and engaging with diverse stakeholders to gather feedback on the societal effects of new technologies. For example, AI developers should involve ethicists, social scientists, and representatives from marginalized communities in the design process to ensure that their products do not unintentionally perpetuate bias or exclusion. (Ada Lovelace Institute, 2022; National Institute of Standards and Technology, 2023)

While national regulations are crucial, the global nature of technology and its rapid spread across borders make international cooperation essential. Automation technologies are being adopted worldwide at different rates, and their effects on labor markets, intellectual property, and corporate accountability vary depending on the region. Therefore, it is essential for countries to work together to establish international agreements and frameworks that address the challenges posed by automation. Such agreements could focus on issues such as cross-border labor mobility, fair wages, and the protection of workers' rights in a globalized, automated economy. This would help ensure that the benefits of automation are shared globally, rather than deepening existing inequalities between developed and developing countries. (Group of Seven, 2023; Council of Europe, 2024; Organisation for Economic Co-operation and Development, 2023)

One of the challenges in establishing international agreements on automation is the differing regulatory approaches and priorities of countries. While some nations may prioritize economic growth and technological innovation, others may be more focused on social welfare and workers' rights. As a result, the negotiation of international frameworks will require compromises and a balance between these varying priorities. Nevertheless, global cooperation will be key to ensuring that automation benefits humanity as a whole, rather than just the wealthiest nations and corporations. (Council of Europe, 2024; Group of Seven, 2023)

In addition to labor and economic issues, global cooperation on automation will also need to address intellectual property rights. As AI systems become more capable of generating new ideas, inventions, and designs, the question of who owns the intellectual property created by machines becomes increasingly complex. For example, if an AI system develops a new drug or creates an innovative product, who should be credited with the invention—the AI, the company that owns the AI, or the programmers who developed it? International agreements on intellectual property must adapt to the realities of automated innovation, ensuring that creators, workers, and consumers are fairly treated. (ISO/IEC, 2023b; Department for Science, Innovation and Technology, 2023)

Another important consideration for global cooperation is the ethical use of automation in areas like surveillance, security, and warfare. As AI systems become more advanced, their use in surveillance and military applications raises serious concerns about privacy, human rights, and the potential for misuse. Governments must establish clear guidelines for the ethical use of these technologies, ensuring that they are not used to violate individual freedoms or escalate conflicts. International agreements could provide a framework for ensuring that automated surveillance tools and military robots are used responsibly, with a focus on protecting civil liberties and preventing unnecessary harm. (European Union Agency for Cybersecurity, 2023; Council of Europe, 2024)

While the focus on ethical guidelines and international cooperation is critical, it is equally important to ensure that public engagement and education play a central role in shaping the future of automation. As automation technologies become more pervasive, it is crucial for the public to have a voice in how these systems are developed and deployed. Governments and businesses should engage in transparent dialogues with communities, labor unions, and advocacy groups to address concerns and foster a broader understanding of automation's impact on society. Public engagement can also help identify potential blind spots in policy and regulation, ensuring that the needs and rights of all individuals are considered in decision-making processes. (U.S. Office of Management and Budget, 2024; UNESCO, 2023)

One way to promote public engagement is through the establishment of advisory boards or committees that include a diverse range of stakeholders. These groups can provide input on the ethical implications of new technologies and offer recommendations for regulatory frameworks that balance technological progress with social responsibility. By involving the public in the conversation about automation, societies can ensure that technological advancements align with shared values and contribute to the common good, rather than benefiting a select few at the expense of others. (Department for Science, Innovation and Technology, 2023; OECD, 2023)

In addition to fostering public engagement, it is essential to invest in education and workforce development programs that help individuals adapt to the changing job market. As automation continues to reshape industries, many workers will need to acquire new skills to remain competitive. Educational institutions, vocational training centers, and employers all have a role to play in equipping the workforce with the skills necessary for an automated future. This might involve offering reskilling programs, emphasizing STEM (science, technology, engineering, and mathematics) education, and promoting lifelong learning to ensure that workers are not left behind in a rapidly evolving labor market. (International Labour Organization, 2023; Organisation for Economic Co-operation and Development, 2023)

Finally, the development of ethical guidelines and regulatory frameworks for automation must be seen as an ongoing process. As technology continues to advance, new ethical challenges will emerge that require adaptive and proactive solutions. By establishing robust and flexible frameworks today, governments, businesses, and technology developers can ensure that the future of automation is not only technologically advanced but also just, equitable, and beneficial for all members of society. Through collaboration, global cooperation, and a commitment to ethical responsibility, automation can become a force for positive change that enhances the well-being of individuals and communities around the world. (European Union, 2024; National Institute of Standards and Technology, 2023; ISO/IEC, 2023a, 2023b)

THE ROLE OF WORKERS IN SHAPING THE FUTURE OF WORK

While much of the conversation around automation focuses on the potential negative impacts it may have on workers, it is important to recognize that workers themselves can play a crucial role in shaping the future of work. Labor unions, workers' organizations, and other advocacy groups have the power to advocate for policies that prioritize the rights and interests of workers as automation technologies are deployed. These groups can serve as a voice for workers in discussions with policymakers, tech companies, and other stakeholders, helping to ensure that their concerns are heard and addressed. Through active engagement and collaboration, workers can influence how automation is integrated into industries, ensuring that its implementation does not come at the cost of fair wages, job security, or ethical treatment. (International Labour Organization, 2023; Organisation for Economic Co-operation and Development, 2023)

One key avenue for worker involvement is through collective bargaining. Labor unions can negotiate with employers to establish fair terms for the integration of automation in the workplace. This might include negotiating job retraining programs, ensuring that displaced workers have access to new opportunities, or advocating for a redistribution of the benefits that automation brings. For instance, unions could push for policies such as profit-sharing or employee stock options, ensuring that workers benefit from the increased productivity driven by automation, rather than simply being replaced or left behind. By being proactive in these discussions, unions can help protect the interests of workers while supporting the responsible implementation of new technologies. (International Labour Organization, 2023; Organisation for Economic Co-operation and Development, 2023)

Beyond collective bargaining, workers must also embrace the need to adapt to the changing landscape of work. As automation continues to transform industries, the types of jobs available may shift, with an increasing demand for specialized skills and technical expertise. Workers who can acquire new skills and adapt to new technological systems will be better positioned to thrive in this evolving environment. This might involve pursuing education in fields such as data science, artificial intelligence, or robotics, as well as learning how to collaborate effectively with automated systems. The ability to work alongside machines rather than being replaced by them will be a critical skill for future workers. (Organisation for Economic Co-operation and Development, 2023; Stanford Institute for Human-Centered Artificial Intelligence, 2024)

In addition to technical skills, workers will also need to embrace new forms of labor and collaboration. For example, the rise of the gig economy, facilitated in part by automation, has already begun to change traditional employment relationships. Remote work, freelance opportunities, and short-term contracts are becoming more prevalent, providing workers with more flexibility but also introducing new challenges, such as lack of job security or access to benefits. As automation enables even greater flexibility and collaboration across borders, workers will need to navigate these new forms of employment, which may require different skills, mindsets, and strategies for career development. For many workers, the future may involve balancing multiple gigs or working with a combination of traditional employers and automated platforms. (Organisation for Economic Co-operation and Development, 2023)

In this changing landscape, workers who take an active role in shaping their professional futures will be better positioned for success. This means not only acquiring new technical skills but also developing a deeper understanding of how automation will impact various industries and how they can contribute to those changes. Workers who engage in continuous learning, participate in professional networks, and stay informed about trends in automation will be better equipped to navigate the challenges and opportunities ahead. Moreover, those who are flexible and open to new forms of work may find opportunities to lead in emerging industries that capitalize on automation, such as remote work coordination, AI oversight, or automation system management. (International Labour Organization, 2023; Organisation for Economic Co-operation and Development, 2023)

In addition to adapting to new technologies, workers may also play a vital role in ensuring that automation is deployed ethically and responsibly. This could involve advocating for transparency in the way automation technologies are used and ensuring that the rights and dignity of workers are respected throughout the process. By joining forces with advocacy groups, labor unions, and policymakers, workers can help push for regulations that protect their interests, such as fair wages, job security, and access to benefits. In this way, workers can be part of a broader movement that seeks to create a future of work where technology and human labor coexist in a mutually beneficial way. (International Labour Organization, 2023; Group of Seven, 2023)

Furthermore, the shift toward automation also presents an opportunity for workers to drive innovation and influence the design of new technologies. Workers who have firsthand experience of the challenges and opportunities presented by automation can offer valuable insights into how these systems should be developed and deployed. By participating in conversations about the design of automated systems, workers can help ensure that these systems are developed with their needs in mind. This could involve advocating for more user-friendly interfaces, ensuring that automation tools are inclusive and accessible, or pushing for safeguards that protect workers from exploitation. Workers, in collaboration with tech developers, can influence the future of automation in ways that benefit both individuals and society as a whole. (Ada Lovelace Institute, 2022; National Institute of Standards and Technology, 2023)

Ultimately, workers must view automation not just as a threat but as an opportunity to shape their future and contribute to the evolution of work. While there are undeniable challenges associated with automation, workers have the potential to influence the development of these technologies and the future of work through active engagement, education, and advocacy. By embracing the changing nature of work and working alongside automation technologies, workers can help ensure that the future of work remains just, equitable, and empowering for all. (International Labour Organization, 2023; Organisation for Economic Co-operation and Development, 2023)

CONCLUSION: BUILDING A FUTURE OF WORK THAT BENEFITS ALL

The future of work in an age of automation presents both significant challenges and immense opportunities. On one hand, automation technologies like artificial intelligence, robotics, and machine learning hold the promise of greatly increasing productivity, reducing operational costs, and enabling entirely new industries to emerge. The ability of machines to perform repetitive, labor-intensive tasks with precision and efficiency could unlock new economic opportunities, potentially improving standards of living globally. However, alongside these advantages come complex ethical concerns that cannot be ignored. The rapid displacement of workers, particularly those in low-wage and manual labor sectors, raises important questions about human dignity, purpose, and identity. Automation risks exacerbating income inequality, with the wealth generated by these technologies concentrated in the hands of a few powerful entities, leaving the majority of workers behind. (International Labour Organization, 2023; Organisation for Economic Co-operation and Development, 2023; Stanford Institute for Human-Centered Artificial Intelligence, 2024)

Addressing these challenges requires thoughtful and proactive policymaking. Governments must develop frameworks that ensure automation benefits society as a whole, rather than deepening the divide between the rich and poor. This includes providing adequate social safety nets, such as universal basic income or job retraining programs, for workers displaced by automation. Furthermore, policies must ensure that technology companies are held accountable for the impact of their innovations, particularly when these technologies have wide-reaching effects on employment and social structures. As automation continues to evolve, it is essential that it does not become a tool that only benefits a select few, but rather is designed and deployed in a way that enhances the overall well-being of all people. (U.S. Office of Management and Budget, 2024; European Union, 2024; Organisation for Economic Co-operation and Development, 2023)

The role of education and workforce development will be central to a future where automation coexists with human labor. In this new era, it is crucial to emphasize lifelong learning and the development of skills that complement automation rather than compete with it. Workers will need to acquire technical expertise in fields like artificial intelligence, robotics, and data analysis, while also focusing on uniquely human skills such as creativity, empathy, and problem-solving. By investing in education systems that are flexible, inclusive, and responsive to the changing demands of the labor market, we can ensure that workers are equipped with the tools they need to thrive in an automated future. This will also require businesses to invest in reskilling and upskilling initiatives to help their employees transition into new roles and industries. (International Labour Organization, 2023; Organisation for Economic Co-operation and Development, 2023)

Ultimately, the future of work in the age of automation hinges on our ability to balance technological progress with a commitment to ethical principles and human well-being. By prioritizing human dignity, social justice, and fairness, we can create a future where automation serves the broader interests of society. This means that, alongside innovation, we must foster a global dialogue about the role of technology in our lives, ensuring that the benefits of automation are shared equitably. With the right combination of regulation, technological development, and international cooperation, it is possible to build a future where automation not only enhances human potential but also empowers workers and contributes to a more just and sustainable global society. (Council of Europe, 2024; Group of Seven, 2023; ISO/IEC, 2023b; International Energy Agency, 2024)

CHAPTER 11 ESSAY PROMPTS

The Future of Work—Ethics and Automation

1. How does automation challenge traditional concepts of work and human dignity? Discuss the ethical implications of machines replacing human labor.
2. In what ways can automation exacerbate income inequality, and what policies could be implemented to mitigate these effects?
3. How should policymakers balance the benefits of automation with the risks of job displacement? What role can universal basic income (UBI) play in this balance?
4. What are the potential ethical issues related to the accountability of autonomous systems, such as self-driving cars or AI in healthcare, and how should they be addressed?
5. How can automation technologies be developed and implemented in a way that ensures the fair distribution of economic benefits? Discuss the ethical challenges of wealth concentration due to automation.
6. Explain how automation can affect human relationships in the workplace. What are the ethical concerns related to the dehumanization of labor through automation?
7. Discuss the role of education and workforce development in preparing workers for an automated future. How can we ensure that workers are equipped with the necessary skills to adapt to automation?
8. How does automation impact marginalized communities and what steps can be taken to ensure they are not left behind in the transition to an automated economy?
9. Analyze the ethical dilemmas that arise when AI systems perpetuate biases, such as in facial recognition or hiring practices. How can these biases be mitigated through better AI design and regulation?
10. What are the ethical considerations for companies when adopting automation technologies? How should businesses balance efficiency and profit with their responsibility to workers and society?

REFERENCES

Ada Lovelace Institute. (2022). *Participatory data stewardship: A toolkit for a rights-respecting data ecosystem*. Author.

Council of Europe. (2024). *Framework Convention on Artificial Intelligence, Human Rights, Democracy and the Rule of Law*. Author.

Department for Science, Innovation and Technology. (2023). *A pro-innovation approach to AI regulation (White Paper)*. HM Government.

European Data Protection Board. (2022). *Guidelines 05/2022 on the use of facial recognition in the area of law enforcement*. Author.

European Union. (2024). *Artificial Intelligence Act*. European Parliament and Council.

European Union Agency for Cybersecurity. (2023). *Artificial Intelligence threat landscape 2023*. ENISA.

Group of Seven. (2023). *Hiroshima AI Process: International Code of Conduct for organizations developing advanced AI systems*. Author.

Information Commissioner's Office. (2023). *Guidance on AI and data protection* (updated). Author.

International Energy Agency. (2024). *Electricity 2024: Data centres and AI analysis*. Author.

International Labour Organization. (2023). *Generative AI and jobs: A global analysis of possible effects on job quantity and quality*. Author.

International Monetary Fund. (2024). *World Economic Outlook: GenAI and the future of work (Chapter 3)*. Author.

ISO/IEC. (2023a). *Information technology—Artificial intelligence—Risk management (ISO/IEC 23894:2023)*. International Organization for Standardization.

ISO/IEC. (2023b). *Artificial intelligence—Management system—Requirements (ISO/IEC 42001:2023)*. International Organization for Standardization.

McKinsey Global Institute. (2023). *Generative AI and the future of work in America*. McKinsey & Company.

Organisation for Economic Co-operation and Development. (2023). *OECD Employment Outlook 2023: Artificial intelligence and the labour market*. OECD Publishing.

Pew Research Center. (2023). *How workers and the public view artificial intelligence in the workplace*. Author.

Stanford Institute for Human-Centered Artificial Intelligence. (2024). *AI Index Report 2024*. Stanford University.

Stockton Economic Empowerment Demonstration. (2021). *SEED: Year two outcomes report*. Author.

U.S. Bureau of Labor Statistics. (2024). *Employment projections, 2024–2034*. U.S. Department of Labor.

U.S. Equal Employment Opportunity Commission. (2023). *Assessing adverse impact in software, algorithms, and AI used in employment selection procedures under Title VII (Technical assistance document)*. Author.

U.S. Office of Management and Budget. (2024). *Memorandum M-24-10: Advancing governance, innovation, and risk management for agency use of artificial intelligence*. Executive Office of the President.

UNESCO. (2023). *Guidelines for the governance of digital platforms: Safeguarding freedom of expression and access to information*. Author.

World Economic Forum. (2023). *The Future of Jobs Report 2023*. Author.

World Health Organization. (2021). *Ethics and governance of artificial intelligence for health*. Author.

Responsible Innovation and Global Challenges

12

ABSTRACT

Chapter 12, "Responsible Innovation and Global Challenges," explores the critical role of ethical practices in the development and deployment of emerging technologies. It emphasizes the need for responsible innovation that balances technological advancement with social, environmental, and economic considerations. The chapter highlights the importance of a multidisciplinary approach, involving experts from diverse fields to address the ethical challenges posed by technologies like AI, automation, and biotechnology. It also discusses the role of governments, businesses, and international organizations in creating regulatory frameworks and policies that ensure the responsible use of technology, promote global cooperation, and mitigate the risks of inequality, bias, and environmental harm. Ultimately, the chapter argues for a future where innovation benefits society equitably, fosters sustainability, and prioritizes human well-being.

INTRODUCTION

In the face of rapid technological advancements, the need for responsible innovation has never been more pressing (Council of Europe, 2024; European Union, 2024; U.S. Office of Management and Budget, 2024). As technologies like artificial intelligence (AI), machine learning, biotechnology, and automation continue to reshape industries, societies, and economies, their ethical implications must be carefully considered (Department for Science, Innovation and Technology,

2023; Organisation for Economic Co-operation and Development, 2023). The responsibility of developers, policymakers, and global organizations extends beyond simply creating new technologies; it is about ensuring these innovations contribute to a better, more equitable future for all (World Economic Forum, 2023; Ada Lovelace Institute, 2022). This chapter explores the role of responsible innovation in addressing complex ethical issues in a tech-driven world, emphasizing the importance of synthesizing knowledge from multiple disciplines to navigate the challenges of our increasingly digital and interconnected age (Stanford Institute for Human-Centered Artificial Intelligence, 2024).

As innovation accelerates, new technologies often emerge faster than ethical frameworks can be developed to guide their use (National Institute of Standards and Technology, 2023; International Organization for Standardization & International Electrotechnical Commission, 2023a). The rapid pace of technological change has outstripped the ability of traditional institutions, including governments and regulatory bodies, to keep up (Council of Europe, 2024; European Union, 2024). This gap between innovation and regulation creates significant challenges in ensuring that emerging technologies are deployed in ways that are both safe and socially beneficial (Department for Science, Innovation and Technology, 2023). Without clear guidelines and robust ethical considerations, technologies can inadvertently perpetuate inequality, bias, and environmental harm (European Data Protection Board, 2022; International Energy Agency, 2024a, 2024b, 2024c, 2024d). The need for responsible innovation has thus become a central issue in the global conversation about the future of technology (World Economic Forum, 2023).

Responsible innovation is not just about mitigating risks, but also about ensuring that technological progress actively contributes to human flourishing (Ada Lovelace Institute, 2022). It requires asking fundamental questions about who benefits from new technologies and who might be harmed (Organisation for Economic Co-operation and Development, 2023). For example, advancements in AI and machine learning promise to revolutionize industries, but without proper oversight, they could also reinforce existing biases, exacerbate social divides, and displace workers (U.S. Equal Employment Opportunity Commission, 2023; International Labour Organization, 2023). The ethical responsibility of innovators is to anticipate these outcomes and design technologies that promote fairness, equity, and human dignity (National Institute of Standards and Technology, 2023).

The ethical challenges posed by new technologies are multifaceted and complex (International Organization for Standardization & International Electrotechnical Commission, 2023b). They encompass issues of privacy, data security, transparency, accountability, and the potential for misuse (European Data Protection Board, 2022). In the field of AI, for example, algorithms that are used to make decisions in criminal justice, hiring, and healthcare have been shown to perpetuate racial, gender, and socioeconomic biases (U.S. Equal Employment Opportunity Commission, 2023; Organisation for Economic Co-operation and Development, 2023). In biotechnology, advancements in genetic engineering and CRISPR technology raise questions about the ethics of altering the human genome and the potential unintended consequences of such changes (Stanford

Institute for Human-Centered Artificial Intelligence, 2024). As these technologies continue to evolve, the ethical implications will only grow more complex and far-reaching (World Economic Forum, 2023).

To navigate these challenges, it is essential that innovation occurs within a framework that integrates ethical considerations from the outset (National Institute of Standards and Technology, 2023). This means involving ethicists, sociologists, environmental scientists, and other experts in the design and implementation of new technologies (Ada Lovelace Institute, 2022). Interdisciplinary collaboration is key to ensuring that technological innovations are not developed in isolation, but with a holistic understanding of their social, economic, and environmental impacts (Organisation for Economic Co-operation and Development, 2023). By bringing together a diverse range of perspectives, innovators can create solutions that are more likely to be beneficial for all sectors of society (World Economic Forum, 2023).

Furthermore, the role of government and regulatory bodies cannot be overstated (Council of Europe, 2024; European Union, 2024). Effective policies and regulations are necessary to ensure that technology development aligns with public interest (U.S. Office of Management and Budget, 2024). However, creating these policies is no simple task. Governments must strike a balance between fostering innovation and protecting public welfare (Department for Science, Innovation and Technology, 2023). Overregulation could stifle technological progress, while under-regulation could lead to unintended consequences, such as the unchecked proliferation of harmful technologies (Group of Seven, 2023). The challenge for policymakers is to create regulatory frameworks that are both flexible enough to allow innovation and robust enough to prevent harm (International Organization for Standardization & International Electrotechnical Commission, 2023a).

One of the greatest challenges in responsible innovation is the issue of global inequality (Organisation for Economic Co-operation and Development, 2023). Technology has the potential to benefit people worldwide, but without careful planning, it could exacerbate existing inequalities (World Economic Forum, 2023). For instance, access to advanced technologies such as AI, biotechnology, and high-speed internet is still limited in many parts of the world, especially in developing countries (Pew Research Center, 2023). As technology continues to advance, it is crucial that global efforts be made to ensure that these innovations do not widen the gap between the rich and the poor (International Monetary Fund, 2024). This requires international cooperation and a commitment to ensuring that the benefits of innovation are distributed equitably (Group of Seven, 2023).

In addition to addressing global inequality, responsible innovation must also tackle environmental sustainability (International Energy Agency, 2024). Technological development has historically been a major driver of environmental degradation (European Union Agency for Cybersecurity, 2023). From the industrial revolution to the rise of consumer electronics, the pace of innovation has often come at the expense of the planet. As we develop new technologies, we must be mindful of their environmental impact and prioritize sustainability (International Energy Agency, 2024). This includes designing products that are energy-efficient, reducing waste in manufacturing processes, and considering the lifecycle of products from creation to disposal (National Institute of Standards and Technology, 2023). Responsible innovation, in this sense, is not just about improving human well-being but also about preserving the planet for future generations (World Economic Forum, 2023).

The ethical implications of technology also extend to the realm of individual rights and freedoms (European Data Protection Board, 2022). As technologies such as AI and big data become more embedded in daily life, there are growing concerns about surveillance, privacy, and the erosion of personal freedoms (Organisation for Economic Co-operation and Development, 2023). In particular, the widespread collection of data by tech companies raises questions about consent, ownership, and the potential for misuse (U.S. Office of Management and Budget, 2024). Innovations that promise greater convenience, such as personalized recommendations and facial recognition, may also come with significant trade-offs in terms of privacy (European Union, 2024). It is essential that ethical considerations be made to protect individuals' rights in an increasingly connected world (Group of Seven, 2023).

The potential for technology to influence political processes is another area where responsible innovation is crucial (World Economic Forum, 2023). From social media algorithms to the use of AI in election campaigns, technology has the power to shape public opinion, influence political discourse, and even sway election outcomes (Stanford Institute for Human-Centered Artificial Intelligence, 2024). While technology can democratize information and empower citizens, it can also be weaponized for propaganda, misinformation, and manipulation (Organisation for Economic Co-operation and Development, 2023). The role of responsible innovation in this context is to ensure that technological tools are used to promote democratic values, transparency, and fair representation rather than undermining them (Council of Europe, 2024).

In the context of business and industry, companies have a responsibility to consider the broader social impacts of the technologies they develop (International Organization for Standardization & International Electrotechnical Commission, 2023b). This means going beyond profit maximization and ensuring that innovation is aligned with ethical principles (National Institute of Standards and Technology, 2023). Corporate social responsibility (CSR) should be a guiding framework for businesses involved in the development of emerging technologies (World Economic Forum, 2023). Companies must assess the potential risks and benefits of their innovations and make decisions that prioritize the welfare of employees, customers, and the broader community (Organisation for Economic Co-operation and Development, 2023).

Education and public awareness also play a critical role in promoting responsible innovation (Pew Research Center, 2023). As technological advancements continue to shape the world, it is essential that individuals are equipped with the knowledge and skills to engage with these changes in an informed and ethical manner (Organisation for Economic Co-operation and Development, 2023). This includes fostering digital literacy, critical thinking, and an understanding of the ethical implications of technology (World Economic Forum, 2023). By educating the public and empowering individuals to make informed choices, societies can better navigate the complex ethical landscape of the digital age (Ada Lovelace Institute, 2022).

Another important aspect of responsible innovation is the need for long-term thinking (International Organization for Standardization & International Electrotechnical Commission, 2023a). Many of the technologies we develop today will have long-lasting impacts that extend far beyond their initial use (National Institute of Standards and Technology, 2023). It is crucial that we anticipate the long-term consequences of new technologies and design them with future generations in mind (International Monetary Fund, 2024). This includes considering the social, economic, and environmental implications of innovation over the coming decades, as well as ensuring that future generations inherit a world that is better, not worse, as a result of technological progress (International Energy Agency, 2024).

Finally, the future of responsible innovation lies in the active involvement of global communities (Group of Seven, 2023). Innovation is no longer confined to one country or region; it is a global endeavor that affects people everywhere (World Economic Forum, 2023). As such, international cooperation will be essential in ensuring that technological progress is harnessed for the greater good (Council of Europe, 2024). This involves creating global standards for ethical innovation, sharing knowledge and resources, and working together to address challenges such as climate change, inequality, and access to technology (European Union, 2024; Organisation for Economic Co-operation and Development, 2023).

In conclusion, the role of responsible innovation in addressing global challenges is both complex and urgent (U.S. Office of Management and Budget, 2024). As technology continues to advance, we must ensure that innovation is guided by ethical principles that prioritize human well-being, environmental sustainability, and social equity (National Institute of Standards and Technology, 2023). By integrating diverse perspectives, fostering global cooperation, and implementing effective regulations, we can create a future where technological progress serves the greater good and addresses the most pressing challenges of our time (World Economic Forum, 2023). Responsible innovation is not a choice; it is a necessity for building a just, sustainable, and equitable world in the face of rapid technological change (Council of Europe, 2024).

THE NEED FOR RESPONSIBLE INNOVATION

At its core, responsible innovation is the practice of designing, developing, and deploying technologies in ways that maximize societal benefits while minimizing potential harm (National Institute of Standards and Technology, 2023). It emphasizes foresight, accountability, and inclusivity, seeking to balance innovation with ethical responsibility (Ada Lovelace Institute, 2022).

Responsible innovation for global challenges: aligning AI, biotech, and automation with human rights, equity, and sustainability

Read through a global-justice lens, the image frames "responsible innovation" as solving shared problems without exporting harm: center human rights in design (consent, privacy, non-discrimination), assess risks to vulnerable groups and ecosystems before deployment, pair breakthroughs with guardrails (impact assessments, transparency, auditability, redress), and share benefits through access, affordability, and capacity-building—especially where resources are scarce—so speed and scale never outrun accountability.

As technology continues to advance, the potential for unintended consequences grows, especially when innovations outpace the development of corresponding ethical frameworks and regulatory measures (Department for Science, Innovation and Technology, 2023). With the speed of technological development increasing exponentially, society faces the challenge of ensuring that new technologies do not cause more harm than good (International Organization for Standardization & International Electrotechnical Commission, 2023a). Responsible innovation, therefore, requires active engagement from developers, policymakers, businesses, and communities to ensure that the benefits of technological progress are shared broadly while mitigating risks (World Economic Forum, 2023).

As technology evolves, the ethical considerations surrounding it become increasingly complex (European Union Agency for Cybersecurity, 2023). For example, the rapid growth of artificial intelligence (AI) and big data analytics has led to groundbreaking developments in fields such as healthcare, finance, and transportation (Stanford Institute for Human-Centered Artificial Intelligence, 2024). In healthcare, AI has the potential to revolutionize diagnoses, personalize treatments, and predict disease outbreaks (National Institute of Standards and Technology, 2023). However, the application of AI in these areas also raises important ethical concerns about privacy, data security, and consent (European Data Protection Board, 2022). With access to sensitive personal information, AI systems may be prone to misuse, such as unauthorized data collection or biased algorithms, that could harm vulnerable populations or undermine trust in healthcare systems (Organisation for Economic Co-operation and Development, 2023). Similarly, in finance, algorithms used to assess creditworthiness can unintentionally perpetuate biases, leading to unequal treatment of marginalized communities (U.S. Equal Employment Opportunity Commission, 2023).

The rise of automation and robotics presents another crucial area of concern (International Labour Organization, 2023). Automation holds the promise of significantly improving productivity, efficiency, and safety in industries such as manufacturing, logistics, and healthcare (World Economic Forum, 2023). Yet, it also brings with it the threat of large-scale job displacement (International Monetary Fund, 2024). While automation can reduce labor costs and enhance performance, millions of workers, particularly in low-skill, routine jobs, risk losing their livelihoods (Organisation for Economic Co-operation and Development, 2023). This displacement could exacerbate existing economic inequalities, as the benefits of automation are often unevenly distributed (World Economic Forum, 2023). Skilled workers in fields such as engineering, programming, and robotics may see increased demand for their expertise, while those without the means to reskill may face long-term unemployment (U.S. Bureau of Labor Statistics, 2024). Responsible innovation in this context involves not only developing technologies that benefit society but also implementing policies that support workers through retraining programs and social safety nets (International Labour Organization, 2023).

The consequences of irresponsible innovation can also be seen in the growing gap between technological haves and have-nots (Pew Research Center, 2023). As new technologies are developed and deployed, there is a risk that they will disproportionately benefit those with the resources and infrastructure to access them, leaving behind entire populations (Organisation for Economic Co-operation and Development, 2023). For instance, access to high-speed internet, advanced healthcare, and cutting-edge educational tools remains limited in many parts of the world (World Economic Forum, 2023). This digital divide has the potential to deepen existing social inequalities, as those without access to the latest technologies may be left further behind in terms of education, employment, and quality of life (International Monetary Fund, 2024). In order to address these global disparities, responsible innovation must prioritize inclusivity, ensuring that the benefits of technology are accessible to all, regardless of geography or socioeconomic status (Ada Lovelace Institute, 2022).

One of the key principles of responsible innovation is foresight (International Organization for Standardization & International Electrotechnical Commission, 2023a). Developers and innovators must anticipate the potential consequences of new technologies before they are widely implemented (National Institute of Standards and Technology, 2023). This foresight involves not only considering the direct impact of technologies but also accounting for unintended long-term effects (International Energy Agency, 2024). For example, the environmental impact of certain technologies, such as the proliferation of electronic waste and energy consumption of data centers, must be considered alongside their immediate benefits (European Union Agency for Cybersecurity, 2023). In this regard, sustainable innovation practices that prioritize energy efficiency, recyclability, and minimal environmental impact are essential (International Energy Agency, 2024). Furthermore, foresight requires a commitment to anticipating and mitigating risks that may not be immediately apparent, such as the ethical implications of AI-driven surveillance systems or the potential for social disruption caused by widespread automation (Council of Europe, 2024).

Accountability is another crucial element of responsible innovation (National Institute of Standards and Technology, 2023). Innovators must be held accountable for the technologies they create and deploy (International Organization for Standardization & International Electrotechnical Commission, 2023b). This accountability extends to developers, companies, and governments who have a responsibility to ensure that technologies are not only safe but also align with societal values and ethical principles (World Economic Forum, 2023). For instance, AI systems used in hiring decisions should be transparent and accountable to ensure that they do not perpetuate discrimination or bias (U.S. Equal Employment Opportunity Commission, 2023). Companies that develop AI tools must take responsibility for how their products are used and must be transparent about the data sources, algorithms, and decision-making processes behind their systems (Department for Science, Innovation and Technology, 2023). Regulatory bodies, too, must enforce ethical standards and hold companies accountable when technologies cause harm (European Union, 2024). Without accountability, there is a risk that the pursuit of profit will outweigh the consideration of human welfare (Organisation for Economic Co-operation and Development, 2023).

Inclusivity is another essential aspect of responsible innovation (Ada Lovelace Institute, 2022). The development of new technologies should involve a broad range of voices and perspectives, particularly those from marginalized or underrepresented communities (Pew Research Center, 2023). This inclusivity helps ensure that technologies are designed to serve the needs of diverse populations, rather than reinforcing existing inequalities (World Economic Forum, 2023). Involving ethicists, social scientists, community leaders, and representatives from vulnerable groups in the innovation process helps to ensure that the potential risks and benefits of new technologies are fully understood (Organisation for Economic Co-operation and Development, 2023). By fostering inclusivity, innovation can better address the concerns of all stakeholders and prevent technologies from exacerbating social, economic, and cultural divides (Council of Europe, 2024).

To be truly responsible, innovation must also be guided by ethical frameworks that are adaptable to the rapidly changing technological landscape (U.S. Office of Management and Budget, 2024). The ethical implications of AI, biotechnology, and automation are constantly evolving, and regulatory bodies must be flexible enough to keep up with these changes (European Union, 2024). Traditional regulatory models, which often operate at a slower pace, may struggle to address the complex ethical dilemmas posed by emerging technologies (Department for Science, Innovation and Technology, 2023). Therefore, the development of dynamic, forward-thinking ethical guidelines is essential (National Institute of Standards and Technology, 2023). These guidelines should be able to respond to new challenges as they arise, incorporating lessons learned from previous innovations and adapting to the unique ethical considerations of each new technology (International Organization for Standardization & International Electrotechnical Commission, 2023b).

Another critical aspect of responsible innovation is public engagement (Ada Lovelace Institute, 2022). The development of new technologies should not be an insular process (World Economic Forum, 2023). It is essential that the public is informed and included in discussions about the ethical implications of technological advances (Pew Research Center, 2023). Public engagement ensures that the development of technologies aligns with societal values and that citizens have a voice in shaping the future of innovation (Organisation for Economic Co-operation and Development, 2023). This engagement can take many forms, from public consultations and debates to collaborative initiatives between tech companies and civil society organizations (Group of Seven, 2023). Through active dialogue and participation, individuals can better understand the benefits and risks of new technologies and hold developers accountable for their actions (Council of Europe, 2024).

The role of education in fostering responsible innovation cannot be overstated (Organisation for Economic Co-operation and Development, 2023). As technology becomes increasingly integrated into daily life, it is essential that future generations are equipped with the knowledge and skills to engage with emerging technologies in an ethical and responsible manner (World Economic Forum, 2023). This includes not only technical skills but also a deep understanding of the social, ethical, and environmental implications of technology (Pew Research Center, 2023). By integrating ethics and social responsibility into STEM education, we can cultivate a new generation of innovators who are committed to designing technologies that prioritize human welfare and environmental sustainability (International Energy Agency, 2024).

Global cooperation is another critical element of responsible innovation (Group of Seven, 2023). Many of the ethical challenges posed by new technologies, such as AI and biotechnology, are global in nature and require international collaboration to address (Council of Europe, 2024). For example, issues related to privacy and data protection are not confined to a single country or region but affect individuals worldwide (European Data Protection Board, 2022). In order to effectively regulate emerging technologies, nations must work together to create international standards and frameworks that promote ethical innovation on a global scale (European Union, 2024). This cooperation can help ensure that the benefits of new technologies are shared equitably across borders and that the risks are managed in a way that respects the rights of all people, regardless of where they live (International Monetary Fund, 2024).

In addition to international collaboration, the responsible innovation framework must be rooted in social responsibility (International Organization for Standardization & International Electrotechnical Commission, 2023b). Innovators must recognize that their work has far-reaching consequences beyond the immediate benefits of technology (National Institute of Standards and Technology, 2023). They must consider the broader impact of their innovations on society, the environment, and future generations (International Energy Agency, 2024). By prioritizing social responsibility, innovators can help ensure that technological progress is aligned with the greater good and does not come at the expense of vulnerable communities or the planet (World Economic Forum, 2023).

In conclusion, responsible innovation is essential for ensuring that technological advancements contribute positively to society (Council of Europe, 2024; U.S. Office of Management and Budget, 2024). As technologies like AI, big data, and automation continue to evolve, they bring with them both significant opportunities and potential risks (Organisation for Economic Co-operation and Development, 2023). By emphasizing foresight, accountability, inclusivity, and adaptability, we can ensure that innovation serves the public good and does not harm the most vulnerable members of society (Ada Lovelace Institute, 2022; National Institute of Standards and Technology, 2023). Responsible innovation requires the active involvement of developers, policymakers, businesses, and the public to create ethical guidelines and regulatory frameworks that ensure technology enhances human well-being, promotes social equity, and safeguards the environment (European Union, 2024; World Economic Forum, 2023). The challenge of balancing technological progress with ethical responsibility is immense, but it is a challenge that must be met in order to build a just and sustainable future for all (International Monetary Fund, 2024).

ETHICAL ISSUES IN A TECH-DRIVEN WORLD

The rapid pace of technological change brings a host of complex ethical dilemmas (World Economic Forum, 2023). As innovations in artificial intelligence (AI), big data, and automation continue to reshape industries and societies, the ethical challenges they present become more pronounced (Organisation for Economic Co-operation and Development, 2023). One of the most prominent concerns is the question of privacy (European Data Protection Board, 2022). With increasing amounts of personal data being collected, stored, and analyzed, the line between individual privacy and collective security is often blurred (U.S. Office of Management and Budget, 2024). Technologies such as facial recognition and surveillance systems raise significant questions about consent, autonomy, and the potential for misuse (European Union, 2024). These technologies can be incredibly beneficial for law enforcement, national security, and public safety, but they also pose risks to individuals' fundamental rights, particularly in terms of surveillance without consent (Council of Europe, 2024). As these tools become more prevalent, the ethical responsibility of designers, policymakers, and corporations in managing personal data becomes paramount (Group of Seven, 2023).

The balance between individual privacy and public safety is a delicate one (Council of Europe, 2024). On the one hand, technological innovations like AI-powered surveillance can help identify and prevent criminal activities, safeguard national borders, and monitor public health (European Union, 2024). On the other hand, the widespread use of such technologies can lead to the erosion of civil liberties and a growing sense of surveillance (European Data Protection Board, 2022). In many instances, individuals are unaware of how their data is being collected or used, leading to a lack of consent and transparency (U.S. Office of Management and Budget, 2024). Moreover, the potential for abuse, such as surveillance used for political control or social manipulation, raises concerns about the concentration of power and the infringement of individual freedoms (Organisation for Economic Co-operation and Development, 2023). Therefore, to ensure responsible innovation, stakeholders must prioritize data privacy and develop technologies that respect individuals' rights while balancing public safety and security concerns (Group of Seven, 2023).

Another pressing ethical issue in the tech industry is bias in AI systems (U.S. Equal Employment Opportunity Commission, 2023). Machine learning algorithms often rely on vast datasets to make decisions, from predicting creditworthiness to assessing job applicants' suitability for employment (National Institute of Standards and Technology, 2023). However, if these datasets are not carefully curated, they can perpetuate and even amplify existing biases related to race, gender, socioeconomic status, and other factors (Organisation for Economic Co-operation and Development, 2023). For instance, biased algorithms used in hiring, lending, or criminal justice decisions can reinforce social inequities, making it more difficult for marginalized groups to access opportunities or resources (U.S. Equal Employment Opportunity Commission, 2023). In the criminal justice system, predictive algorithms used to assess the likelihood of reoffending can disproportionately target communities of color, leading to racial discrimination in sentencing (World Economic Forum, 2023). Similarly, AI-powered hiring tools may overlook qualified candidates from underrepresented groups if the training data reflects biased hiring practices from the past (Pew Research Center, 2023).

To address these challenges, it is critical for technologists to design fair, transparent, and accountable AI systems (National Institute of Standards and Technology, 2023). Ethical AI design requires a commitment to diversity in data collection, ensuring that datasets are representative of the populations they are meant to serve (Organisation for Economic Co-operation and Development, 2023). This can help prevent the inadvertent reinforcement of stereotypes or exclusionary practices (World Economic Forum, 2023). Additionally, transparency in algorithmic decision-making is essential for building trust in AI systems (U.S. Equal Employment Opportunity Commission, 2023). When algorithms make decisions that affect individuals' lives, they should be explainable, allowing users to understand how and why decisions were made (National Institute of Standards and Technology, 2023). Regular audits and ongoing monitoring of AI systems are also necessary to identify and rectify biases as they emerge, ensuring that AI systems remain fair and equitable over time (International Organization for Standardization & International Electrotechnical Commission, 2023b).

While AI and machine learning hold immense promise, their potential for harm must not be underestimated (Organisation for Economic Co-operation and Development, 2023). As these technologies become increasingly integrated into daily life, the need for responsible design and ethical oversight becomes even more critical (World Economic Forum, 2023). This is especially true when it comes to ensuring that AI systems are not only technically proficient but also socially responsible (National Institute of Standards and Technology, 2023). For instance, AI applications in healthcare, such as diagnostic tools, must be designed with careful attention to avoid reinforcing existing disparities in healthcare access and outcomes (Stanford Institute for Human-Centered Artificial Intelligence, 2024). If AI systems are trained on data that predominantly reflects the experiences of certain demographic groups, they may struggle to accurately diagnose conditions in underrepresented populations, exacerbating health inequalities (Pew Research Center, 2023).

The ethical issues surrounding AI bias and data privacy are not limited to any one sector or industry but are pervasive across a wide range of domains (Organisation for Economic Co-operation and Development, 2023). In the field of education, for example, AI tools that assess student performance or predict academic success may unintentionally favor students from affluent backgrounds who have greater access to resources and support (Pew Research Center, 2023). Similarly, in finance, automated credit scoring systems could penalize individuals with less traditional financial histories, further entrenching financial inequalities (World Economic Forum, 2023). In each of these cases, the ethical responsibility of developers and regulators is to ensure that AI systems are equitable and just (National Institute of Standards and Technology, 2023). This may require ongoing efforts to adjust algorithms and ensure they do not perpetuate historical injustices (U.S. Equal Employment Opportunity Commission, 2023).

Addressing these ethical issues is not solely the responsibility of technologists but also requires input from ethicists, sociologists, and other stakeholders who can bring a broader perspective to the challenges posed by AI and emerging technologies (Ada Lovelace Institute, 2022). Collaborative efforts that involve interdisciplinary teams can help identify potential ethical pitfalls early in the design process, allowing for the development of more inclusive and socially responsible technologies (World Economic Forum, 2023). Policymakers also have a role to play in setting regulations that hold developers accountable for the impact of their technologies, while also ensuring that innovation is not stifled by overly restrictive rules (European Union, 2024a, 2024b). Effective regulation should promote the responsible use of technology while encouraging continued innovation and progress (Department for Science, Innovation and Technology, 2023).

In addition to bias and privacy, another critical issue in responsible innovation is the environmental impact of emerging technologies (International Energy Agency, 2024). The development and deployment of advanced technologies, particularly AI and big data, often require significant energy consumption and resource use (European Union Agency for Cybersecurity, 2023). Data centers that power AI systems, for instance, consume vast amounts of electricity, contributing to global carbon emissions (International Energy Agency, 2024). As the demand for these technologies increases, so too does their environmental footprint (Stanford Institute for Human-Centered Artificial Intelligence, 2024). To ensure that innovation is truly responsible, technologists and policymakers must consider the environmental implications of new technologies and work to mitigate their impact (National Institute of Standards and Technology, 2023). This could involve developing more energy-efficient algorithms, utilizing renewable energy sources for data centers, and prioritizing the creation of sustainable technologies (International Energy Agency, 2024).

Moreover, as automation and robotics continue to reshape industries and labor markets, questions about the future of work and economic equity become even more urgent (International Labour Organization, 2023). While automation has the potential to improve efficiency and productivity, it also carries the risk of widespread job displacement (International Monetary Fund, 2024). Millions of workers, particularly in low-wage or routine jobs, may find themselves replaced by machines (World Economic Forum, 2023). In order to ensure responsible innovation, society must take steps to manage the economic and social impact of automation (Organisation for Economic Co-operation and Development, 2023). This may include providing workers with the skills and training necessary to transition into new roles, as well as implementing social safety nets to support those who are displaced by technological change (U.S. Bureau of Labor Statistics, 2024). Without such measures, automation could exacerbate inequality and create social unrest (International Labour Organization, 2023).

At the heart of responsible innovation is the need for ethical leadership in the tech industry (World Economic Forum, 2023). Developers, business leaders, and policymakers must take a proactive role in shaping the future of technology by prioritizing ethical considerations in their decision-making processes (National Institute of Standards and Technology, 2023). This includes not only addressing issues such as bias, privacy, and environmental impact but also ensuring that technological advancements contribute to the common good (Council of Europe, 2024). By fostering a culture of responsibility and accountability, the tech industry can help guide the development of technologies that promote social justice, environmental sustainability, and human dignity (International Organization for Standardization & International Electrotechnical Commission, 2023b).

Ultimately, the responsibility for ensuring that innovation serves the greater good lies with all stakeholders, including governments, corporations, and individuals (Group of Seven, 2023). While technological advancements hold tremendous potential, they also present significant ethical challenges (European Union, 2024). By addressing these challenges head-on and prioritizing responsible innovation, society can harness the power of emerging technologies to create a more equitable, sustainable, and just world (World Economic Forum, 2023). However, this requires ongoing dialogue, collaboration, and a commitment to ensuring that the benefits of innovation are shared broadly, while the risks are minimized and managed carefully (Organisation for Economic Co-operation and Development, 2023). Only through this holistic approach can we ensure that technology continues to serve humanity in ways that align with our shared values and aspirations (U.S. Office of Management and Budget, 2024).

GLOBAL CHALLENGES AND TECHNOLOGICAL IMPACT

As technology continues to expand across borders, its global impact becomes more pronounced (World Economic Forum, 2023). The proliferation of advanced technologies, such as artificial intelligence, automation, and digital platforms, is reshaping economies and societies at an unprecedented rate (International Monetary Fund, 2024). While these innovations bring immense potential for growth and efficiency, they also highlight deep-seated inequalities, particularly between developed and developing countries (Organisation for Economic Co-operation and Development, 2023). Technological innovation is not inherently neutral—its effects depend heavily on how it is developed, deployed, and regulated (Council of Europe, 2024). The benefits of new technologies are often unevenly distributed, with wealthy nations and corporations reaping the greatest rewards, while many low-income countries struggle to keep pace (Pew Research Center, 2023). This unequal distribution of technological resources and capabilities threatens to deepen existing global disparities and create new forms of inequality (World Economic Forum, 2023).

In developing countries, the digital divide remains a significant barrier to equitable participation in the global economy (Organisation for Economic Co-operation and Development, 2023).

Many nations in the Global South face obstacles such as inadequate infrastructure, limited internet connectivity, and a lack of technical expertise, which prevent them from fully engaging with the digital economy (Pew Research Center, 2023). While developed nations may benefit from the latest innovations in fields like healthcare, education, and finance, many low-income countries are left behind, unable to access or utilize these technologies (World Economic Forum, 2023). For instance, telemedicine, which has revolutionized healthcare delivery in wealthier nations, remains out of reach for millions of people in developing regions who lack reliable internet access or basic healthcare infrastructure (International Monetary Fund, 2024). Similarly, the rise of online education platforms has the potential to transform global learning, yet millions of students in rural or underserved areas still face barriers to accessing digital education resources (Organisation for Economic Co-operation and Development, 2023).

This disparity in access to technology leads to a widening gap in various aspects of human development (World Economic Forum, 2023). The unequal availability of technologies further exacerbates existing global inequalities in areas such as healthcare, education, and economic opportunities (Organisation for Economic Co-operation and Development, 2023). For example, countries with limited access to advanced technologies are often unable to offer their citizens the same quality of education or healthcare services as those in more technologically advanced nations (Pew Research Center, 2023). As a result, individuals in developing countries may face diminished opportunities for social mobility, reinforcing cycles of poverty and limiting their ability to participate in the global economy (International Monetary Fund, 2024). The lack of digital infrastructure also limits the ability of entrepreneurs in low-income countries to tap into global markets, leaving them disadvantaged in a world where digital platforms and e-commerce play an increasingly central role (World Economic Forum, 2023).

Furthermore, automation, while offering opportunities for efficiency and growth, poses a major challenge to global labor markets (International Labour Organization, 2023). As robots, artificial intelligence, and machine learning systems take over routine and low-skill jobs, workers in many sectors face the threat of job displacement (International Monetary Fund, 2024). This shift is particularly pronounced in developing countries, where large portions of the workforce are employed in manual labor, agriculture, and manufacturing (World Economic Forum, 2023). As advanced technologies replace human labor in these sectors, many workers will find themselves without viable alternatives, especially in regions with limited access to retraining and upskilling opportunities (Organisation for Economic Co-operation and Development, 2023). In some cases, automation may also lead to the outsourcing of jobs to countries with cheaper labor markets, further exacerbating inequality and economic instability in both developed and developing regions (U.S. Bureau of Labor Statistics, 2024).

The displacement of workers due to automation presents a complex challenge for governments and policymakers worldwide (International Labour Organization, 2023). In regions where jobs are already scarce or difficult to secure, automation may lead to widespread unemployment and social unrest (World Economic Forum, 2023). This shift in labor markets requires a concerted effort from governments, businesses, and global organizations to create policies that foster economic resilience and help workers adapt to the changing job landscape (U.S. Office of Management and Budget, 2024). Without adequate policies in place, the rapid pace of technological advancement could leave millions of people without access to stable employment, pushing vulnerable populations further into poverty (International Monetary Fund, 2024). As such, there is a growing need for international cooperation to ensure that the benefits of automation are shared equitably and that the risks of job displacement are mitigated (Council of Europe, 2024).

One of the primary strategies for addressing the challenge of job displacement is retraining and upskilling initiatives (Organisation for Economic Co-operation and Development, 2023). These efforts can help workers transition from industries that are being automated to new roles that require advanced technical skills (World Economic Forum, 2023). Governments and businesses must work together to invest in education and training programs that equip workers with the skills needed to thrive in a tech-driven world (U.S. Bureau of Labor Statistics, 2024). In addition to technical training, these programs should emphasize soft skills such as problem-solving, creativity, and adaptability, which are increasingly valued in the modern workforce (World Economic Forum, 2023). By providing workers with the tools they need to succeed in emerging industries, societies can mitigate the social and economic disruptions caused by automation (International Labour Organization, 2023).

Global organizations also have an important role to play in addressing the challenges posed by automation (International Labour Organization, 2023; World Economic Forum, 2023). International bodies can help facilitate dialogue and cooperation between governments, businesses, and civil society (Group of Seven, 2023). These organizations can advocate for policies that promote fair labor practices, ensure access to education and training, and support workers who are displaced by automation (Organisation for Economic Co-operation and Development, 2023). In addition, global cooperation can help bridge the digital divide by ensuring that developing countries have the resources and infrastructure they need to access and benefit from emerging technologies (International Monetary Fund, 2024). By working together on a global scale, nations can create a more inclusive and sustainable future for all (Council of Europe, 2024).

While the automation of labor presents significant challenges, it also offers opportunities for innovation and economic growth (International Monetary Fund, 2024). If managed properly, automation can drive productivity and efficiency, creating new industries and business models that can generate new jobs (World Economic Forum, 2023). For example, the rise of the renewable energy sector, the growth of digital platforms, and the expansion of industries such as robotics and AI all have the potential to create new employment opportunities for workers in both developed and developing countries (Stanford Institute for Human-Centered Artificial Intelligence, 2024). However, the success of these efforts will depend on how governments and businesses manage the transition (U.S. Office of Management and Budget, 2024). Policies that promote sustainable development, support innovation, and prioritize workers' welfare will be essential in ensuring that the benefits of automation are broadly shared (Organisation for Economic Co-operation and Development, 2023).

Another important aspect of addressing the challenges of automation is ensuring that workers have access to social safety nets and protections (International Labour Organization, 2023). As automation disrupts traditional employment structures, many workers may face periods of unemployment or underemployment as they transition to new industries or careers (U.S. Bureau of Labor Statistics, 2024). Governments must invest in robust social safety nets that provide financial support, healthcare, and retraining opportunities for displaced workers (International Monetary Fund, 2024). Universal basic income (UBI) is one proposal that has gained attention in recent years as a way to provide economic security in an increasingly automated world (World Economic Forum, 2023). By providing a guaranteed income, UBI could help ensure that workers are able to meet their basic needs while they transition to new roles or develop new skills (Organisation for Economic Co-operation and Development, 2023).

The impact of automation on global labor markets also underscores the need for a more inclusive approach to innovation and development (World Economic Forum, 2023). Rather than allowing technology to exacerbate existing inequalities, we must prioritize policies that ensure the benefits of innovation are distributed more equitably (Council of Europe, 2024). This includes addressing issues such as income inequality, access to education, and the digital divide (Pew Research Center, 2023). By ensuring that all individuals have the opportunity to participate in the digital economy, we can create a more just and sustainable future for all (Organisation for Economic Co-operation and Development, 2023). This will require not only technological innovation but also a commitment to social responsibility, ethical development, and global cooperation (Group of Seven, 2023).

Finally, the global nature of technological innovation calls for a more holistic and interconnected approach to policy development (Council of Europe, 2024). As technologies transcend borders, the policies that govern them must also be global in scope (European Union, 2024). National policies alone will not be sufficient to address the challenges of automation and technological disruption (International Monetary Fund, 2024). International cooperation and multilateral agreements will be necessary to establish global norms and standards for the responsible development and deployment of technologies (Group of Seven, 2023). By working together, nations can ensure that technological progress benefits everyone and that the risks associated with automation and digital innovation are mitigated in a way that promotes global equity and sustainability (World Economic Forum, 2023).

In conclusion, while technology has the potential to drive economic growth and improve living standards across the globe, its benefits are not automatically shared equally (Organisation for Economic Co-operation and Development, 2023). As technological innovation continues to expand, it is crucial that we address the digital divide, mitigate the risks of automation, and ensure that workers in all regions have the skills and protections they need to thrive in a rapidly changing world (International Labour Organization, 2023; U.S. Bureau of Labor Statistics, 2024). This will require collaborative efforts from governments, businesses, and global organizations to develop policies that promote inclusive growth, foster economic resilience, and prioritize social responsibility (Group of Seven, 2023; Council of Europe, 2024). Only through these concerted efforts can we create a future in which technology serves the common good and benefits all people, regardless of where they live (World Economic Forum, 2023).

SUSTAINABLE DEVELOPMENT AND TECHNOLOGICAL INNOVATION

One of the key areas where responsible innovation can make a significant impact is in the context of sustainable development. As the world faces growing environmental challenges, including climate change, resource depletion, and ecosystem destruction, the role of technology in addressing these issues cannot be overstated (IPCC, 2023; United Nations, 2024). Innovations in renewable energy, sustainable agriculture, and waste management offer promising solutions to global environmental crises. However, these technologies must be developed and deployed with sustainability in mind, ensuring that they do not inadvertently create new environmental or social problems (UNEP, 2024a; IEA, 2024a). In order to effectively tackle climate change and other environmental issues, it is critical that technological solutions be integrated into broader social, economic, and political contexts to maximize their benefits while minimizing unintended consequences (IRENA, 2024; WEF, 2024).

For example, renewable energy technologies, such as solar, wind, and hydroelectric power, have emerged as essential tools in the fight against climate change. By reducing reliance on fossil fuels, these innovations offer a way to curb greenhouse gas emissions and promote a more sustainable energy future (IEA, 2024a; IRENA, 2024). However, the production and deployment of renewable energy technologies also pose their own challenges. For instance, the manufacturing of solar panels and wind turbines requires significant amounts of raw materials, such as rare earth and other critical minerals, which can lead to environmental degradation if not sourced responsibly (IEA, 2024c; European Union, 2024b). Similarly, large-scale renewable energy projects, if not carefully planned, can have negative impacts on local ecosystems and communities (WMO, 2024; United Nations, 2024). Responsible innovation in renewable energy requires developers to consider not only the technological feasibility of these solutions but also their environmental and social ramifications (UNEP, 2024a; WEF, 2024).

The same principles of responsible innovation apply to the adoption of electric vehicles (EVs), which are widely seen as a crucial step toward reducing greenhouse gas emissions from the transportation sector. While EVs offer significant environmental benefits by reducing dependence on fossil fuels, the environmental impact of lithium and other battery mineral extraction, which is required for

EV batteries, must also be considered (IEA, 2024b; World Bank, 2024). Lithium extraction can result in water scarcity and ecosystem degradation, particularly in regions where water resources are already limited (IEA, 2024c). Additionally, the energy demands of EV charging infrastructure must be addressed to ensure that the transition to electric vehicles does not place an unsustainable burden on energy grids or exacerbate environmental problems (IEA, 2024a; Khandekar et al., 2024). In this case, responsible innovation means carefully evaluating the entire lifecycle of EV technologies, from the extraction of raw materials to the disposal and recycling of batteries, to ensure that these solutions do not create new ecological harms (IEA, 2024c; UNEP, 2024a).

Another promising area of responsible innovation in sustainable development is in sustainable agriculture. Agricultural practices are a major contributor to environmental degradation, including deforestation, soil erosion, and the depletion of water resources. However, technological innovations in precision farming, vertical farming, and genetically modified or improved crop varieties offer the potential to make agriculture more sustainable and efficient (FAO, 2023). For instance, precision farming techniques use sensors, drones, and data analytics to optimize the use of water, fertilizers, and pesticides, reducing waste and minimizing the environmental impact of farming practices (FAO, 2023; United Nations, 2024). While these technologies hold great promise, they must be developed and deployed with an eye toward inclusivity and fairness, particularly in regions where small-scale farmers may not have access to the necessary resources or training to implement these innovations (United Nations, 2024; OECD, 2023).

The development of waste management technologies also plays a crucial role in sustainable development. As global populations grow and urbanization continues, the amount of waste generated by human activity increases, placing significant strain on ecosystems and natural resources (UNEP, 2024b). Innovations in waste recycling, composting, and waste-to-energy technologies offer solutions for managing waste in more sustainable ways. For example, waste-to-energy technologies convert organic waste into energy, reducing the need for landfills while simultaneously providing a source of renewable energy (UNEP, 2024b). However, the environmental and social impacts of waste management technologies must be carefully considered. For instance, the incineration of waste to generate energy can release harmful pollutants into the air, and the long-term storage of toxic materials can pose risks to human health (UNEP, 2024b). Responsible innovation in waste management requires that these technologies be designed with safety, environmental impact, and social equity in mind, alongside e-waste strategies to handle rapidly rising electronic waste streams (ITU & UNITAR, 2024).

The integration of AI into environmental sustainability efforts is another area where responsible innovation is crucial. AI-powered solutions have the potential to revolutionize how we monitor and manage environmental issues. For instance, AI can be used to create predictive models for natural disasters, helping communities prepare for and respond to events such as hurricanes, wildfires, and floods (WMO, 2024). These models can also be used to optimize energy consumption, reduce waste, and improve agricultural practices (IEA, 2024d; FAO, 2023). However, the use of AI in these contexts must be approached with caution. Predictive models must be designed with an emphasis on fairness and inclusivity, ensuring that they account for the needs and vulnerabilities of marginalized communities, particularly those in developing regions who are most at risk from climate change and environmental degradation (United Nations, 2024; OECD, 2023).

Furthermore, the development and deployment of AI-powered solutions for climate change and environmental sustainability must be transparent and accountable. AI models can sometimes produce biased or inaccurate predictions if they are trained on incomplete or unrepresentative data (OECD, 2023). For instance, climate services or adaptation tools that fail to account for the specific challenges faced by low-income or rural communities may inadvertently perpetuate environmental injustice (United Nations, 2024). Responsible innovation in AI requires that developers ensure that these technologies are not only technically effective but also socially responsible and inclusive, particularly in vulnerable regions that may face the worst effects of climate change (UNEP, 2024a; United Nations, 2024).

In addition to the environmental and social implications of technology, the economic feasibility of sustainable innovations must also be considered. While many of the technologies discussed—such as renewable energy, electric vehicles, and sustainable agriculture—hold great potential for addressing environmental challenges, their widespread adoption often requires significant investment and infrastructure development (IEA, 2024a; WEF, 2024). In many cases, these technologies are more expensive in the short term than their conventional counterparts, which can present a barrier to their adoption, particularly in developing countries or economically disadvantaged regions. Responsible innovation in this context involves finding ways to make sustainable technologies economically viable, ensuring that they are accessible to all, regardless of socioeconomic status. This may require policy interventions, such as subsidies, carbon pricing, or incentives, to promote the adoption of green technologies and ensure that they are financially sustainable in the long term (UNEP, 2024a; World Bank, 2024).

At the heart of responsible innovation in sustainable development is a commitment to long-term thinking and foresight. While technological solutions may offer immediate benefits, it is important to consider their long-term impacts on both the environment and society. For instance, the widespread adoption of electric vehicles could lead to a reduction in greenhouse gas emissions, but the environmental impact of battery disposal and the energy demands of charging infrastructure must also be accounted for (IEA, 2024b; Khandekar et al., 2024). Similarly, the development of AI-powered solutions for environmental monitoring must be accompanied by ongoing research and oversight to ensure that these technologies continue to meet the needs of communities in a changing world (WMO, 2024; IEA, 2024d). Responsible innovation requires a commitment to evaluating the long-term impacts of technological advancements and making adjustments as necessary to ensure that they align with broader social and environmental goals (Greenhouse Gas Protocol, 2023; IRENA, 2024).

In conclusion, responsible innovation in the context of sustainable development is essential for addressing the global environmental challenges of our time. While technology holds immense promise in tackling issues such as climate change, resource depletion, and ecosystem destruction, it is crucial that these innovations be developed and deployed with careful consideration of their environmental, social, and economic impacts. By prioritizing sustainability, fairness, and inclusivity, responsible innovation can help create a future in which technology contributes to a more just, sustainable, and equitable world (IPCC, 2023; United Nations, 2024). It is up to policymakers, businesses, and innovators to ensure that the technologies of tomorrow not only address the challenges of today but also promote a healthier and more sustainable planet for future generations (UNEP, 2024a; WEF, 2024).

COLLABORATION AND MULTIDISCIPLINARY APPROACHES

To address these complex ethical issues, it is essential to adopt a multidisciplinary approach to responsible innovation. As technology rapidly evolves, its impacts span various aspects of human life and the environment, making it imperative to draw on expertise from a range of fields (United Nations, 2024). A multidisciplinary approach brings together specialists from diverse areas such as technology, ethics, law, sociology, economics, and environmental science, among others, to collaborate in analyzing and addressing the challenges posed by new technologies (OECD, 2023; IRENA, 2024). Each of these disciplines offers unique perspectives that are critical in evaluating the social, ethical, and environmental consequences of technological developments, ensuring a comprehensive and well-rounded approach to innovation (UNEP, 2024a; WEF, 2024).

For instance, when developing artificial intelligence (AI) systems, the contributions of ethicists are essential to ensure that the technology aligns with fundamental human values, such as fairness, accountability, and transparency. AI systems often process vast amounts of data, and without careful ethical oversight, these systems can unintentionally perpetuate or exacerbate societal biases (OECD, 2023). Sociologists can help assess the broader social impacts of AI, considering how it might affect different groups within society, particularly marginalized communities (United Nations, 2024). Legal experts play a critical role in ensuring that AI technologies comply with existing laws and regulations and are subject to appropriate governance structures (European Union, 2024a). By integrating these varied perspectives, the development of AI can be steered in a direction that is more likely to respect human rights and minimize harm, fostering public trust in the technology (United Nations, 2024; UNEP, 2024a).

Similarly, in the context of addressing global challenges such as climate change, a multidisciplinary approach is crucial. Environmental scientists are well-placed to evaluate the ecological consequences of technological innovations, ensuring that the benefits of new technologies do not come at the expense of environmental sustainability (IPCC, 2023; WMO, 2024). They can assess the long-term impacts of emerging technologies on ecosystems and help to design solutions that avoid unintended environmental degradation. Economists bring an understanding of the costs and benefits of different technological solutions, helping to determine the economic viability of innovations (WEF, 2024; IEA, 2024a). Their input is essential in ensuring that sustainable technologies are affordable and accessible, particularly for low-income regions that may lack the resources to implement such solutions (United Nations, 2024; ITU, 2024a). Policymakers, meanwhile, are instrumental in creating the regulatory frameworks that guide technological development, ensuring that innovations align with broader sustainability goals and that their benefits are distributed equitably across society (European Union, 2024a, 2024b).

In addition to these academic and professional fields, it is important to engage stakeholders from various sectors, including business, government, and civil society, in the conversation about responsible innovation. Business leaders, for example, play a crucial role in implementing ethical standards and ensuring that innovations are developed in ways that align with corporate social responsibility goals (WEF, 2024). Government representatives are essential for creating and enforcing regulations that safeguard the public interest and ensure that technological innovations are aligned with national and global policy objectives (United Nations, 2024). Civil society organizations, including advocacy groups and non-governmental organizations, provide a voice for communities that may be affected by new technologies, ensuring that their concerns and needs are considered in the innovation process (UNEP, 2024b; ITU & UNITAR, 2024). By involving these diverse stakeholders, a more inclusive and participatory approach to responsible innovation is achieved (OECD, 2023).

Furthermore, the integration of these different disciplines and stakeholders helps to foster innovation that is not only ethically responsible but also socially inclusive. Technologies have the potential to either alleviate or exacerbate existing social inequalities, and it is crucial that innovation processes prioritize inclusivity (United Nations, 2024). Involving sociologists, community leaders, and representatives from marginalized groups in the innovation process can ensure that technological solutions are accessible to all, particularly those who are most vulnerable (OECD, 2023). This approach can help prevent the digital divide from widening and ensure that technological benefits are distributed more equitably (ITU, 2024a).

Education also plays a key role in promoting responsible innovation. Incorporating multidisciplinary approaches into education and training programs can help cultivate the next generation of innovators who are equipped with the knowledge and skills to address complex ethical issues in their work (United Nations, 2024). Universities and research institutions have a critical responsibility to foster cross-disciplinary collaboration and to provide students with opportunities to engage with real-world ethical challenges (WEF, 2024). By emphasizing the importance of ethics and social responsibility alongside technical expertise, educational institutions can help create a workforce that is prepared to innovate in ways that benefit society while minimizing harm (UNEP, 2024a).

A multidisciplinary approach also encourages ongoing dialogue and adaptation. As new technologies emerge and societal values evolve, it is important for innovation processes to be flexible and responsive to changing circumstances (OECD, 2023). Regular collaboration between experts from different fields can help ensure that technological developments remain aligned with evolving ethical standards and societal needs (United Nations, 2024). This approach fosters a dynamic, iterative process of innovation, where feedback and reflection are integral to refining and improving technologies over time (IRENA, 2024; IEA, 2024a).

The role of global cooperation is also crucial in the context of responsible innovation. Many of the challenges that arise from technological advancements, such as climate change, data privacy, and AI ethics, are global in nature (United Nations, 2024). Solutions to these problems require coordinated efforts across borders and sectors. International collaboration allows for the sharing of knowledge, resources, and best practices, ensuring that innovations are developed with a broader global perspective in mind (UNEP, 2024a; IRENA, 2024). Global partnerships also provide opportunities to harmonize regulatory frameworks, creating consistency and ensuring that ethical standards are upheld worldwide (European Union, 2024a; ITU, 2024a).

Ultimately, adopting a multidisciplinary approach to responsible innovation helps to create technologies that are not only effective and efficient but also ethical, equitable, and sustainable (WEF, 2024; United Nations, 2024). By involving experts from a wide range of fields and ensuring that all relevant perspectives are considered, we can ensure that technological advancements contribute positively to society and the environment. This approach fosters innovation that is mindful of the long-term impacts on humanity and the planet, promoting a future where technology serves as a force for good (IPCC, 2023; UNEP, 2024a).

In conclusion, the complex ethical issues raised by technological innovation require a coordinated, multidisciplinary approach to address effectively. By bringing together expertise from diverse fields and engaging multiple stakeholders, we can ensure that innovations are developed in ways that maximize their benefits while minimizing harm (United Nations, 2024; OECD, 2023). This collaborative approach fosters a more holistic, inclusive, and responsible innovation process that prioritizes ethical responsibility and social good, ultimately contributing to a more just and sustainable future (IRENA, 2024; WEF, 2024).

THE ROLE OF POLICYMAKERS AND REGULATION

Governments and policymakers have a critical role to play in fostering responsible innovation by implementing regulations and policies that ensure the ethical development and use of technology (United Nations, 2024; UNEP, 2024a). As technological advancements continue at an accelerated pace, they bring both immense opportunities and significant risks. While the pace of technological development often outstrips regulatory frameworks, it is crucial for governments to anticipate the potential risks and challenges associated with emerging technologies and establish guidelines to address them (OECD, 2023; European Union, 2024a). Without proper oversight, new technologies can lead to unintended consequences, such as exacerbating social inequalities, violating privacy, or undermining public trust. Thus, proactive policymaking is essential to guide the ethical deployment of new technologies (WEF, 2024; United Nations, 2024).

One area that requires careful regulation is the use of artificial intelligence (AI) in sensitive fields such as healthcare, criminal justice, and education. In healthcare, AI can be transformative, enabling better diagnostic tools, personalized treatment plans, and improved patient outcomes. However, it also raises concerns about data privacy, algorithmic bias, and transparency in decision-making (OECD, 2023; European Union, 2024a). For example, AI algorithms may inadvertently reinforce existing healthcare disparities if they are trained on biased datasets, or if they fail to account for the social determinants of health. To address these challenges, regulators must create frameworks that enforce transparency in AI decision-making processes, ensure that data used for training is diverse and representative, and establish robust mechanisms for data protection and patient privacy (European Union, 2024a; United Nations, 2024).

Similarly, in criminal justice, AI tools are increasingly being used for predictive policing, sentencing recommendations, and risk assessments. However, the use of AI in these areas can perpetuate or even worsen existing racial and socio-economic biases (OECD, 2023). For example, predictive policing algorithms may disproportionately target marginalized communities, while risk assessment tools could unfairly influence sentencing decisions based on biased historical data. Policymakers must establish clear rules and guidelines to ensure that AI systems in criminal justice are used in ways that promote fairness and equity, preventing discriminatory outcomes (European Union, 2024a; United Nations, 2024). This may involve requiring audits of AI systems, implementing safeguards against bias, and ensuring that humans remain in control of critical decisions that impact people's lives (OECD, 2023).

In education, AI has the potential to revolutionize personalized learning, but it also presents risks related to data privacy, surveillance, and the loss of human interaction. Educational institutions must carefully consider how AI technologies are used to monitor student performance, track behavior, and recommend learning paths (United Nations, 2024). Policymakers can help by creating standards for data collection and usage, ensuring that students' personal information is protected and that AI systems are used to support, rather than replace, human educators (OECD, 2023). Ensuring that these technologies are used ethically will require collaboration between educators, technologists, and lawmakers to create policies that uphold students' rights and foster a supportive learning environment (United Nations, 2024).

Moreover, policies promoting the ethical use of emerging technologies can encourage companies to prioritize social good, invest in sustainability, and ensure that their innovations are used for the benefit of all, rather than just for profit (WEF, 2024). Corporate social responsibility initiatives and regulations can incentivize companies to adopt ethical practices, such as reducing carbon footprints, ensuring fair labor practices, and investing in technologies that address social challenges (UNEP, 2024a). Governments can offer tax incentives or subsidies to companies that integrate sustainability into their business models, and impose penalties for those that engage in unethical practices, such as exploiting vulnerable populations or contributing to environmental harm (IEA, 2024a; European Union, 2024b). By aligning innovation with societal values, policymakers can help foster a business environment where technological advancements contribute positively to both the economy and society (United Nations, 2024).

In addition to creating regulatory frameworks, governments can play an essential role in fostering public awareness and ensuring that citizens are informed about the implications of new technologies. Public engagement initiatives, such as consultations, workshops, and educational campaigns, can help ensure that society's diverse voices are heard in discussions about emerging technologies (United Nations, 2024). These initiatives also provide an opportunity for governments to build trust with the public, ensuring that citizens feel confident in the ethical oversight of new technologies (UNEP, 2024b). A well-informed public can advocate for responsible innovation and hold both private and public entities accountable for the societal impact of their technological decisions (OECD, 2023).

Finally, international cooperation is necessary to address the global challenges posed by emerging technologies. The implications of AI, automation, biotechnology, and other advanced technologies extend beyond national borders, requiring global standards and agreements (United Nations, 2024). Governments must collaborate with international organizations to create harmonized regulations and frameworks that ensure the ethical use of technology worldwide (UNEP, 2024a; IRENA, 2024). This could include agreements on issues such as data privacy, intellectual property, and the regulation of emerging technologies, ensuring that innovations are deployed in ways that benefit all people and mitigate harm across global contexts (European Union, 2024a; ITU, 2024a). By working together, governments can help ensure that technological advancements contribute to a more equitable and sustainable global future (WEF, 2024; IPCC, 2023).

GLOBAL COOPERATION FOR ETHICAL INNOVATION

The challenges posed by technological innovation are inherently global in nature, and therefore, they require global cooperation (United Nations, 2024). As technological advancements continue to transcend national borders, their impact is no longer confined to a single country or region. Innovations in artificial intelligence, biotechnology, renewable energy, and automation, for example, have the potential to affect all aspects of society, including economies, security, and human rights. Given these far-reaching implications, it is essential for countries to collaborate on creating shared ethical standards and regulatory frameworks that can guide the responsible development and deployment of emerging technologies (UNEP, 2024a; IRENA, 2024). Without such international cooperation, there is a risk that technological advancements could exacerbate existing global inequalities or result in unintended harmful consequences (ITU, 2024a; World Bank, 2024).

International agreements can play a critical role in ensuring that emerging technologies are developed and deployed responsibly. These agreements provide a platform for countries to set common ethical standards and regulatory policies that prioritize safety, human rights, sustainability, and equity (United Nations, 2024; European Union, 2024a). For example, agreements could be established to regulate the use of AI in sensitive areas such as healthcare, criminal justice, and military applications, ensuring that these technologies are used in ways that protect privacy, reduce bias, and uphold human dignity (OECD, 2023). Moreover, international cooperation could help address the ethical concerns surrounding issues like data ownership, surveillance, and the militarization of technology, ensuring that technology serves the collective good and does not disproportionately benefit powerful nations or corporations (WEF, 2024; UNEP, 2024a).

International organizations like the United Nations (UN) and the World Economic Forum (WEF) are uniquely positioned to facilitate dialogue among governments, businesses, and civil society groups. These platforms bring together diverse stakeholders from around the world to discuss the challenges and opportunities presented by new technologies (United Nations, 2024; WEF, 2024). Through multilateral discussions, these organizations can promote the development of global ethical guidelines that take into account regional differences and local contexts, ensuring that technology is developed in ways that are culturally appropriate and socially responsible (UNEP, 2024b; IRENA, 2024). By fostering collaboration and creating shared frameworks, these organizations can help bridge the gaps in technological access, ensuring that the benefits of innovation are distributed equitably (ITU, 2024a).

The digital divide is one of the most pressing global challenges related to technological innovation. While developed nations often lead the way in adopting and deploying new technologies, many developing countries still struggle with limited access to the internet, infrastructure, and skilled labor (ITU, 2024a). This divide creates an uneven playing field, where only a small portion of the global population benefits from technological advancements. International cooperation is essential to address this divide by creating programs that support technology access and digital literacy in underserved regions (United Nations, 2024; ITU & UNITAR, 2024). By pooling resources, expertise, and funding, developed nations and international organizations can work to ensure that developing countries are not left behind, enabling them to participate in the digital economy and harness the benefits of emerging technologies (WEF, 2024).

In addition to addressing the digital divide, global cooperation is necessary to ensure that the benefits of technological innovation are shared equitably. As technologies such as automation and

AI transform industries and job markets, it is essential that the benefits of these advancements are not concentrated in the hands of a few powerful corporations or wealthy nations (OECD, 2023). International agreements can help ensure that the global workforce is equipped with the skills needed to thrive in a rapidly changing labor market, supporting policies that promote retraining, reskilling, and job creation (United Nations, 2024). Moreover, these agreements can set standards for fair labor practices, including ensuring that the deployment of automation and AI does not lead to the exploitation of workers or the widening of income inequality (WEF, 2024; UNEP, 2024a).

Finally, global cooperation is critical for fostering innovation that is sustainable and aligned with long-term goals such as climate action and social well-being. The development of technologies to address global challenges, such as climate change, resource depletion, and public health crises, requires a collective effort (IPCC, 2023; IEA, 2024a). International partnerships can help share knowledge, resources, and technologies that contribute to sustainable solutions (IRENA, 2024; UNEP, 2024a). By working together, countries can pool their expertise to create technologies that have a positive environmental impact, such as clean energy solutions, sustainable agriculture practices, and waste management systems (FAO, 2023; UNEP, 2024b). Moreover, through global cooperation, nations can align their efforts to address technological challenges in a way that benefits all people, not just the wealthy or technologically advanced (United Nations, 2024; ITU, 2024a). This collaborative approach to innovation will be key to ensuring a more equitable and sustainable global future (WEF, 2024; World Bank, 2024).

CONCLUSION

Responsible innovation is not just a guiding principle but an essential framework for addressing the profound ethical challenges that emerge in our increasingly tech-driven world (United Nations, 2024; UNEP, 2024a). As technology evolves at an unprecedented pace, the need for thoughtful, inclusive, and ethically sound innovation becomes more urgent (OECD, 2023; IPCC, 2023). Innovations in artificial intelligence, biotechnology, and automation, among others, have the potential to reshape our societies in fundamental ways. However, to fully realize their benefits, these technologies must be developed and implemented with a keen awareness of their societal, environmental, and ethical implications (European Union, 2024a; WEF, 2024). The process of responsible innovation requires a multidisciplinary approach, where knowledge from diverse fields—such as ethics, law, sociology, technology, and environmental science—comes together to ensure that emerging technologies promote social good, protect human rights, and advance sustainability (IRENA, 2024; United Nations, 2024).

Collaboration among stakeholders, including developers, policymakers, civil society, and the global community, is critical in fostering responsible innovation (UNEP, 2024a; ITU, 2024a). By working together, we can build systems that not only push the boundaries of technological progress but also safeguard human dignity and promote equity (United Nations, 2024; WEF, 2024). This means ensuring that technology does not exacerbate existing inequalities or create new forms of discrimination. Ethical considerations, such as bias in algorithms or the environmental impact of new technologies, must be central to innovation practices (OECD, 2023; IEA, 2024d). Through ongoing dialogue and cooperation across borders, industries, and disciplines, we can anticipate the consequences of emerging technologies and craft solutions that serve the collective interest, rather than the interests of a few (UNEP, 2024b; IRENA, 2024).

While the challenges of responsible innovation are substantial, they are not insurmountable. With proper foresight, regulation, and the willingness to adapt to changing circumstances, the world can navigate the complexities of technological development (European Union, 2024a; United Nations, 2024). Government regulations, international frameworks, and industry standards must evolve to keep pace with innovation, providing the necessary oversight and guidance to ensure that technology is used ethically and responsibly (WEF, 2024; OECD, 2023). Moreover, the rapid pace of technological change should not be viewed as a barrier to responsible innovation but as a catalyst for change. By embracing the challenges that come with new technologies, we can create an environment where ethical considerations are integrated into every stage of innovation, from research and design to deployment and beyond (UNEP, 2024a; IPCC, 2023).

The role of global cooperation cannot be overstated. In our interconnected world, technological challenges are not confined to national borders, and solutions must therefore be approached from a global perspective (United Nations, 2024). The digital divide, access to technology, and the environmental impact of new innovations are issues that require collective action (ITU, 2024a; World Bank, 2024). International cooperation, through agreements, partnerships, and shared frameworks, is essential for ensuring that emerging technologies benefit everyone, not just the wealthiest or most technologically advanced nations (European Union, 2024a; IRENA, 2024). By addressing global challenges together, we can ensure that innovation contributes to sustainable development, reduces inequalities, and provides equitable access to the benefits of technology (WEF, 2024; UNEP, 2024b). This global approach will help bridge the gap between different regions, creating a more inclusive and just future (United Nations, 2024; IPCC, 2023).

The future of innovation hinges on our ability to shape technology in ways that are aligned with ethical principles and the broader well-being of society. This means developing technologies that not only solve today's problems but also anticipate the needs of tomorrow (IRENA, 2024; WEF, 2024). By prioritizing long-term sustainability, fairness, and inclusivity, we can ensure that the technologies of the future contribute to a world where all people can thrive (United Nations, 2024; UNEP, 2024a). This is the essence of responsible innovation: a commitment to creating technologies that serve humanity's best interests while preserving the planet for future generations (IPCC, 2023). By doing so, we can ensure that innovation remains a force for good—driving progress, enhancing lives, and contributing to a more equitable and sustainable global society (United Nations, 2024; IEA, 2024a).

In conclusion, responsible innovation is more than just an ideal; it is a crucial approach to ensuring that technology serves humanity in ways that are ethical, inclusive, and sustainable. As we continue to navigate the complexities of a tech-driven world, the collaboration of diverse stakeholders, the development of thoughtful policies, and the commitment to global cooperation will be key in addressing the challenges and harnessing the potential of emerging technologies (UNEP, 2024a; European Union, 2024a). Through responsible innovation, we have the opportunity to address some of the world's most pressing issues—climate change, social inequality, healthcare disparities, and more—while safeguarding the well-being of future generations (IPCC, 2023; United Nations, 2024). This vision of a responsible, ethical, and sustainable future is within reach, provided we remain committed to shaping technology in ways that benefit all of humanity (WEF, 2024; IRENA, 2024).

CHAPTER 12 ESSAY PROMPTS

Responsible Innovation and Global Challenges

1. How can responsible innovation address the ethical challenges posed by rapidly advancing technologies like AI, automation, and biotechnology?
2. Discuss the role of interdisciplinary collaboration in responsible innovation. How can expertise from fields such as ethics, law, and environmental science improve technological development?
3. What are the ethical implications of emerging technologies, and how can they be mitigated through responsible innovation? Provide examples.
4. Explain the concept of the "digital divide" and its implications for responsible innovation in developing countries. How can global collaboration address these disparities?
5. How does the rapid pace of technological development pose challenges to traditional regulatory frameworks, and what strategies can governments use to ensure ethical technology deployment?
6. Analyze the role of international organizations, such as the UN or World Economic Forum, in promoting responsible innovation. What are their strengths and limitations in influencing global technological standards?
7. What are the risks of technological innovations reinforcing existing social biases, and how can responsible innovation address issues like algorithmic bias and fairness in AI systems?
8. In the context of sustainable development, how can responsible innovation contribute to tackling global challenges such as climate change and resource depletion? Provide examples of technologies that balance innovation with sustainability.
9. How can responsible innovation promote economic resilience in the face of technological disruptions, such as automation and job displacement? What role do policymakers play in ensuring a smooth transition for affected workers?
10. How can responsible innovation be applied to the development of global health technologies, ensuring that innovations are accessible, equitable, and beneficial to all populations, particularly in low-income regions?

REFERENCES

Ada Lovelace Institute. (2022). Participatory data stewardship: A toolkit for a rights-respecting data ecosystem. Author.

Council of Europe. (2024). Framework Convention on Artificial Intelligence, Human Rights, Democracy and the Rule of Law. Author.

Department for Science, Innovation and Technology. (2023). A pro-innovation approach to AI regulation [White paper]. HM Government.

European Data Protection Board. (2022). Guidelines 05/2022 on the use of facial recognition in the area of law enforcement. Author.

European Union. (2024a). Artificial Intelligence Act. Author.

European Union. (2024b). Critical Raw Materials Act. Author.

European Union Agency for Cybersecurity. (2023). Artificial intelligence threat landscape 2023. ENISA.

Greenhouse Gas Protocol. (2023). Land sector and removals guidance. Author.

Group of Seven. (2023). Hiroshima AI Process: International Code of Conduct for organizations developing advanced AI systems. Author.

International Energy Agency. (2024a). World energy outlook 2024. Author.

International Energy Agency. (2024b). Global EV outlook 2024. Author.

International Energy Agency. (2024c). Critical minerals market review 2024. Author.

International Energy Agency. (2024d). Electricity 2024: Data centres and AI analysis. Author.

International Labour Organization. (2023). Generative AI and jobs: A global analysis of possible effects on job quantity and quality. Author.

International Monetary Fund. (2024). World economic outlook: GenAI and the future of work (Chapter 3). Author.

International Telecommunication Union, & United Nations Institute for Training and Research. (2024). The global e-waste monitor 2024. Authors.

Intergovernmental Panel on Climate Change. (2023). AR6 synthesis report: Climate change 2023. Author.

Khandekar, A., Gorsche, C., McLaren, J., Zhou, E., & Neubauer, J. (2024). Electric vehicle charging demand profiles for retail buildings (Report No. NREL/TP-6A20-90638). National Renewable Energy Laboratory.

National Institute of Standards and Technology. (2023). Artificial Intelligence Risk Management Framework (AI RMF 1.0) (NIST AI 100-1). U.S. Department of Commerce.

Organisation for Economic Co-operation and Development. (2023). OECD employment outlook 2023: Artificial intelligence and the labour market. OECD Publishing.

Pew Research Center. (2023). How workers and the public view artificial intelligence in the workplace. Author.

Stanford Institute for Human-Centered Artificial Intelligence. (2024). AI Index report 2024. Stanford University.

United Nations. (2024). The Sustainable Development Goals report 2024. Author.

United Nations Environment Programme. (2024a). Emissions gap report 2024. Author.

United Nations Environment Programme. (2024b). Global waste management outlook 2024. Author.

U.S. Bureau of Labor Statistics. (2024). Employment projections, 2024–2034. U.S. Department of Labor.

U.S. Equal Employment Opportunity Commission. (2023). Assessing adverse impact in software, algorithms, and AI used in employment selection procedures under Title VII [Technical assistance document]. Author.

U.S. Office of Management and Budget. (2024). Memorandum M-24-10: Advancing governance, innovation, and risk management for agency use of artificial intelligence. Executive Office of the President.

World Bank. (2024). Commodity markets outlook: Critical minerals and the energy transition. Author.

World Economic Forum. (2023). The future of jobs report 2023. Author.

World Meteorological Organization. (2024). State of the global climate 2024. Author.

www.ingramcontent.com/pod-product-compliance
Lightning Source LLC
LaVergne TN
LVHW080015210926
848481LV00011B/3031